AF881055

SOTHIS PRESS

Sothis Press aims to completely reimagine the public domain book with an emphasis on universal access to captivatingly designed print editions. We believe that technology will enable every single book ever published to be perpetually accessible in beautiful print form. Sothis is beauty in ubiquity.

www.sothis.press
1 Waterhouse Square, London, ECIN 2ST

Appendix to Railway Practice

Containing a Copious Abstract of the Whole of the Evidence Given Upon the London and Birmingham, and Great Western Railway Bills, When Before Parliament, Properly Digested and Arranged, With Marginal Notes

S. C. Brees

SOTHIS PRESS

ISBN 978-1-0365-2612-2

Appendix to Railway Practice
is in the public domain.

This edition is published by
 The company
is registered in England and Wales.
The Company Number is 14900389.

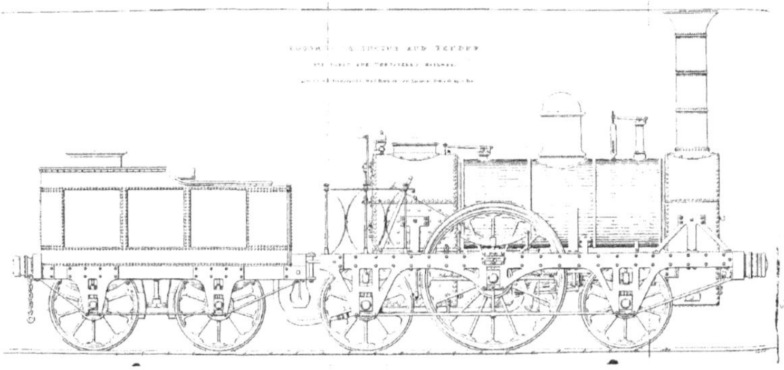

APPENDIX

TO

RAILWAY PRACTICE,

CONTAINING

A COPIOUS ABSTRACT OF THE WHOLE OF THE

EVIDENCE

GIVEN UPON THE

LONDON AND BIRMINGHAM, AND GREAT WESTERN

RAILWAY BILLS,

WHEN BEFORE PARLIAMENT,

PROPERLY DIGESTED AND ARRANGED, WITH MARGINAL NOTES.

*** Among the several Witnesses examined will be found the following eminent Civil Engineers:—

GEORGE STEPHENSON, ESQ.
ROBERT STEPHENSON, ESQ.
I. K. BRUNEL, ESQ.
J. U. RASTRICK, ESQ.
JOSEPH LOCKE, ESQ.
CHARLES VIGNOLES, ESQ.
HENRY R. PALMER, ESQ.
GEORGE W. BUCK, ESQ.
HENRY H. PRICE, ESQ.
DR. DIONYSIUS LARDNER.
GEORGE LEATHER, ESQ.
WILLIAM C. MYLNE, ESQ.
FRANCIS GILES, ESQ.
COL. G. HENDERSON.
THOMAS CABREY, ESQ.

&c. &c. &c.

TO WHICH IS ADDED,

A GLOSSARY OF TECHNICAL TERMS,

USED IN

CIVIL ENGINEERING,

EXPLAINING AND ILLUSTRATING EVERY WORD IN ORDINARY USE;

AND

THE DETAILS OF HAWTHORNE'S CELEBRATED LOCOMOTIVE ENGINE,

FOR THE PARIS AND VERSAILLES RAILWAY.

BY S. C. BREES, C.E. &c.

LONDON:
JOHN WILLIAMS, LIBRARY OF SCIENCE AND ART, 106, GREAT RUSSELL STREET.
1839.

LONDON:
PRINTED BY J. DAVY, QUEEN STREET, KING STREET, LONG ACRE.

PREFACE.

The valuable nature of the Engineering Evidence, lately given in the Committee Rooms of the Houses of Parliament, being generally admitted, no apology is required for an Abstract or Digest of the same.

The Author begs merely to remark, that he is fully sensible that the importance of the subject required a more experienced hand; and although he was in some measure aware of the difficulties of the task when he undertook it, (from a knowledge that several professional gentlemen had commenced arranging the Evidence, and afterwards abandoned it) yet they have far exceeded his anticipations, notwithstanding he trusts the result of his labors will not prove unacceptable to the Profession and the Public generally.

A Glossary of Technical Terms used in Civil Engineering is added, containing a short explanation of every word in ordinary use, and general remarks upon the subject connected therewith.

And the Details of the celebrated Locomotive Engine, invented and manufactured by Messrs. R. and W. Hawthorne, for the Paris and Versailles Railway* conclude the volume, which will not be found the least interesting portion of it.

* Kindly communicated to Mr. Williams, the publisher, by the Inventors.

INTRODUCTION.

One of the principal exhibits of a Civil Engineer's talent and resources is displayed in the Committee Rooms of the Houses of Parliament, in his examination as a Witness to prove the practicability or the contrary of proposed Public Works, comprising the numerous Railways, Canals, &c. throughout the kingdom.

And as the same individual has frequently to advocate and support totally opposite systems and contingencies, upon different undertakings, much ingenuity is consequently displayed on these occasions:—— for instance, where an Engineer appears as a Witness in favor of a Line of Railway with very favorable Gradients, his answers to the questions are always full and explicit, and he states boldly, without fear of contradiction, the great advantages of a level railway, compared with an undulating line, containing long and steep inclinations thereon; but he is not so communicative respecting the means which are taken, or the sacrifices which are made, to obtain this advantageous run of levels.

The next line he may chance to be examined upon may have unfavorable Gradients,—when his mind becomes enlarged, and his memory awakened, he takes a more extensive view of this case, and calls to recollection many other circumstances requiring equal consideration as the Gradients; and as the latter (upon the line in question) are quite available for the practical working of Locomotive Engines, he doubts the policy of running the Company into ruinous expenses merely to acquire an assumed desideratum: the best line must certainly be that which is the most feasible and practicable.

Why not, he remarks, leave some parts of the line for the future Management to improve from time to time, when the funds of the Company can afford it; who can say that the country which has produced men whose works have excelled and rendered secondary upon a level plane mankind's best servant "the noble horse," proudly outstripping all comparison; * who dare say it does not or will not possess talent that shall vanquish him in the chace even "o'er hill and dale:" therefore, it may not be imprudent to leave something for the rising generation to overcome, particularly as it will benefit the present.

The Witness next proceeds to prove that the slopes upon the whole line run in proper juxta position with each other, and explains that as none of it is upon a level, but the whole disposed and arranged at a corresponding rate of clivity, and rendered equally advantageous for a line of transit both ways, there cannot therefore be any waste of power upon it. To prove which, he delivers in a long table of the power gained in passing down the slopes by gravity, by which he obtains a set off against the assistance required in getting up them: upon which the learned Counsel reminds the Committee of the even balance of mechanical power and gravity upon the line. He concludes by stating, that as they are the best levels the country can afford, they must suffice.———

The system pursued in the Committee Rooms of the Houses of Parliament, in point of fact, amounts to this:—the Evidence given on one side explains all the advantages of an undertaking, with the several adjuncts and details, say of a proposed line of railway or canal; it also justifies the many inconveniences which may arise out of the same: while the other party state all the disadvantages of the line of railway or canal in question, accompanying it with a long list of grievances which the Public will suffer, and perhaps conclude with a suggestion of a better direction for a line of communication; and every argument that can be brought to bear upon the case is made available, *pro. et con.*, until at length the subject becomes exhausted.

It is therefore for the Reader, after having carefully studied the many bearings of any particular measure described in the Evidence, to judge of the expediency of the same, and to weigh all the various objections, discriminating between those which are real and such which are imagi-

* It is not improbable that he would here introduce a passing tribute to the name of "Stephenson," which will not prove the least injurious to his cause.

nary, separating mere local and private interests from public benefit, and attaching whatever degree of faith to the several parties that he may think proper.

The Author has considered that any comment of his upon the opinions contained in the Evidence wholly unnecessary and uncalled for. He has confined his efforts entirely to the explanation of the subject, which he has endeavoured to render plain and intelligible; and it is the matter elicited from the several Witnesses in the course of debate which is interesting to the Profession, rather than the point at issue.

The Evidence contains answers to many important engineering questions, of a most essential and valuable nature, which were broached to the most distinguished of the Profession, and the only men capable of answering them.

The narrative style of composition is employed throughout the Evidence, in preference to a mere report of the same, as it is the most concise; and the Author has endeavored in all cases to ascertain the true opinion of the several Witnesses upon any point of dispute, which having ascertained, he has condensed and reduced to as few words as possible, employing the same idioms, comparisons, and style of the individuals, as far as convenient and consistent: He has been obliged occasionally to modify it, in order to render the meaning clear, as there are some phrases the sense of which depend partly upon the emphasis with which they are delivered, and when written down precisely as spoken, admit of an obscure and doubtful construction: and this is more particularly the case with argumentative discourses, there being many parts of the Evidence very confused and intricate, independant of typographical errors—it is frequently found necessary to glance over several pages, in search of circumstances to confirm the judgment, to discover the true bearings of the case.

To conclude, the Author hopes he has not misunderstood, or misrepresented the opinions or assertions of any gentleman, either from among the professional or general Witnesses examined, and he will be most happy to correct any such errors, upon the nature of the same being communicated to him.

LONDON AND BIRMINGHAM RAILWAY.

ABSTRACT OF EVIDENCE,

Given before a Committee of the House of Lords, June 1832.

COUNSEL FOR THE BILL.

Mr. Serjeant Merewether.
Mr. Follet.
Mr. Alexander.
Mr. Rushton.

COUNSEL
FOR THE
OPPOSITION TO THE BILL.

Mr. Harrison.
Mr. D. Pollock.

WITNESSES, (Engineering).

Mr. Robert Stephenson.
Mr. Francis Forster.
Mr. George Hennet.
Mr. Thomas Gooch.
Mr. J. U. Rastrick.
Mr. Henry R. Palmer.
Mr. Joseph Locke.
Mr. J. Copeland.
Mr. Francis Wedge.
Mr. Wm. W. Gardner.
Mr. Layton Cook.
Mr. Philip Hardwick.

Mr. H. Booth.
Mr. Hardman Earle.
Mr. Joseph Pease.
Mr. Thomas Lee.
Mr. John Hart.
Mr. Fred. Clements.
Matthew Holman.
Mr. John Norton.
Mr. Thomas Norton.
Mr. John Shackell.
Mr. Wm. Partridge.
Mr. Wm. Shore.
Mr. John Swainson.
Capt. Richard Moorsom, R.N.
Mr. Peter Lecount, R.N.
Mr. Wm. Meade Warner.
Mr. Charles Whitworth.
Mr. John Sharpe.
Mr. Robert Attenborough.
Mr. Oliver Mason.
Mr. Edward T. Moore.
Mr. Fred. Barry.
Mr. John Travers.
Mr. Henry Hemsley.
Mr. Fred. Barnes.
Mr. Richard Purkess Westall.
Mr. John Moss.
Mr. Henry Cheetham.
Mr. Thomas Badger.
Mr. John Cheetham.
Mr. Richard Whitmore.
Lieut.-Gen. Sir. J. W. Gordon, Bart. K.C.B.
Mr. Augustus Godby.
Mr. Wm. Kay.
Mr. James Marshall.
Mr. James Forster.
Mr. John Dillon.
Mr. Richard Creed.

*** This Committee came to the conclusion that the promoters of the Bill had not made out a sufficient case to warrant the forcing of the proposed Railway through the lands and property of so large a portion of dissentient land owners and proprietors.

Ex. ROBERT STEPHENSON, ESQ. C.E.

Warrington Railw. Leicester & Swanington Railway. Route of proposed line to Birmingham. Goes near Watford.

I was Engineer of the Warrington Railway, which is 5 miles long; and of the Leicester, which is 16 miles long; it is between Leicester and the coal field at Swanington, near Ashby de la Zouch, and is not yet completed.——I was likewise engaged on the Liverpool and Manchester Railway under my father.——I have been nearly two years examining the country for the proposed line to Birmingham, which begins at Oxhey Lane and goes from thence to the South end of Watford, through the Colne valley, and passes the parks of Lords Clarendon and Essex, avoiding the same by a tunnel and an acute curve, which would be attended with no inconvenience; we then pass over the chalk ridge near Ivinghoe, and descend into the vale of Buckinghamshire, passing the Ouse; we then proceed towards Brockwell, and in a direct line to Blisworth, and so on to Weedon, where there is a considerable bend in the line, in order to avoid a couple of crossings over the Grand Junction Canal and a very high bank on the grounds of Mr. Thornton; whose park we avoid, but we still cross some of his land, but mostly in a tunnel. We then go towards Kilsby, where we surmount much high ground by means of a tunnel; from thence to Rudworth, in nearly a direct line towards Coventry, where we have Meriton Ridge to surmount, which is very high; from thence we descend into the vale of Blyth, which runs towards the Trent; thence to Birmingham to Novia Scotia Gardens.—— ——Primrose Hill tunnel will be in London clay, also another at Oxhey Lane; at Northchurch and Watford there are tunnels in chalk. The next considerable tunnel we come to is at Kilsby, which lies in the clay formation; as all clays are very easy to tunnel, unless they have much sand in them, we shall find no difficulty. The next tunnel is a short one at Brockhall. ——(I have just finished a tunnel one mile long at Leicester.)—— —— Nearly all the cuttings near the London road are of the same slope.—— All considerable cuttings have a slope of 2 to 1, the very small cuttings vary from 8 to 10 feet, the inclination of the embankments are invariably 2 to 1, that is a base of 2 feet to 1 foot perpendicular; and to the best of my judgment these slopes will be sufficient, provided they are carefully made, and the water is kept out, which is a most important point; if it is not, they will not stand at any slope. Generally speaking, when the cuttings are upon an equality, the water can easily be taken off by a

Blisworth and Weedon. Coventry. Tunnels. Soils, &c. they go through. Clay a good soil to tunnel unless much mixed with sand. Tunnel at Leicester 1 mile long. Slopes of cuttings mostly 2 to 1. Embankments all 2 to 1. Necessity of keeping the water out of the slopes.

surface drain, and arrangements can be made to prevent the water getting on the bank while the embankments are being formed, also for keeping it off after they are formed.—— ——The excavation exceeds the embankments by about 1,250,000 cubic yards.

DETAILS OF ESTIMATE.

Estimate.

EVIDENCE as to SLOPES, BRIDGES, and TUNNELS.

Contents of Excavations and Embankments.

	Excavations.	Embankments.
From London to Oxhey Lane - - - - - -	975,084	975,084
From Oxhey Lane to South End of Watford Tunnel -	901,813	836,475
From Watford Tunnel to Box Moor - - - - -	554,120	554,120
From Box Moor to Ivinghoe - - - - - -	444,838	444,838
From Ivinghoe to Leighton Buzzard - - - -	1,480,434	774,381
From Leighton Buzzard to Bletchley - - - -	570,651	570,651
From Bletchley to Castlethorpe - - - - -	1,237,147	1,237,147
From Castlethorpe to Gayton - - - - -	997,530	997,530
From Gayton to London Road Tunnel at Weedon, No. 1. -	454,987	454,987
From London Road Tunnel to Buckley Wharf - -	694,827	654,827
From Buckley Wharf to Kilsby - - - - -	347,643	277,300
From Kilsby Tunnel to Clifton and Rugby Road - -	387,575	387,575
From Clifton and Rugby Road to Church Lawford Road -	275,013	218,769
From Church Lawford to Road from Brandon to Coventry -	467,813	357,155
From Brandon Road to Warwick Road - - - -	509,489	509,489
From Warwick Road to Fletchampstead - - -	116,932	—
From Fletchampstead to End of the River Blythe Embankment	684,391	467,158
From Blythe Embankment to Marston Green Embankment	250,440	250,440
From Marston Green to Birmingham - - - -	730,389	730,389
Cubic Yards - - -	12,081,116	10,698,315

This will be proved by { Mr. ROBERT STEPHENSON, Mr. RASTRICK, Mr. PALMER.

ABSTRACT OF ESTIMATE.

	Estimate proved in the House of Commons.
	£
Excavations and Embankments - - - - - -	179,000
Tunnelling - - - - - - - -	250,286
Masonry.—This Item is increased in consequence of an Agreement with the Commissioners of the Metropolitan Roads to add to some of our Bridges in Width and Height, and also an Agreement with the Trustees of the Radcliffe Library Estates to increase the Number of Arches in the Wolverton Viaduct, and also an Addition of Two Bridges over the Avon near Brandon, to avoid the Diversion of the River - - -	350,574
Rails, Chairs, Keys, and Pins - - - - - -	212,940
Blocks and Sleepers - - - - - -	102,960
Ballasting and laying Rails - - - - - -	102,960
Fencing at 740*l.* per Mile - - - - - -	76,032
£	1,874,752
Land - - - - - - - - -	250,000
Six Water Stations at 500*l.* - - - - - -	3,000
Six intermediate Pumps - - - - - -	600
Offices, &c. requisite at each End of the Line, for Convenience of Passengers, &c., and Walling for enclosing the Space for Depôt	16,000
Forty Locomotive Engines, 1,000*l.* - - - - -	40,000
300 Waggons at 30*l.* - - - - - -	9,000
Sixty Coaches at 200*l.* - - - - - -	12,000
£	2,205,352
Contingencies - - - - -	294,648
£	2,500,000

My estimate allows two lines of railway the whole distance, including a siding for every 5 miles——I made this calculation on my own experience, having been employed in making bridges and works of a similar description, and I have put the prices in some places considerably higher than usual.——The engines are not to be exclusively furnished by the Company, but approved of by them; neither are they intended to be entirely supplied by my father, but any efficacious engine will be adopted.——I have included bridges for every existing road, except where two roads intersect, as by a slight diversion of the road, one bridge would be sufficient. I stated in the Commons, that I made the number of occupation bridges average about the same as works of a

A siding every 5 miles.

Engines.

Particulars of occupation bridges.

similar kind, as the Grand Junction Canal, which goes about 50 miles in the same district. I thought that a good criterion.———The first class of bridges are those for turnpike roads, which are twenty-two in number, at £2500 each, making £55,000. Over the important parish roads there are fifty-five, at £1000. each; over the inferior roads one hundred and twenty-seven, at £500. each, the accommodation bridges are two hundred, at £300. each; there are seven bridges over canals, at £3200. each; there are two or three skew bridges over canals, for which distinct estimates have been made, amounting to £3500. each. I estimate the Primrose Hill, the Watford, the Watling Street, the Brockhall, the Kilsby, and the Becknell tunnels at £32. per running yard, the shorter ones at £26. only; (The tunnel under Islington, which is much smaller than ours, in the proportion of 3 to 5, was £32. per lineal yard.) as in the longer tunnels there is provision made for ventilating them, which in the shorter is unnecessary. A small drift is driven along the top of the tunnel, to communicate with the shafts, and any vapour that may arise from the engine ascends into the upper drift.———They are all in 18 inches brickwork. The tunnel at Liverpool was in exceedingly bad material; at Leicester we had loose running sand, and it was only 14 inches. If they are turned in three bricks, the expense would be nearly 5*s*. per yard extra.—— ——I have estimated brick facing to all the tunnels.———I consider two lines of railway quite sufficient, notwithstanding the collateral travelling. The width of the railway on a level is 30 feet, and 6 feet on each side, for the fence and ditch, making 42 feet, as it is considered advantageous to make our sidings (which will require a little more space) where the railway is upon a level. We are empowered to take 20 yards all the way. ———Where we have stone we employ it to make the blocks, (where the rock is not suitable we shall employ wood). I have allowed 2*s*. 6*d*. for each block, I never knew them to exceed that; on the Liverpool line most of them were 1*s*. and 1*s*. 2*d*. I believe the average was 4*s*. per yard for blocks and sleepers, which I consider about the same price. I intend the sleepers to be 8 or 9 inches by 4 inches; upon the Liverpool and Manchester Railway they are not so large, I do not know of any that cost more than 5*s*. The motion of the coaches on wooden sleepers is easier than on stone, and the noise less; perhaps the wood would be cheaper than stone, as the line goes through a great quantity of timber. Mr. Walker calculates them at 4*s*. or 5*s*. per square foot, and as each sleeper contains 2 cubic feet, he makes the expense 11*s*. or 12*s*. for each sleeper. I am not aware of a single sleeper being laid down at such a cost.

Price of turnpike road bridge £2,500.
Parish road ditto, £1,000.
Inferior road ditto, £500.
Accommodation ditto, £300.
Canal do. £3,200.
Skew do. £3,500.
Expence of long tunneling, £32. per yard.
Short ditto, £26.
Cost and Description of the tunnel under Islington.
To be in 18-inch brick work.
L. and M. line in the same.
Tunnel at Leicester in 14-inch.
If in 3 bricks it will be 5s. per cubic yd. more.
Space allowed for sidings.
Blocks and sleepers.
2s. 6d. price of a block.
L. and M. line, 2s.
The motion on sleepers easier than on stone blocks.
Mr. Walker's calculation of sleepers.

Water getting upon the rails causes them to sink, and get out of place.

Ballasting.

Wells good criterions to judge of the soil.

Sand stone and free stone perpendicular.

Lias $1\frac{1}{2}$ and 1 to 1. Chalk $\frac{3}{4}$ to 1.

Primrose Hill borings,

Meriton Ridge.

—— ——If water gets in upon the rails they invariably sink, (as they stand unequal) and get out of place; therefore, I have provided a proper bed, 2 feet thick, which also prevents the necessity of cutting a drain, which is upon the surface. The material for making this bed (on which the blocks and sleepers are laid) must be hard dry stuff, such as gravel, broken red sand stone, or chalk, all of which we shall find along the line. —— ——Regarding the soil, I enquired of people who had bored or sunk wells in the immediate neighbourhoods, and judged of their accuracy by borings I had taken. In some places we have the sand stone rock and free stone perpendicular; the slope of lias is generally $1\frac{1}{2}$ to 1, and sometimes 1 to 1; the slope I propose in chalk is $\frac{3}{4}$ to 1, there are instances of this slope to be seen. Our boring at Primrose Hill is 56 feet deep, which is the bottom of the tunnel; it is 1 foot of vegetable mould, 7 feet of moist yellow clay, 13 feet of tough brown clay, and 30 feet of London clay.——Supposing that Meriton Ridge (which is $\frac{1}{4}$ to 1, part of the bottom would be perpendicular) was in marl, with loose thin lamina sand stone, about 2 or 3 inches thick, and the cutting would not stand at 2 to 1, I should most decidedly make a short tunnel, as it would be much cheaper. In the former case, the expence of excavating and tunneling would be nearly the same, as all the cuttings are priced at 3*s*. per yard; taking the average width of an excavation at 30 feet, and the depth at 60, gives an area of 200 yards, which makes it £30. per yard, the expence of tunneling being £32. If we should resort to tunneling, instead of open cuttings, we should save the expense of bridges, &c., as the communications with the land would be left open. The expense of tunneling would likewise be greater than a 50 feet excavation, with the sides perpendicular, at 3*s*. per cubic yard.——At Ashton parish, near Birmingham, the cuttings are nealy 50 feet deep, through red marl, with red clay on the top, and in one part of the cutting there is a covering of yellow sand; I make the slopes of this 1 to 1. In the same parish, which is 4 feet gravel and soil, 9 feet yellow lime stone, and 15 feet blue shale, I intend making it perpendicular at the bottom, until we come to the loose stratum of alluvial earth.——Marl is a sort of loose clay, a red greasy soil.——Blue shale is the soil that iron stone lies in; it melts into clay when exposed to the air, but as long as the water is kept from it, it will stand perpendicular. (There is a deep cutting through which the Grand Junction Canal goes in this same stuff, it has a slope of 4 to 1; but I do not consider that a depth of 50 feet in blue shale would stand perpendicular in this situation, as there is no rock covering to keep the water off.)——Upon the Liver-

Comparison of the expence of passing the above in a tunnel and open cutting.

Red marl, red clay, &c. 1 to 1.

Descriptn. of marl.

Ditto of blue shale.

Slope upon the G. J. Canal.

pool and Manchester line, when we met with a soft stratum, the rock and the other part of the cutting was employed to build a wall, (*i. e.* merely a dry wall) to prevent frost, and keep the water from it.——Wherever we find a soft stratum of shale, we shall underset the rock by a wall, for which ample price has been allowed; as the whole of the cutting is estimated at 3*s.* per cubic yard, and the excavating the shale will not cost more than 1*s.* or 1*s.* 3*d.*——Upon the whole line we have four or five ridges of hills to pass over. If we had gone near Willesden, taking a different direction for about 10 miles out of London, as stated by the Counsel in opposition to the Bill, we should have added another summit to our railway, and caused a considerable length of tunneling at Willesden Hall, also increased the expense £30,000., for which we should only shorten the line 1¼ mile. Since the Bill was in the House of Commons, we have been obliged to alter the level of the line at that part next London, in consequence of a communication from the Metropolitan Road Commissioners; instead of having an inclination of about 1 in 320, we are obliged to make it about 1 in 150,* which increases the expense £10,000. proceeding until we get nearly on a level. On the road from Paddington to Edgware it is 1 in 304, which is not an objectionable slope.——We go behind Primrose Hill. My object in going round this high hill is to reduce the quantity of excavation in London clay, as it is considered objectionable. The waste of power in going round this hill is equal to surmounting 16 feet of perpendicular height; I consider 16 to 20 feet of elevation equal to a mile, (I am quite sure that 26, the number suggested by the Counsel, is not a right calculation) but it depends much upon the friction of the waggons employed, as they vary from 6 to 9 lbs. per ton, I consider that 8 lbs. is amply sufficient for the friction on a level; that is, you may overhang a pulley, and it will draw a ton. In a Report of Mr. James Walker to the Manchester and Liverpool Directors, on the difference of locomotive and stationary engine power, (he went to the North to examine circumstances connected with the same,) he took the friction at 12½, but he cannot produce a single instance of the same; I answered his Report, and I put it at 1 to 200 on a plain surface, which is the friction actually existing on the coal waggons in the North of England; although, by comparison, it is a bad railroad. I am sure there is not a waggon on the Liverpool and Manchester line, moving with a friction of 8 lbs. per ton, which would be 1 to 280. The reason I put it at 1 in 200 was this, that the waggons on railroads in the North of England have axletrees of

Where soft shale is met with it will be underset.

Excavating and building the above 3s. per cubic yard.

There are 4 or 5 summits to pass over.

Proposed deviation for 10 miles out of London.

Levels.

1 in 304 not an objectionable slope.

16 to 20 ft. elevation is equal to a mile.

Friction from 6 to 9 lbs. per ton on a level.

Explanation of the above.

Mr. James Walker made it 12½.

Particulars of same, &c.

Liverpool and Manchester line—the friction almost equal to 8 lbs. or 1 in 280.

* In other parts, these are stated to be 230 and 152.

considerably larger size than those on the Liverpool and Manchester; where the bearings are put on the outside of the wheels, and the size of the axletrees reduced to 1¾ inches; the diameter of the axletrees of coal waggons in the North of England is full 3 inches, which causes a greater degree of friction.—— ——The height where the railroad commences at Maiden Lane, is I think about 9 feet above the sea, and exactly 20 feet above the surface of the ground; the passengers will walk up the same, as they do at Manchester, (where the elevation is a little higher,) by means of stairs; the goods will be raised up by machinery. I estimated the expense of this at each end of the line at £8000.—— ——From London to Oxhey Lane the heavy cuttings are 2 to 1; there are some smaller ones, near Harrow, 1½ to 1, which slope extends to Watford Heath, where we enter the chalk part of the excavation at the mouth of Watford tunnel, it is taken at 1 to 1, and on the part nearest the tunnel ¾ to 1. Thence we go to the Tring summit, where they are all ¾ to 1; the slope at Box Moor in chalk is 1 to 1. In some of the cuttings we expect to find a great quantity of gravel, which is sloped 1½ to 1; in the deep cuttings we expect to find chalk, there we slope it 1 to 1; at Leighton Buzzard, at the South end of the tunnel, we slope it 1½ to 1; there is a small portion, 1 to 1, at the other end of the tunnel; where the rock extends perpendicular it is 1½ to 1, and continues so during the distance of the sand; the excavations thence to Bradwell are all 1½ to 1; the slopes of Blisworth are ¼ to 1, (the entrance of the tunnel of the Grand Junction Canal, at Blisworth, is of the same slope as I have got it. The clay would stand at a considerable inclination if the water was kept off, for which no provision seems to be taken; as it is running over in all directions, which causes the clay to slip, and as it slips they remove it, which increases the evil,) and continue so to Weedon; at Weedon they are ½ to 1; to Churchlawford there is a small piece 1 to 1; thence to Coventry it would be 1½ to 1; at Coventry it is rock ¼ to 1, and continues so to Berkswell; there it is 1½ and 1 to 1 up to Birkenhill, where it is 1½, the last part of the Blythe, and thence to Birmingham, has the same slope.—— ——I consider a communication can be made at any part of the line with Derby, as the ground is very favourable. I likewise surveyed part of the country with a view of extending the railway to Warwick; it opens into an excellent country. The line passes about 4½ miles from Northampton, from which there might be a branch, which should join us at Blisworth; from this point it is about 40 or 50 miles to Derby, from which it would be very easy to make a communication with Nottingham, as it is an excellent country, nearly a dead level.

Depôt at Manchester.

Do. upon proposed line.

Slope of cuttings.

Gravel 1½ to 1.

Chalk 1 to 1.

Slopes upon the Grand Junction Canal at Blisworth.

Communication with Derby.

Do. Warwick.

Do. Northampton.

Do. Nottingham.

Ex. Mr. FRANCIS FORSTER.

I have made borings for the proposed line, which are as follows:

At Oxbey Lane in the Parish of Watford—

Yellowish clay - . - -	3 feet.
Darkish clay - - - - -	18 feet.
Yellowish sandy loam - - -	5 feet.
Black clay, very hard - - -	7 feet.
	33 feet.

The borers were stopped at this depth; it is supposed by flints.

At Watford Heath a boring was made of 27½ feet in depth, and there were found—

Rubble - - - - - -	4 feet.
Clay - - - - - - -	8 feet.
Blue clay - - - - - -	8½ feet.
Chalk and marly chalk - -	7 feet.
	27½ feet.

Account of the Borings.

About half a mile toward Birmingham beyond this boring, the chalk may be observed in situations on the side of the lane from Watford to Watford Heath, and there is a chalk pit in the field No. 17, in Bushy Parish.

On the banks of the River Colne, near Otter's Poole, the strata were found by six borings to consist of gravel covered by brown clay, varying from 4 feet to 7½ feet in thickness.

In the line of the tunnel in Cashio Township, the strata consists of chalk and chalk flints, with an irregular covering of gravel. This may be seen in a gravel pit in field 8, in Leavesden Township, and in a chalk pit about twenty chains to the South. There are also several wells in Leavesden Lane which have been sunk through gravel and into the chalk, many feet below the level of the tunnel.

In a boring made in the road from Ashbridge to North Church there was found—
White chalk without flints - - 33 feet.

In two borings made in the line from Stokehammond to Winslow there were found—

In 1st boring,	blue clay - -	28 feet.
In 2d do. -	blue clay - - -	18 feet.

In a boring on the summit, near the crossing of the London Road, between Fenny Stratford and Stoney Stratford, there were found—

Brownish yellow clay, with fragments of chalk -	13 feet.
Grey, blue, and blackish shale - - -	17 feet.
	30 feet.

At Denbigh Hall Public House, in field No. 48, Township of Fenny Stratford, a well has been sunk and bored to the depth of 73 feet, through dark coloured shale, without finding water.

In a boring at the summit in field No. 9, Woolverton Parish, there were found—

Mixed brown earth - - - -	16 feet.
Dark blue clay - - - -	29 feet.
	45 feet.

By several borings at the crossings of the River Ouse and Tow, the strata were found to consist of yellow and blue clay, varying from 7 to 10 feet in thickness, and resting on a sub-stratum of gravel.

Several borings were made on the line of railway at the great cutting near Blisworth; at the South end of the cutting borings were made to the rock in four different places down the side of the hill, and the limestone rock was found at the depth of from 4 to 6 feet.

In a boring in the centre of the Blisworth cutting, in the lane between Nos. 50 and 55, in Roade Parish, there were found—

Stony yellow clay - - -	11 feet.
Blue shale - - - - -	14 feet.
Do. passing into rock - -	7 feet.
	32 feet.

Account of the Borings.

In a quarry crossed by the line in No. 2, Courteenhall Parish, near the summit of Blisworth ridge, the limestone rock is marked; and the section shows—

	Feet.	In.
Loose shivery limestone - - - -	2	9
Yellow marl - - - - - - -	1	6
Shivery limestone - - - - -	1	6
Limestone - - - - - - - -	8	0
Yellow marl, stated by the quarry man to be	7	0

And to be underlaid by a bed of colitic freestone, of unknown thickness.

At the North end of the tunnel there were found by boring—

Yellow clay - - - - - - -	8 feet.
Blue clay - - - - - -	25 feet.

As the basil of the limestone rock makes its appearance in the lane from Roade to Blisworth, very near to this boring, the above-mentioned clay is considered to be a diluvial mass, reposing against the end of the hill.

In a boring at Stowe Hill, made on the 24th April, on the side of the road from London to Weedon (commencing 33 feet below the level of the crossing of the railway line), there were found—

Yellow clay - - - - - -	14 feet.
Blue shale - - - - - -	8 feet.
	22 feet.

In this road, near the crossing of the railway line, there is a cutting through the hill in blue and blackish shale, the height of which is 25 feet, and the slope one and one third to one.

In a well by the side of the canal at Stowe Hill gate (a well has been sunk for water, but without success), there were found—

Yellow clay - - - - - -	14 feet.
Blue shale - - - - - -	32 feet.
	46 feet.

In a boring in a lane at Dodford, made 23d April, near the proposed tunnel under the road from Weedon to Daventry, commencing at 30 feet below the crossing of the railway line, there was found—

	Feet.	In.
Brownish blue clay, with frebbles -	4	6
Do. very strong -	7	6
Strong blue clay - - - - -	9	6
Soft yellow sandstone - - -	1	6
Sandstone - - - - - -	6	0
	29	0

At the proposed tunnel near Watling Street, the strata have been proved by a well at Norton Lodge, 35 feet deep, in which was found—

Soil and clay - - - - -	7 feet.
Gravelly clay - - - - -	14 feet.
Blue clay - - - - - -	14 feet.

The only water in this well comes from the top.

In a well at a farm house on the side of Watling Street, a well has been sunk to the depth of 45 feet through—

Gravel - - - - - - -	25 feet.
Yellow clay - - - - - -	5 feet.
Blue clay - - - - - -	3 feet.
Sand - - - - - - -	12 feet.
	45 feet.

Account of the Borings.

In confirmation of the strata sunk through in this well, there is a gravel pit in the field No. 15, Norton Parish, and another large one near the bottom of the same field.

Bored at Kilsby, first in the road from Ashby to Watford, at the South-east end of the proposed tunnel, and found—

	Feet	In.
Soil - - - - - - - -	2	6
Yellow clay - - - - - - -	5	6
Brown clay - - - - - - -	2	6
Stone - - - - - - - -	0	6
Blue shale - - - - - - -	17	0
Blue marl - - - - - - -	19	0
	47	0

Or at 4 feet below the level of the tunnel.

At the North-west end of the tunnel, near the village of Kilsby, was found—

	Feet.	In.
Soil - - - - - - - -	5	6
Yellow and blue clay - - - -	9	0
Blue shale - - - - - - -	6	6
Brown stone - - - - - - -	1	8
Blue marl, with thin beds of stone -	22	10
	45	6

Or about 3 feet below the bottom of the proposed tunnel.

In the same parish a boring was made at the Western end of the tunnel (No. 11 on plan), 66 feet deep; which proved—

18th, 19th June	Yellow clay - - - - -	18 feet.
	Blue clay - - - - - -	20 feet.
	Dark blue clay - - - - -	28 feet.
		66 feet.

In the Parish of Watford, County of Herts, a boring was made at Oxhey Lane (No. 5 on the plan), 48 feet deep; which proved—

20th, 21st, and 22d June	Yellow and brown clay - -	20 feet.
	Sand, with a little water in it -	5 feet.
	Yellow and blue clay - -	2 feet.
	Red, yellow, and variegated clay -	9 feet.
	Sand (dry) - - - - -	1 foot.
	Variegated clay, very hard, and marly chalk - - -	11 feet.
		48 feet.

Account of the Borings.

At Watford Heath, Parish of Watford (No. 39 on plan), a boring was made; proving—

22d, 23d, 25th, and 26th June	Yellow and brown clay - -	12 feet.
	White sand, with water in it -	1 foot.
	Yellowish and blueish clay -	17 feet.
	Yellow and green sand, very hard	6 feet.
		36 feet.

In the lane from Watford to Watford Heath, about half a mile from this last hole, the chalk appears at about 5 feet from the surface, and (in No. 49) near the same place a chalk pit is now working.

There is also a chalk pit in (No. 17 on plan) Bushey Parish, where the chalk is only a few feet from the surface.

The Tring cutting was proved thus: In the lane, No. 1 on the plan, Tring Parish, the chalk was found at 18 inches from the surface; and in the corner of No. 7 there is a chalk pit of some depth, where the chalk is only 1 foot from the surface; and in No. 6 on the opposite side of the line, the chalk also shows itself near the surface.

At the lane between the Counties of Herts and Buckingham a boring was made, proving—

27th, 28th, 29th, and 30th June.	Soil - - - - - -	2 feet.
	Chalk - - - - - -	43 feet.
	Grey chalk - - - - -	5 feet.
		50 feet.

Near the line of railway in Pitstone Parish there are a few gravel pits, in one of which, at 6 feet from the surface, a boring was made, which proved chalk at 6 inches, and thus continued for 10 feet.

Ex. Mr. THOMAS GOOCH.

I have made borings on the proposed line, which are as follows:

Boring in lane, No. 53, Roade Parish—

	Feet.	In.
Soil and clay - - - - -	2	0
Stony yellow clay, with loose stones	9	0
Blue clay - - - - - -	14	0
Yellow limestone - - -	16	0
Soft yellow limestone, with a little water	1	6
Hard yellow limestone - -	0	9
Blue shale - - - - - -	0	9
Strong blue limestone, very hard -	2	0
Stronger shale - - - -	1	6
Very hard blue limestone - -	4	3
Stony blue shale - - - -	4	3
	58	0

Boring in lane, No. 39, Roade Parish—

	Feet.	In.
Sand and loam - - - - -	5	0
Clay - - - - - - -	2	0
Yellow limestone - - - -	14	0
Blue limestone, very hard, mixed with thin beds of shale - -	13	0
Blue shale, similar to that in the above hole - - - - - -	5	0
	39	0

Account of the Borings.

A small quantity of water oozed into the hole.

Boring in road from Blisworth to Roade, No. 1, Courtenhall, on the plan—

	Feet.	In.
Sand and loose Stones - - -	3	0
Yellow Limestone - - - -	13	0
Soft limestone - - - -	0	6
Yellow limestone - - - -	2	0
Strong brown clay - - -	4	0
Yellow limestone, with a three-inch parting (a little water) - -	4	0
Blue shale - - - - - -	14	6
Very hard blue limestone, with a soft stratum of two inches - -	6	9
Very dark blue shale - - -	2	9
Brown sandy loam, with water at the bottom - - - - -	4	0
Rather soft green coloured stone, containing hard strata - -	3	6
	58	0

Account of the Borings.

Boring on road from Ashton to Roade, No. 10, Ashton Parish, on plan—

	Feet.	In.
Gravel and soil - - - - -	4	0
Yellow limestone - - - - -	9	6
Blue shale - - - - - -	13	6
	27	0

No water.

Boring made in quarry in No. 30, Ashton Parish—
N.B. The line passes over this quarry.

		Feet.	In.
Quarry	Soil and loose stones - - - -	3	0
	Yellow limestone - - - - -	11	0
	Yellow limestone - - - -	6	6
	Blue shale, getting very hard towards the bottom - - - - -	10	6
		31	0

Ex. Mr. JOHN RASTRICK, C.E.

I have had some experience in the construction of Railroads; I was employed to obtain information for the Liverpool and Manchester.——— ——— I have been twice over the country of the proposed line to Birmingham, and as far as my observation have gone, I think it the best line for a railroad that can be got, that is, the best line within a limited distance. I was engaged about a fortnight upon it; (I certainly did not make observations with a view to look for another line.) I have also examined Mr. Stephenson's Estimate, and I consider the works may be executed for less than he has stated.———I went by the Plan and Section furnished me, and from the Book of Reference I found to whom the several fields belonged.———Regarding the Occupation Bridges, when a field was severed belonging to one proprietor I put a bridge, where there are several fields lying together one or two bridges would suffice for all of them. Upon this principle I made one hundred and ninety-eight Occupation Bridges, but I have taken them at two hundred; which number is independent of all other roads at present existing. I also made the necessary calculation of the number of bridges for the Turnpike roads. The line passes over the Holyhead Road, and we allow 20 feet head-way; in no instance do we pass under it by a bridge, but we pass three times under in a tunnel.———In all public roads we pass, I calculate, an inclination of 1 in 15 to 1 in 20 and 1 in 25, according

Opinion upon the proposed line.

And estimate.

Data in forming occupation bridges.

200 in the proposed line.

Inclination of roads 1 in 15 to 1 in 25.

to the nature of the roads.—— ——Regarding the slopes, generally speaking, I agree with Mr. Stephenson; perhaps there are instances in which it may be prudent to increase them. I made a calculation of them at 3 to 1, instead of what they are stated at by Mr. Stephenson, which made an addition of £17,303.; but I should not be inclined to make them as much as 3 to 1, (I have not had experience in London clay). Where the cuttings are shallow 2 to 1, or less, is quite sufficient, and there are few cuttings of great importance along the line. Mr. Stephenson makes his deep cuttings 2 to 1, and his shallow cuttings $1\frac{1}{2}$ to 1. One of the principal cuttings is at Primrose Hill, there is another, rather deep, where he crosses the London Road to Harrow, perhaps it may be necessary to increase these slopes to $2\frac{1}{2}$ to 1, the expense of which would be under £5000.——— In works of this nature slips will occur, even in the best material.——— Although I made allowance for greater slopes, my estimate was considerably under Mr. Stephenson's, independent of contingencies it amounts to £1,875,527.; I have added the sum of £374,473. for contingencies, which brings it to £2,500,000. (Mr. Stephenson's estimate); this does not include the purchase of land, which amounts to £250,000.———In the course of my experience, I have had pass through my hands, the estimates of all the Engineers of the day that have come before Parliament, and I think Mr. Stephenson's is 30 per cent. higher, than any I remember. In general, when I have been engaged to support a Bill in Parliament, on looking over the estimates I have been obliged to consider what part of the work could be left out, to bring it within the amount stated.———The soil about Coventry is principally rock, in some parts of the fields it is to be seen upon the surface, also by the roadside, and places may be observed where quarries have been dug; there is a quarry a little deeper than this room (speaking of the Committee room) within 150 yards of the line, which is at Beachwood; the quarry has been worked, and is good hard stone; the stone for the Locks on the Warwick Canal was got from it. ———Supposing the cutting at Meriton Ridge is marl, with a strata of rock, the intermediate parts may be blocked up by the rock which is got out of the cutting.——Marl is of such a nature that, when mixed with sand, it will slope at 1 to 1.—— ——I have made some experiments to determine the friction of carriages upon the Liverpool and Manchester Railway, and the results were various. I reduced the friction to 6 lb. per ton upon some of the carriages, but the average was under 8 lbs., which would average nearly 20 feet in a mile.———I spent several days upon it; in fact, at different times, I have been two or three months upon this

Mem. regarding the Slopes.

Risk of Slips.

Estimate.

Parliamentary Estimates generally made too low.

Good Stone found on the line.

Meriton Ridge.

Marl and Sand will stand at 1 to 1.

Friction upon a Railway nearly 8 lbs. per ton, which is nearly 20 feet in a mile.

Description of Old Carriages L. & M.

Friction on these 12½ lbs.

Do. present Carriages.

Friction as low as 5 lbs. upon one of them.

railway, trying different carriages which I had constructed for that purpose, &c. &c.——The carriages originally used were constructed upon the old fashioned plan, the axletrees of which were 3 inches in diameter, and the bearings on the inside; they ran generally upon cast iron; I took the friction of them at about $12\frac{1}{2}$. The carriages that are now used have their bearings on the outside, and are reduced to $1\frac{3}{4}$ and $1\frac{1}{2}$ diameter; these are the carriages with which I tried the experiments. There was one carriage the friction of which came to about 5 lbs. In the first instance the whole weight was placed on the axle, the wheels being upon the outside. The friction has been in proportion to the reduction of the diameter of the axle.

Ex. HENRY ROBINSON PALMER, ESQ. C.E.

Engineer to the London Docks.

Estimate of Proposed Line

Instance of Clay standing at ½ to 1.

Slopes of Line.

The advantage of keeping Chalk perpendicular.

Instance of Iron Sandstone standing perpendicular.

I have had considerable expreience in surveying and estimating canals, railroads, and other public works, but I have not executed or had the management of a railway.——I have examined the proposed line to Birmingham, and I estimate the cost of the works, exclusive of contingencies, at £1,893,788.; this amount includes all the Bridges, &c. I have added £356,000. for Contingencies, to bring it to Mr. Stephenson's Estimate, which is £2,500,000.; this sum does not include the Land. I consider the prices I have allowed quite sufficient.——I had in an excavation at the London Docks very similar Clay to that of Primrose Hill, and there was a slope which stood at $\frac{1}{2}$ horizontal to 1 perpendicular, and remained nearly two years without any artificial assistance. I consider that a slope of 2 to 1 will be sufficient for the cutting on the proposed line, provided the Soils are similar to the borings.——I have known Chalk stand quite perpendicular, and even overhanging; it would be advantageous to make it perpendicular, as it would be less exposed to the action of the weather. Upon the Holyhead Road, near Dunstable, there is some Chalk with a slope of 1 to 1; but I presume it has been cut for the purpose of obtaining material, and not because it would not stand at a steeper inclination.——I observed near Leighton Buzzard 30 to 35 feet of Iron Sandstone standing perpendicular, and I should prefer this stratum perpendicular rather than sloping. I think that a Soil which will stand alone is capable of bearing all the shaking that the motion of the Carriages is likely to produce.

Ex. Mr. JOSEPH LOCKE, C.E.

I have had experience in the formation of railroads for the last ten or twelve years.———I was the Resident Engineer upon the Liverpool and Manchester Railroad. (Chat Moss upon this railway is 4 miles long, and before the latter was made, was a mere uncultivated common).—— ——My estimate of the Tunnels, fronts, and shafts upon the proposed line to Birmingham amounts to £249,979. I allowed 18 inches thickness brickwork, which I consider sufficient, whether the boring be hard or soft. ———The tunnel at Liverpool is in two bricks, except such part where there are houses over it, and the material is bad; there it is at 2½. Some part of the 18 inch work goes through wet Shale. It is 2250 yards long. The distance in Shale is perhaps 300 or 400 yards, and about 150 to 200 yards in Clay; the rest is chiefly in Sandstone Rock. There was some quick sand mixed with the clay, which we stopped up with timber and straw; and when we left the tunnel over Sunday the sand and water would run through, and leave a cavity above the roadway, but the brickwork did not give way. A trifling settlement also took place, which affected one or two of the houses over it; but although the surface of the tunnel (*i. e.* the earth work) fell in, in consequence of the weight of earth and houses upon it, yet the brick work stood firm.———There is a centre drain made in it. (As a principle it would be better to prevent the water getting into a tunnel by making it water tight.)—— ——On the same railway both larch and oak Sleepers are used, the price of which varies from 2*s.* to 3*s.*; oak is generally the most expensive. Mr. Stephenson's estimate for wooden sleepers is 5*s.* Larch sleepers are used upon the Warrington line, (Mr. Robert Stephenson was the Engineer of the Warrington line, both under the Act of Parliament, and for the execution of the same) and they are better than Stone Blocks (which I once tried) for embankments.—— ——I have gone through Mr. Stephenson's estimate, and I think many of his prices too high. Upon the Liverpool and Manchester the Occupation Bridges are in the proportion of rather more than one in a mile; there are eight under embankments, which average £238. each: Mr. Stephenson allows £300. for such.—— ——I consider a rise of about 18 to 20 feet in practice would be about equal to a mile of distance, although some carriages would warrant the supposition that 16 feet would

Chat Moss 4 miles long.
Estimate of the tunnels on the pro. line.
18 inches thickness for tunnel sufficient.
Tunnels L. and M. 2 bricks thick.
Where there are houses over it 2½.
It is 2250 yds. long.
300 or 400 yards in Shale.
150 or 200 in Clay.
Rest in Sandstone.
Difficulties encountered in same.
Methods used in surmounting them.
Drainage of a Tunnel.
Larch and oak Sleepers.
Price of the same on the L. and M. & Birmingham, &c.
Warrington line.
Sleepers better than Blocks for Embankments.
Mr. Stephenson's estimate too high.
Occ. Bridges L. and M. rather more than 1 to the mile, and averaging £238. each.
Mr. Stephenson allows £300. for the same.

A rise of 18 to 20 Feet equal to a Mile of distance.

be equal to a mile: my opinion is founded upon experiments I have made. —— ——The length of the Cuttings and Embankments upon this Proposed Line, are of nearly the same proportions as those upon the Liverpool and Manchester, and I believe that from end to end of the latter there is not

All the L. and M. is either in Embankment or Cutting.
Description of and amount of same.
Slopes L. and M.
Method employed to prevent the escape of water from same.
Slopes upon do.
Slip of the Sankey Viaduct.

a mile that runs on the surface of the ground; it is all either in Embankment or Cutting, which are nearly of equal length, the Cutting upon it amounts to about 3,000,000 cubic Yards.———I believe the Slopes have stood very well, there were some cases in which the water came out, which we remedied thus: a few stakes were driven in, and binders put upon them, the expense of which was very trifling; it was through clay and sand mixed together. Where it is Clay without Sand it stands very well at $1\frac{1}{2}$ to 1. There was a Slip upon the Sankey Viaduct, owing to the wetness of the clay; the Slope was made $1\frac{3}{4}$ to 1 in consequence of it, but it is $1\frac{1}{2}$ to 1 at the other parts, the colour of this Clay varies from

The width of Embankments been increased.
Mounds to Parapets and Embankments.
Description and execution of do.
Breaks to Carriages L. and M.
Description and application of the above.

blue to yellow. Since the Railway has been opened the Embankments have been increased to 25 feet width.—— ——I do not consider Mounds placed against the Parapets of Bridges of any use; they will cost perhaps 6*d*. per lineal Yard. The expence of having these Mounds on each side of the Embankment would be about £88. per mile; they are more fancied security than any thing else.—— ——The Carriages on the Liverpool and Manchester are supplied with a Break, which presses on the wheel; it projects before the end of the carriage, thereby preventing the others coming in contact with it; if any thing prevents the first carriage going forward, the next consequently comes against it, and the impetus of the concussion acts upon the break, the third acts in a similar manner upon the second, and so on, which decreases their momentum without producing any sudden

Steepest Plane on the L. & M. 1 in 96.
Method of working it.
Steepest on the Birmingham 1 in 330.
L. and M. crossed roads at a level.
Precaution taken with same.

shock.—— ——On the Plane of 1 in 96 of the same Railway, (which is the steepest on the line) we have what we call a Stationary Locomotive to push the Train up, and delays may sometimes occur upon it.———The Steepest Inclination on the Birmingham is 1 in 330.———On the Liverpool and Manchester some Parish Roads are crossed upon a Level; we have Gates on each side of the road, and an Attendant, sometimes there are two Attendants. I have not heard of any accident arising from them since the opening of the railroad.

Ex. Mr. JAMES COPELAND, *Contractor.*

I executed 12 miles of the Liverpool and Manchester, and I have recently completed a contract upon the Leicester and Swanington Railroad, where there is a tunnel which passed through 500 yards of loose and dry running Sand; to overcome which we were obliged to make a wooden tunnel, before we could turn the brick work, which increased the expense; the remainder of it was in Clay.——I am willing to enter into a contract for the Tunnels on the Birmingham line for the same price that was paid me for the former, or £31. 10*s*. per Yard for tunnels where there are ventilators, and £26. without ventilators.——I have made allowance for any difficulties that might arise. If the tunnels should be in hard dry clay the expense will be less.——The tunnel at Leicester is smaller than the above are intended to be, but men can work to greater advantage in a large tunnel.

Tunl. at Leicester.

Part of it is in dry running Sand. Method used to overcome same.

Est. for Tuns. on the Birming. line.

£31. 10*s*. with Ventilators. £26. without Ventilators. If hard dry Clay they would be less.

Ex. Mr. FRANCIS WEDGE, *Land Surveyor.*

I have surveyed the proposed line from Woolverton to Birmingham, which is 60 miles, and allowing for two lines, I make 710¾ acres; this Land I estimate at £88,436; the Houses upon it I estimate at £11,440. more (I have not allowed any thing for goodwill); this includes the station at Nova Scotia Gardens, which covers 4¼ acres (the separate valuation of this is £725.). The average number of Years' purchase I have taken is thirty-five; the usual number obtained for property in that district is about twenty-seven or twenty-eight. The proportion of building land is about 5 acres, which is in the vicinity of Birmingham, which I have calculated at twenty-five years' purchase, (*i. e.* I have considered the value of the land necessary to rent, and have allowed twenty-five years purchase in that value, at so much per running yard). The sum above stated does not include timber, or compensation for injury that may be done, but this is kept separate.

Land required from Wool. to Bir. 710¾ acres. Valued at £88,436. And the houses at £11,440.

At 35 years' pur.

Particulars of the above.

Ex. Mr. WILLIAM WELLS GARDENER,

Land Surveyor at Biggleswade.

Land from Tring to Wool. 228 acres.
Valued at £16,165.
Ave. £7. per Acre.

I valued the Land from (where Mr. Wedge left off) Tring to Woolverton, a distance of 22 miles. 228 acres are required for the two lines, and I valued the same at £16,165., which averages upwards of £7. per acre. I have put it at £10. per cent. beyond the fair occupation value, which I have calculated at thirty years' purchase, adding three years' purchase for Contingencies, making thirty-three years' Purchase. (The quantities were given me by the Engineer.) It is all agricultural land, and there is but one cottage on it.

Ex. Mr. LAYTON COOK,

Chairman of Committee of the Agricultural Society of Roads.

Land from Kilburn to Tring 345 acres.
Valued at £62,794.
Particulars of same.

I have valued the Land on the Proposed line, from Kilburn to Tring, at £62,794., which comprises 345 acres. About thirty buildings are included in this Valuation, four or five of which have business carried on in them. I have included the goodwill of these houses.

Ex. Mr. P. HARDWICK.

Land from Kilburn to London 43 acres.
Land valued at £30,606.
Buildings 35,235.
£65,841.
Particulars of same.

I have valued the Land and Buildings on the proposed line from Kilburn to London, which comprises 43 acres, at 65,841, the Land £30,606. the Buildings (including goodwill) £35,235., which is I consider a full and liberal price. I have put it at a greater rate of Purchase, than ever it was taken at before, as it is generally found to exceed an estimate. Goodwill in London, when brought before a jury, is rarely found correct, as it is a very difficult subject to estimate.

Ex. Mr. HENRY BOOTH.

I am Treasurer of the Liverpool and Manchester Railway, the Act of which was obtained in 1826, and it opened in 1830.——From the middle of September, 1830, to the middle of last month, which includes twenty-one months, 669,211 Passengers have been booked at our office, and we take up about one-sixth of that number more on the road, making 780,000; so that the average per month is 37,000, or about 1200 per day.——The first class of Carriages generally takes less than $1\frac{1}{2}$ hours, and the second 2 hours for the journey, which is 30 miles. The fare by the first class 5*s*., the second class 3*s*. 6*d*. We take the mail three times a week each way, and our charge is 1*d*. per mile, which is two-thirds of the price it used to be; the fare by this coach is extra. Only one fatal accident has occurred during that period; a man in the second class insisted on jumping out, he did so, was lamed, and died. There has not been a single instance of the railway being stopped by winter, even in very severe weather.——There are twenty-two regular daily Coaches, and six extra during the summer months; the greatest number they could hold is seven hundred, they average four hundred and fifty, which is somewhat more than one-third the number conveyed by the railway. The average fare by Coaches is 10*s*. inside, and 6*s*. out, they occupy from 4 hours to $4\frac{1}{2}$ hours. I attribute the increase of Passengers to the cheapness, expedition, and ease of conveyance by the railway.——The first 6 months last year we carried 42,000 tons of goods, the next half-year 65,000, and for $5\frac{1}{2}$ months, up to the 14th of this month, 65,000, which gives a very great increase.——Cotton is the principal article conveyed. Coffee, Rum, Corn and Flour, are also conveyed in very considerable quantities, and almost every description of merchandise; the conveyance of goods is on the increase: we let them follow the train of coaches, that they may not be in the way; they are 2 hours on the road.——The charge by Canal was 15*s*. per ton, it is now 10*s*., which is the price by the Railway; Sugar, which was 10*s*., is now 9s. These remarks apply to the generality of Canals; the reduction in other goods is not so great.——The Passage of goods by Canal occupies but 12 hours now, considerable improvements having been made upon it since the railway was opened; but their passage is occasionally interrupted, owing to the Tides, which they are obliged to suit their departure to.

Description of L. and M.
Opened in 1830.
No. of passengers.
Average 1200 per day.
Time of transit.
The mail.
Accidents.
Fares.
Railway not impeded by frost.
No. of Coaches.
Goods conveyed by the Railway.
Amount of, &c.
Charge for goods by Canal and Railway.
Time of transit by Canal.

Value of land increased by the Railway.

——The value of Land along the line has increased since the Railway opened.——We take large quantities of Bacon, Corn, Eggs, Flour, and fresh Butter (from Ireland), from Liverpool to the Manchester market; the same provisions were formerly transmitted by Canal, but I do not think there were many eggs before the railway was opened.——We took a Regiment of eight hundred Soldiers, and a large quantity of baggage, which were necessary to be shipped to Ireland; from the time they mounted the Carriages until they embarked in the ships at Liverpool little more than 3 hours elapsed, of which 2 hours was occupied on the journey.——A man has commenced a dairy farm by way of trial, and we convey his milk 15 miles.——Seven or eight hundred Persons are constantly employed on this Railway, and they had no previous experience on the subject. The gross Receipts for the 12 months, ending the 21st of December, was £155,502., the Disbursements £84,504., leaving a Profit of £70,097.; the last half-yearly dividend is 4½ per cent. a share, which gives 9 per cent. per ann: the £100. shares are now selling for £200., although the novelty of the undertaking increases the expenses of the Company.——Both our Receipts and Expenses have very much exceeded the anticipations formed when we obtained the Bill, not that the Profit has been greater, but it has been on a larger scale. We had much to learn at that time, and from the results we have obtained experience.——The amount of Poors' Rates distributed by the Company to the different Parishes through which the line passes is about £4000. annually.——The Estimate upon which the Bill was obtained amounted to £500,000., and upon such portions which this Estimate covered they expended £800,000.; this amount does not include the Warehouses, and Stations for the Engines and carrying part, which amounted to £200,000. more. No doubt we incurred Expenses by Experiments, on account of the novelty of the undertaking, Locomotive engines, &c. We had great engineering difficulties on the line, particularly Chat Moss, but it was a matter of money more than any thing else.——We have got an Act to make a more complete entrance into the town of Liverpool; the Estimate amounts to £130,000., the money for which may be borrowed, or more shares can be made. We are not authorised to levy more Tolls, all we expect is an extension of business. ——There was a Bridge over the Irwell that was not included in the first estimate; therefore, added to the expenditure, it cost more than was expected, namely, £20,000.: it was made in consequence of the alteration in the Terminus. We at present pay no Duty; coaches, of course, pay duty in various ways.

Conveyance of Provisions by railway.

Do. Soldiers.

Do. Milk.

Men employed.

Receipts and Disbursements.

Pays 9 per cent.

Receipts and Expenses greater than was expected.

Amount of Poors' Rates paid by railway.

Parliamentary estimate £500,000.

And cost £800,000.

And Warehouses, Stations & Engines £200,000. more.

Difficulty of Chat Moss, &c.

Estimate of the improved Entrance into Liverpool £130,000.

No Duty at present upon Railways.

(The Witness delivered in the last Report on the Liverpool and Manchester Railway, which was as follows :)

LIVERPOOL AND MANCHESTER RAILWAY.

Sixth Annual Meeting.

REPORT.

Liverpool, 28th March, 1832.

Pursuant to the Provisions of the Act, the Directors have prepared a detailed Statement of the Accounts of the Concern for the Half Year ending 31st December last, the general Results of which are as follows :— Half-yearly Report 1832—L. and M.

	Tons.	
The Tonnage of Merchandise conveyed between Liverpool and Manchester, from 1st July to the 31st December 1831, amounts to	52,224	Tonnage.
Road Traffic - - - - - - -	2,347	
	54,571	
Between Liverpool and the Bolton Junction - - -	10,917	
Coal from Huyton, Elton Head, and Haydock Collieries, brought by the Company's Engines - - - - -	7,198	
Coal from Hulton, brought by the Bolton Engines - -	1,198	Coal.
	8,396	
Number of Passengers booked at the Company's Office -	256,321	Passengers.

Number of Trips of 30 Miles.

With Passengers - - - - -	2,944	
Merchandize - - - - -	2,298	
Coal (300), or, equalling 30 Miles -	150	Number of Trips.
Total - -	5,392	

RECEIPTS.

Receipts.					£.	s.	d.
	Coal Department				58,348	10	0
	General Merchandize	£30,764	17	8			
	Coal Department	695	14	4			
					31,460	12	0
					£89,809	2	0

EXPENCES.

		£.	s.	d.			
Expenses.	Office Establishment	902	3	10			
	Coal Disbursements	60	15	5			
	Petty ditto	110	0	5			
	Cart ditto	60	17	8			
	Maintenance of Way	6,599	12	6			
	Charge of Direction	297	19	0			
	Coach Office Establishment	589	5	9			
	Locomotive Power	12,203	5	6			
	Advertising	59	3	4			
	Interest	2,737	7	3			
	Rent	900	5	3			
	Compensation (Coaching Department)	156	7	5			
	Engineering Department	625	0	0			
	Carrying Disbursements	10,450	12	3			
	Taxes and Rates	2,763	5	1			
	Stationary Engine Disbursements	269	4	7			
	Coach Disbursements	6,709	7	11			
	Waggon ditto	979	19	8			
	Compensation (Carrying Department)	786	8	2			
	Police Establishment	1,490	14	1			
	Law Disbursements	98	9	10			
	Bad Debts	175	13	6			
					49,025	18	5
	Net Profit from 1st July to 3st December 1831				£40,783	3	7

The foregoing Particulars, when classed under the different Heads of Disbursement, will exhibit per Ton and per Passenger as shewn in the following Table.

1st July to 31st December, 1831.

DISBURSEMENTS apportioned under the different HEADS of EXPENDITURE.

—	Per Passenger booked.	Per Ton of Merchandize, Liverpool and Manchester.	Per Ton of Coals.	Per Ton on Bolton Tonnage	Coaching Department.	Merchandize Department.	Coal Department.	Bolton Tonnage.	Total.
	s. d.	s. d.	s. d.	s. d.	£. s. d.	£. s. d.	£. s. d.	£. s. d.	£. s. d.
Disbursements in the Merchandize Department, consisting of Porterage, Salaries, Carting, Stationary Engine Disbursements, &c. &c.	- -	4 6½	- -	0 4	- - -	12,442 8 2	- - -	183 4 0	12,625 12 2
Disbursements in the Coaching Department, comprising Porterage, Salaries, Repairs, and including 3½d. per Passenger for Omnibusses	0 7	-	- -	- -	7,455 1 1	- -	- - -	- - -	7,455 1 1
Porterage, &c. in the Coal Department	- -	- -	0 2¾	- -	- - -	- - -	97 1 5	- - -	97 1 5
Locomotive Power Account, proportioned according to the Number of Trips of 30 Miles in each Department, comprising Repairs of Engines, Wages, Coke, &c.	0 6¼	1 11	0 11¼	- -	6,664 13 6	5,119 4 6	339 7 6	- - -	12,203 5 6
Sundry Disbursements, proportioned according to the Receipts as between the Coaching and Merchandize Departments, and according to the Number of Tons and Miles conveyed as between the Liverpool and Manchester and Bolton Trade, comprising Maintenance of Way, Police Establishment, General Office Establishment, &c. &c.	0 6¼	1 1½	0 1	0 9	6,722 5 0	3,072 8 2	39 12 9	409 14 9	10,244 0 8
Rates and Taxes, Interest on Loans and Chief Rents, proportioned according to the Amount of Profit in each Department, calculated exclusively of these Items of Disbursement	0 4¾	0 4⅝	0 4¾	0 3½	5,088 1 6	1,127 4 0	29 14 7	155 17 6	6,400 17 7
Total Disbursements	2 0¼	8 0	1 3¾	1 4½	25,930 1 1	21,841 4 10	505 16 3	748 16 3	49,025 18 5
Net Profit	2 6¼	2 7½	0 4¼	1 9¾	32,418 8 11	7,181 11 10	189 18 1	993 4 9	40,783 3 7
Gross Receipts	4 6½	10 7½	1 8	3 2½	58,348 10 0	29,022 16 8	695 14 4	1,742 1 0	89,809 2 0

Scale of Disbursements.

In conformity with the resolution of the last general meeting, the Directors have caused a series of returns to be made of the number of persons employed on salaries and on days wages, constituting so material a part of the current expenditure. From the returns of six fortnights it appears the average number of men in the Company's employ receiving days wages is 633, and of agents and clerks on salaries, 73. The Directors have also prepared a statement in explanation of the system of management pursued in the general business of the concern, and which they propose to read to the present meeting. It has been their anxious endeavour to introduce a safe and judicious economy into every branch of the expenditure.

No. of Men 633.

Do. Agents and Clerks 73.

The Directors are happy in being able to state, that the business of the Company is gradually increasing. In the half year ending 31st December last the average quantity of merchandize conveyed per railway for a computed period of six weeks, was 15,112 tons; in the six weeks just past, ending 23d of March, the quantity conveyed was 16,632 tons.

Business.

The receipts in the coaching department in the twelve weeks ending on Friday last, the 23d March, was 15,496*l.* 11*s.* 3*d.* In the corresponding twelve weeks last year, 13,965*l.* 6*s.* 11*d.*

Increasing.

Since the general meeting on the 5th January, a trade in coal by the railway to Manchester has been commenced, which promises to become of considerable importance to the concern. Since the 23d January last, 3,615 tons of coal have been brought per railway to Manchester from the Haydock and Hulton collieries.

Coals.

The Directors will embrace this opportunity of submitting a few observations on a change in the arrangement of the coaching department, which is about to be adopted. In order to avoid the various objections and anoyances attendant on the employment of omnibusses on the plan hitherto pursued, by which a heavy expence has been entailed on the Company, with very inadequate advantages to the public, the Directors have considered it expedient to discontinue providing omnibusses on the Company's account; and to obviate the inconvenience which might otherwise arise, the Directors have made arrangements for the establishment of independent omnibusses to run from the railway station to different parts of the town, by which persons wishing to avail themselves of the accommodation will have the opportunity of doing so at a reasonable rate. Passengers and parcels will continue at the Company's coach office as usual, while the confusion and inconvenience of a number of crowded omnibusses arriving at or starting from the Dale Street booking office will be avoided.

Co.'s Omnibusses.

To be discontinued.

The Directors, having observed that during last summer great inconvenience resulted from the large and crowded trains consequent on a too limited number of departures, have determined that in the ensuing season the departures, both from Liverpool and Manchester, shall be more frequent, by which the trains will be kept of moderate size, the journey be more punctually performed, and the regularity and good order of the undertaking be more effectually preserved.

The Trains to start oftener.

The Bill for the construction of the new tunnel the Directors are happy to state has passed the Commons House of Parliament, and they have reason to believe the Act will be obtained without opposition. By means of this tunnel the proprietors are aware it is proposed that passengers shall be conveyed to the very centre of the town.

The new Act

(Tunnel)

By the additional number of locomotive engines and carriages that will be required for the increased nnmber of departures, and especially by the outlay of capital for the construction of the new tunnel, and the unavoidable cost of warming, lighting, and working the same, the Company will incur an increased annual expenditure, which will be very inadequately compensated by the saving of the charge for omnibusses. The Directors however rely confidently for remuneration on that increase of business which additional facilities and improvd arrangements never fail to create.

will increase the Expenses.

Remuneration anticipated,

Considering it important that proprietors should know before-hand at what Periods of the year the dividends will take place, the Directors beg leave to recommend that the Company's accounts be made up twice a year, namely, to the 30th June and 31st December; that in future a half-yearly meeting be held between the 21st and 31st of July and of January in each year, for the purpose of declaring a dividend for the previous half-year, to be payable in cash between the First and Tenth of the following month, namely, of August and February in each year. Time of Dividends.

It may not be improper at this time to state shortly the result of the working of the concern from the commencement to the 31st December last, as detailed in the several statements of accounts which have in their due order been laid before the proprietors.

	£.	s.	d.	
The profits of the Company from the opening of the Railway on the 16th September to the 31st December 1830 were -	14,432	19	5	Profits of the Company from the commencement.
Ditto - for the half year ending 30th June 1831	30,314	9	10	
Ditto - for the half year ending 31st December 1831	40,783	3	7	
	£85,530	12	10	

The Dividends have been as follows:—

	£.	s.	d.	£.	s.	d.	
In respect of Profits to 31st Dec. 1830, 6,375 (Shares of £100) @ 2*l.*	12,750	0	0				Do. Dividends.
Ditto - ditto 30th June 1831, 6,375 @ 4*l.* 10*s.* } Ditto - ditto (£25. Shares) 6,375 @ 9*s.* = 1,593 @ 1*l.* 16*s.* }	31,556	5	0				
Ditto - ditto 31st Dec. 1831, 7,968¼ @ 4*l.* 10*s.*	35,859	7	6				
				80,165	12	6	
Leaving a Balance in the hand of the Treasurer of				£5,366	0	4	Balance.

to meet those contingencies to which the working of every extensive and new undertaking may be considered more or less liable.

The expenditure on capital account, that is, in the formation and completion of the railway and works, the proprietors are aware has been kept altogether distinct from the disbursements appertaining to the traffic or working of the way; these two branches of expenditure being separate and independent of each other.

The outlay of capital, as shown by the accounts, is now drawing to a close. In the expenditure of the large sums which during the last six years have been entrusted to their management the Directors have endeavoured to keep constantly in view the substantial and permanent interest of the concern.

(Signed) Charles Lawrence,
Chairman.

MEMORANDUM explanatory of the General Balance Sheet.

Capital Account.

The Treasurer Dr as follows:

	£.	s.	d.	£.	s.	d.	
To total Amount of Calls and Loans - - - -				1,024,375	0	0	
Less by Amount paid back the Exchequer Loan Commissioners -				5,800	0	0	Capital
				1,018,575	0	0	Account.
Add Premium on Exchequer Bills, &c. -	£1,667	19	8				
Amount of sundry Travelling relinquished	60	7	3				
Amount of over Payments on account of Calls	282	19	0				
				2,011	5	11	
Carried forward				1,020,586	5	11	

	Brought forward				1,020,586	5	11
	The Treasurer C^r.						
	By Amount expended in Completion of the Ways and Works -				992,054	3	6
	Account of Funds remaining on Capital Account - -				28,532	2	5
	Namely, Arrears of Calls - -	£22,453	18	7			
	In the Bank - - -	6,078	3	10	28,532	2	5

Annual or Working Account.

	The Treasurer D^r as follows:						
Working	To Profit on Six Months ending 31st December 1831 -				40,783	3	7
	To Balance remaining above the Amount of the Second Dividend				441	4	3
	To Amount not yet paid of Second Dividend - -				1,664	11	0
					42,888	18	10
Account.	Add Amount in hand on Capital Account, as above -				6,078	3	10
					£48,967	2	8
	Provided for thus:						
	In Bank - - - -	£51,130	14	3			
	In Treasurer's Hand - -	85	6	9			
		51,216	1	0			
	Less Balance D^r of Ledger Accounts:						
	viz. Debits - £16,278 5 0						
	Credits - 14,029 6 8	2,248	18	4			
					48,967	2	8

ABSTRACT OF THE RETURNS.

From the different Departments, of the total Number of Agents, Clerks, Overlookers, and Workmen receiving Salaries or Day Wages in the Service of the Company, for the Week ending 16th March 1832.

——	No. of Agents, Clerks, and Overlookers.	No. of Engine Men, Guards, Labourers, &c.	Salaries and Wages, per Week.	Total per Week.	Total.
Carrying Disbursement.			£. s. d.	£. s. d.	£. s. d.
Manchester End, pr Mr. Green -	22	129	151 6 7		
Do. - Mr. Dixon -	- -	6	5 10 8		
				156 17 3	
Liverpool End, pr Mr. Comber -	14	70	91 19 10		
Do. - Mr. Allcard -	- -	12	9 13 8		
				101 13 6	
Stationary Engine Disbursement.					258 10 9
Liverpool End, pr Mr. Allcard -	- -	11	- - -	- - -	12 4 6
Coach Disbursement.					
Manchester End, pr Mr. Dixon -	- -	4	4 18 0		
Do. - Mr. Green -	5	14	24 5 7		
				29 3 7	
Newton Bridge, pr Mr. Dunn -	1	2	- - -	2 16 0	
Liverpool End, pr Mr. Ellwood -	2	2	6 3 1		
Do. - Mr. Williams -	3	21	30 7 7		
Do. - Mr. Allcard -	1	20	25 13 11		
				62 4 7	94 4 2
Maintenance of Way.					
Liverpool End, pr Mr. Allcard -	4	83	- - -	87 0 5	
Manchester End, pr Mr. Dixon -	4	74	- - -	70 4 2	
					157 4 7
Locomotive Power.					
Liverpool End, pr Mr. Allcard -	2	78	- - -	106 10 9	
Manchester End, pr Mr. Dixon -	1	55	- - -	70 13 8	
					177 4 5
Waggon Disbursement.					
Liverpool End, pr Mr. Allcard -	- -	11	- - -	12 8 9	
Manchester End, pr Mr. Dixon -	- -	3	- - -	3 12 6	
					16 1 3
Coal Disbursement.					
Liverpool End, pr Mr. Allcard -	- -	4	- - -	- - -	3 15 6
Engineering Department.					
Manchester End, pr Mr. Dixon -	1	- -	- - -	7 13 10	
Liverpool End, pr Mr. Allcard -	1	- -	- - -	5 15 5	
Do. - Mr. Stephenson	1	- -	- - -	9 12 4	
					23 1 7
Police Establishment.					
Liverpool End, pr Capt. Brooke -	5	48	- - -	- - -	54 1 0
General Office Establishment -	6	1	- - -	- - -	25 7 8
	73	648	- - -	- - -	821 15 5

Salaries for Agents, Clerks, Overlookers, and Workmen.

Ex. Mr. HARDMAN EARLE,

Merchant of Liverpool for the last Twenty Years.

Director L. and M.

Dealers in Cotton visit Liverpool.

Fares by Coach.

Time occupied by ditto.

Steatland Estate.

Effect of the Railway to Persons residing near.

Collateral Travelg.

Carriages conveyed by Railway.

Time of Transit.

Chat Moss.

Railways increase the value of the Land.

Manure carried by Railway.

I have been a Director of the Liverpool and Manchester Railway for the last three years. The Dealers and Consumers of Cotton are in the habit of coming to Lancaster to make their Purchases; they went to Liverpool before the Railway was made generally once a week. The Fare was 10*s.* and 1*s.* for the Coachman. It was possible for a man rising at four in the morning to go and return in the same day, but he could not get home until half-past nine or ten o'clock at night.——There is a large Mansion in the neighbourhood of Liverpool, called Steatland, within 500 or 600 paces of the Railway; it belongs to my mother; (the House and Grounds cost £12,000.) they have not experienced any inconvenience arising from the Railway, or through the people infesting the grounds to look at the Trains. It is an object of interest to persons residing near, as they can hear the Carriages arrive, although there is no smoke. My mother petitioned against the Bill, and I appeared as evidence, but my opinion is entirely changed from what I have seen.——Collateral Travelling has increased in consequence of the Railway. Passengers come from Southgate from the North to Liverpool to go to Manchester.——Gentlemen's Carriages are taken by the Railway on a Truck.——I have passed from end to end of the Railway in the space of One Hour and Ten or Twelve Minutes.——As an instance of a Railway improving Land I may mention Chat Moss, where the Value of the Land was increased. The other day they were taking manure from Manchester to Chat Moss, they were however unable to get people to it; but where there is any thing like a Station it is much improved in value. We see advertisements recommending a site on account of the Liverpool and Manchester Railway passing near to or through it.——I am not aware of any instance in which Land has been depreciated by the Railway, or of people quitting their houses in consequence of it. I should be glad to purchase land near the Railway for building.——There have been a few instances of Manure being carried by the Railway.——I consider the London and Birmingham Railway will be of public utility.

Ex. Mr. JOSEPH PEASE, *of Darlington.*

I am a Director of the Stockton and Darlington Railway, which has been in operation the last seven years; having been formed in 1822. Only a single line was originally intended, but a double line was laid down; it was constructed principally for Coals; there is no Canal passing parallel to it. I am likewise a considerable owner of Canal property. (Carts and waggons with narrow wheels are injurious to a road.)——A considerable portion of the traffic is carried on by Horses; most of the Passenger Carriages are conveyed by them, as there are not a sufficient number of Locomotives.——The greatest Height of Embankment is 52 feet.——We paid great Sums to Landowners for Gravel, Timber and Stone, taken out of their Estates, for making Bricks and a variety of other purposes.——I am not aware of any Landowner of the present day who considers his Land injured by the Passage of the Railway through it.——As regards the Turnpike Roads, I consider all of them in a much better state of Repair, and the Funds much better, than they were prior to the construction of the Railway, owing to the diminution of that kind of traffic which injures them. I was acting Commissioner of a Trust in which there was a Road from Stockton to Barnard Castle; there were two roads running nearly parallel with the Railway, and another intersected them. The Trustees petitioned against the measure, on the ground that the Money lent on Mortgage would be hazarded by making the Railway, but their fears were not realized.——We are assessed on the amount of our nett Income for Parochial and other Rates: more than Half the Rates of some Parishes are borne by the Company.——We had one Complaint as to Injury done to Game, but we remedied it. We have also had Complaints made of Coals having been pilfered along the Line.——I have a small Estate which is intersected by the Railway, and I have been benefited by it in respect to Drainage; (I observe that where there are Cuttings, Landowners make use of them as Drains). I let that Farm subject to its being given up when the Railway was made, and I have since received one-fifth additional Rent.——In reference to Farms, (not to small Parcels of Land) I do not know of any instance of the Rent being reduced in consequence of the Railway passing through them.——The Company never objected to pay 50 per cent. more for the Land, in consideration of the increased Value arising from the Railway.——The

Desn. of the Stockton and Darlington.

Made principally for Coals.

Passenger's Train worked by Horses.

Greatest Height of Embankment.

Advantage of Railway to Land Prop.

Turnpike Roads improved by the Railway.

Heavy Parochial Rates paid by the Railway.

Complaints against the Railway.

Farms increased in Value by the Railway.

The Railway occupies 300 or 400 Acres.

Land occupied by the Railway, taken at a guess, is near 300 or 400 Acres. ——I do not know of any instance in which more Allowance was made to the Tenants than the value of a way going Crop.——Before the Railway was established there was but one public Conveyance, three times a week, in which there was from three to five Passengers. Three Carriages traverse the Railway daily, conveying an average of 80 to 100 Passengers, and the traffic of late has very much increased. The Outside Fares, according to the old charges, for the 12 Miles from Stockton to Darlington, was 3*s.*; it is now 1*s.*, and in the ratio 1*s.* 6*d.* and 4*s.* 6*d.* Inside. The Railway with its Branches is 42 Miles long.

Allow. to Tenants. No. of Passengers, Coaches, and Fare. Ditto and Fares by the Railway. The Railway is 12 Miles long, and 42 with the Branchs

---

Ex. Mr. THOMAS LEE,

Architect and Surveyor, of Chitty Mill, near Manchester.

I superintend the Estates of several Gentleman in that neighbourhood. ——I am aware of Property being improved in an agricultural point of view, in consequence of the Liverpool and Manchester Railway passing through it; it is the case with the Estates of Mr. Tafford and Colonel Lee: their Land has been let at higher Rents in consequence of it; the Farmers have likewise been benefited by it. I took about three acres myself for building purposes at double the original rent, which made it 50*s.* per acre. ——Farmers send their Produce to Liverpool and Manchester; Milk goes regularly, and I think they have got an advance of one farthing per quart upon it.——Some part of the Land has been taken by Market Gardeners.——Previous to the Railway being made Colonel Lee's Coal was sent from Newton to Liverpool by Canal; he had a Railway about three-quarters of a mile long to it; and could get it taken from his Colliery for about 8*s.* per Ton (I pay 5*s.* for Timber, we used to pay 8*s.* 4*d.*). There have been more Coal Fields opened in consequence of the Railway, and considerable quantities have been taken both ways, which has decreased the price of same.——On the same Gentleman's Property a very large Hotel has been built in consequence of the Railway; he could have sold it for £15,000.——The Poor Rates at Newton amount to 6*d.* in the pound, producing £95., and the Railway pays £20. 7*s.* out of every Rate, of which there are eight in the year. I therefore conceive that the Railway will produce great Advantages to the Land Proprietors.—— There was no communication formerly with such parts of Chat Moss which

Instances of Property being improved by the L. and M. Railway. Farmers are likewise benefited by it. Farmers send their Produce by Railwy. Milk, &c. Market Gardens. Prices by Canal from Newton to Liverpool. Fresh Coal Fields opened through it. Improves. made by the Railway. Poor Rates paid by the Railway at Newton. Consequent advantages to Landowner. Chat Moss.

are now improving; (on which there were some very excellent Crops of Wheat last year;) but a great part of it was capable of Cultivation before the Railway was made.——The principal Improvement arises from the Railway giving Frontage with a facility for building, and the Merchants of Liverpool build Villas, where they can get good land, to which they can retire in three-quarters of an hour, and have the benefit of the air. Manufactories are likewise being built: I have built one Foundry myself, and another is in operation.

Advantages of the Railway to Merchants & Manuftrs.

Ex. Mr. JOHN HART,

Coach Proprietor of Birmingham for the last Seventeen Years.

Fourteen Coaches and two Mails leave Birmingham daily, and the same number leave London for Birmingham. The average number of Passengers by Coach is Three Inside, and Six Out, and by the Mails Three Inside, and Two Out. They perform the Journey from 11½ Hours to 13. The average Fare is £2. 6*s*. Inside, and £1. 6*s*. Outside, including Coach and Guardmen, and the Parcels average £2. per Day each Coach.

Number of Coaches and Mails. Average Passengers by same. Ditto Time of Journey. Ditto Fares.

Ex. Mr. FRED. CLEMENTS, *Surveyor, of Barnet.*

Forty-nine Coaches pass through Barnet up and down daily; 13 of them rest one day in the week, two on Monday, the rest on Sunday; 39 of these are four-horse Coaches. 17 turn off the road at Barnet, three stop at Barnet, one stops at Daventry, 16 turn off at Hockliffe, two stop at St. Alban's, and the remaining 10 go to Birmingham. On an average, 20 to 25 Pair of posting Horses leave Barnet daily, 150 to 170 Pair in the busy time, and 10 in the slack, it sometimes amounts to but three or four.

Stat. of Coaches on the Birm. Road. Ditto Posting.

Ex. MATTHEW HOLMAN, *Ostler at the White Horse Hockliffe.*

Seven Vans pass down and eight up between London and Birmingham; they go on to Manchester. A Van weighs 3½ Tons. The Posting at Hockliffe averages 20 Pair a Day; of course they come from and go to various quarters. 34 Waggons pass in the course of the Week up and down.

Traffic of the Line. Vans — A Van weighs 3½ Tons. Posting. Waggons.

Ex. Mr. JOHN NORTON, *of Daintry.*

Traffic by Canals. Burden of a Fly Boat 10 or 11 Tons. Slow Boat's 28 or 30 Tons.

I was employed upon the Grand Junction Canal at Brunston for 14 Days, from six in the morning until six at night; the Number of Fly Boats that passed during that time was 209, (the average Weight of a Fly Boat is 10 or 11 Tons) and 80 Slow Boats, (some are of 28 and some 30 Tons Burden) also 138 Coal Boats. The Branston Pound is a part of the Grand Junction Canal where several other Canals meet, viz. the Oxford, the Warwick and Knapton, the Warwick and Birmingham, and the Paisley and Coventry. I merely took those that came from Oxford, and entered the Junction right up the Line from Coventry, Wolverhampton, and so on; I took all those that went towards London, whether they came from Coventry, the Oxford, or elsewhere.——I was also stationed at Daintry in July last for 12 Days, from six o'clock in the morning until six at night. 12 Carriages and Postchaises passed during that time with four Horses, and 141 ditto with two Horses, 84 Gigs, 112 Drove of Cattle, 247 Drove of Sheep, 72 Drove of Pigs, all of which were going to London.

Traffic on the Rd. Cattle, &c.

Ex. Mr. THOMAS NORTON.

Traffic on the Road during the Night.

I was stationed at Daintry last July 14 Nights, to observe the Vehicles that passed, from six at night until six the next morning. Six Gentlemen's Carriages and Postchaises with four Horses, 61 ditto with two Horses, nine Gigs, 17 Stage Waggons with five or six Horses each, and under 187 Head of Sheep passed during that time.

Ex. Mr. JOHN SHACKELL, *Carrier, of Birmingham.*

8 Waggons between B. and L. Weight of Loads. Time occupied. Charges by same. Intermediate Traffic very great.

Eight Waggons leave Birmingham for London and return every Week, occupying 50 or 60 Hours for the Journey; and carry on an average 3 Tons up and 2½ Tons down, the Charge is 3*s.* per Cwt. down and 5*s.* up; and 6*s.* or 7*s.* for light and valuable articles. The average Quantity sent per Year amounts to 2288 Tons, the Charge for which is £936. The intermediate Traffic carried on by these Waggons is very considerable, amounting to two-thirds the Traffic of the Road, and Goods taken from these intermediate places pay considerably more; I calculate them at £6240., which makes a Total of £9360.——There are three Waggons from Northampton to Birmingham; they carry 30 Cwt. up, and 10 Cwt. down, or about 3012 Tons annually; the aggregate Charge is 2*s*. per Cwt., which gives £624. Besides the Goods mentioned considerable Quantity of Grain is sent from Northampton to Birmingham annually by Land Carriage. ——There are two Waggons from Rugby to Daintry, which go eight times a Week.——There are two Waggons from Towcester to London, and they go twice a Week.——There are two Waggons up and down from Northampton to London direct.——The Waggons from London to Daintry carry about 5000 up at 3*s*., and 3000 down at 2*s*., which amounts to £1092., and the London and Northampton Waggons carry the same. The reason of the Difference in the Charges between carrying Goods up and bringing Goods down is this: Goods brought from London, being principally raw Materials, will not bear so great a Price as manufactured Goods.——There are three Waggons or Carts from Coventry to Birmingham weekly, and the Goods conveyed by them amounts to £370. 10*s*. ——The Total conveyed by all the Waggons, exclusive of Road Goods, amounts to 4060 Tons, the Charge for which is £19,870. 10*s*.; (this amount does not include the Conveyance of Goods by Waggon when the Canal is interrupted). There has been rather a decrease lately, owing to the badness of trade; in 1825 or 1826 there was an increase.

Total Amount and Charge of Goods sent by Waggons.

Traffic between Northamp. & Bir.

Amount of intermediate Traffic.

Reason of Carr. up being dearer than Carr. down.

Traffic between Coventry and Birm.

Total of Goods conveyed by Waggons £19,870. 10*s*.

Ex. Mr. WILLIAM PARTRIDGE,

Canal Carrier of Thirty-four Years Experience.

I have been engaged 20 years in trading between Birmingham and London, (and from Birmingham, Worcester, Shrewsbury, and Bristol). There are three Routes by Canal to London, one by Worcester and Birmingham, one by Stratford and Worcester, and another round by Coventry. The Fly Boats go the shortest route, and are three days and nights on the journey; the Slow Boats are six or seven days, and they seldom travel at night.——The shortest route to London is taking the whole line of the Worcester and Birmingham Canal, which goes into the Warwick and Knapton, and only seven miles on the Oxford; the Heavy Boats generally go by the Fazeley, which is the longest route, as the Birmingham Canal Company allow them to pass on to Fazeley, to a certain place where the Oxford

3 Routes by Canal.

Fly Boats 3 Days and Nights.

Slow Boats 6 Days.

Shortest Route by Canal.

The longest.

Canal charges the same for going 40 miles as the others do for going seven, which makes it cheap.——The number of Fly Boats which start from Birmingham every week is 25, and the average tonnage of each boat is about 15 tons up and 8 tons down; they sometimes carry from 18 to 19 tons.——The number of Slow Boats that start from Birmingham weekly is about 30, and the average number of tons conveyed by them is 23 up and 5 down.——The Cargoes conveyed to London, consist of all the Manufactured Goods of the neighbourhood, as Nails, Vices, Anvils, Chains, Agricultural Implements. These are charged 40*s.* a Ton, which is a lower Rate of Tonnage and Freight than the other part of the Cargo, which consists of Locks, Coach Pins, Screws, Sadlery, Ironmonger's and Drysaltery Goods, Copper Furniture and Nails, Wire, and Wire manufactured Goods, Iron and Paper Trays, Fenders, Fire Irons, Guns, Swords, and Army Stores, Glass Lamps, Bronze Goods, Steel and other Ornaments, Ivory and Bone Toys, Plated Goods, Carpets, &c. The Freight of these light Goods is 55*s.* The heavy goods are put at the bottom of the Boat, and the light goods on the top. The Cargoes from London to Birmingham consist principally of Wines and Spirits, Grocery, Saltpetre, Tallow, and Mercery Goods, and the principal portion of it is Colonial Produce. The average price of the above is 40*s.* per Ton (the steerage is generally calculated on the back carriage). The Slow Boats convey from Birmingham to London Iron Work, Water and Gas Pipes, Grain, at an average Rate of 22*s.* 6*d.* per Ton. The Cargoes brought from London to Birmingham consist of Timber, Grain, and Foreign Iron for the manufacture of Steel; the average Freight of which is 26*s.* 6*d.*——Goods are likewise conveyed by Canal from the intermediate places, and large supplies of Coals are thrown on the Canal, some at Warwick and Fazeley, on the different routes, and some at Branston. By these routes there is an immense quantity of Hardware Goods, Earthenware, and Pottery Goods conveyed; also, Salt from the Cheshire Salt Works, Cheese out of Cheshire and Derbyshire; also, Manchester and Yorkshire Goods.——A Fly Boat occupies about three minutes in passing a Lock; there are many Locks on this line of Canal. The Canals are generally stopped about 14 Days in the Winter, and about a Week or 10 Days for Repairs at Whitsuntide.——Taking the average of light Goods only by Fly Boats, I make it amount to £68,250. up, and to £29,121. down, making altogether £97,370.——I think that the whole of the Goods that go by Fly Boats will go by the Railway, and the heavy Goods, such as Pig and Bar Iron, Timber, &c. will go by the Canal as they do now; so that the Canal would lose £97,000. entirely.

Number of Fly Boats, & Tonnage.

Do. of Slow Boats.

Description of Fly Boat Cargoes up.

Freight on heavy Goods 40*s.* per Ton.

&c. &c.

Ditto light Goods 55*s.* per Ton.

Description of Cargoes down.

Freight 40*s.* per Ton.

Slow Boat Cargoes 22*s.* 6*d.* up, 26*s.* 6*d.* down.

Intermediate Traffic by Canal.

Locks.

Canal Stoppages.

Amount of Goods by Fly Boats £68,250. up 29,121 down

£97,370. Which would be conveyed by the Railway.

Ex. Mr. WILLIAM SHORE, *of Birmingham.*

There are three Lines of Canal between London and Birmingham, viz. the Coventry Canal, by the way of Fazeley, which is 177 miles long; the Oxford Canal, by Warwick and Knapton, which is 152 miles long, and the Grand Junction Canal, by the Worcester and Stratford Canal, which is about 155 miles long.——The number of Locks on the Fazeley Line are 150, on the Warwick 173, and on the Stratford 161. A Fly Boat occupies four minutes in passing the Boat Locks, and five minutes the Barge Locks, and Slow Boats pass in five minutes; so that 11½ hours are occupied by the Fly Boats, and 14 hours by the Slow Boats, in passing all the Locks on the Fazeley Line. There are six Tunnels on the Fazeley Line, occupying a distance of four miles; on the Worcester there are six, making about 4¼ miles; on the Stratford there are six, making 4½ miles. The Fly Boats are occasionally delayed in passing the Tunnels, owing to the heavy Boats getting in before them.——Canals are generally stopped 14 Days upon an average during the Winter; (persons who trade in heavy articles, such as Coals and Iron, are obliged to provide against the same by laying in a large stock) they are also frequently stopped during the Summer Months, from Saturday Night until Monday Morning or Night. Fly Boats complete the Journey in about three Days and three Nights, and Slow Boats in about six or seven Days. The Tonnage on heavy Goods, Coals and Iron, is 1*d.* per Ton, and on general Goods 1½*d.* per Ton per Mile; in addition to which they charge 8*d.* per Ton for passing the Blisworth Cutting, also 4*d.* for going over the Grass Road Valley, and 10*s.* for Permission to pass a Pair of Boats at Night, and 10*s.* is charged for the same upon the Grand Junction at Branston, where they also charge 1*s.* 6*d.* for dragging the Boats through the Tunnel, which is done by two people, and 2*s.* is charged for the same at Blisworth. On the Warwick and Knapton Canal there is 17½ Miles of the Junction to Oxford, for which they pay extra; the Grand Junction charge 18*s.* for passing a single Boat from Brunston to Paddington, and 21*s.* to the Thames at Brentford. I can judge from the draught of water what weight a Boat carries.

3 Canals to Birmingham.
The Coventry.
The Oxford.
The Grand Junctn.
Description of same
Description of Locks, &c.
4 or 5 min. occ. to pass same.
No. of Tunnels.
Description of same
Stoppages.
Time of a Journey.
Tonnage, and Fees charged.

Rates of Freight to London.

	s.	d.	
Hardwares generally, in Casks or Cases	3	0	per Cwt.
Anvils	2	9	—
Vices	2	9	—
Chains	2	9	—
Frying Pans	2	9	—
Iron Hollow Wares	2	9	—
Smith's Bellows	2	9	—
Muskets, in Cases	2	9	—
Swords, in Ditto	2	9	—
Matchets, in Ditto	2	9	—
Nails, in Casks or Bags	2	6	—
Heavy manufactured Goods, generally	2 6 to 2	9	—
Ale, per Barrel of Thirty-six Gallons	2 3 to 2	6	—

N.B. Goods may be reckoned generally 3s. per Cent. cheaper from thence to London.

Rates of Freight.

Rates of Freight from London.

	s.	d.	
Wines, in Cases or Hampers	2	9	per Cwt.
Spirits, in Ditto	2	9	—
Tea	2	9	—
Coffee and Grocery, generally	2	6	—
Sugar	2	3	—
Soap	2	3	—
Drugs, generally	2 3 to 2	6	—
Oils, Resin, &c. &c.	2	3	—
Dry Saltery, generally	2 0 to 2	3	—
Porter	2	3	—
Hops	2	6	—
Seeds	2	6	—
Candles, in Cases	2	6	—
Perfumery	2	6	—
Nuts, Oranges, &c.	2	6	—
Tobacco and Snuff	2	6	—
Drapery, generally	2	6	—
Hats	2	6	—
Wool	2	9	—
Pearl Shell, in Casks, &c.	2 0 to 2	3	—
Barilla	2	0	—
Grain from Brentford, for Boat Loads	1 0 to 1	3	—

Ex. Mr. JOHN SWAINSON,

Clerk in the Stamp Office, in the Stage Coach Department.

17 four-horse Coaches between L. & M.

According to the Returns furnished by the Stamp Office, there are 17 four-horse Coaches on the Road between London and Birmingham.

Ex. Captain RICHARD MOORSOM, R.N.

I am Secretary of the proposed Railway, and hold 20 Shares in it. I have made a Calculation of the probable Amount of Passengers, as follows:

Calculation of Passengers.

Calculation of Passengers on the London and Birmingham Railway.

1.	2.	3.	4.	5.	6.	7.	8.
PLACES.	Weekly Journies.	Average Number of Persons in each Coach.	Weekly Passengers.	Number of Miles now travelled, not extending beyond Birmingham.	Places at which Passengers are assumed to join or leave the Railway.	Number of Miles by Railway.	Aggregate Number of Miles assumed to be travelled weekly on Railway.
London and Birmingham -	124	9	1,116	112	- - -	112	124,992
— Chester M. -	14	5	70	112	Birmingham -	- -	7,840
— Dudley -	14	9	126	112	- - -	- -	14,112
— Holyhead M. -	14	5	70	108	- - -	- -	7,840
— Kidderminster -	6	9	54	112	- - -	- -	6,048
— Liverpool -	54	- -	486	108	- - -	- -	54,432
— Manchester -	108	- -	972	108	- - -	- -	108,864
— Shrewsbury -	42	- -	378	108	- - -	- -	42,336
— Worcester -	64	- -	576	108	- - -	- -	64,512
— Aylesbury -	14	- -	126	40	Near Tring -	32	4,032
— Ampthill -	12	- -	108	45	Leighton Buzzard	40	4,320
— Banbury -	12	- -	108	76	Wolverton -	50	5,400
— Bedford -	12	- -	108	52	- - -	- -	5,400
— Chesham -	12	- -	108	29	Berkhempstead -	26	2,808
— Derby -	12	- -	108	108	Rugby -	82	8,856
— Glasgow M. -	14	5	70	108	- - -	82	5,740
— Halifax -	14	9	126	108	- - -	82	10,332
— Hemel Hempstead	8	- -	252	25	Hemel Hempstead	22	5,544
— Kettering -	12	- -	108	76	Blisworth, near Northampton -	60	6,480
— Leighton Buzzard	12	- -	108	40	Leighton Buzzard	40	4,320
— Leicester -	14	- -	126	98	Rugby -	82	10,332
— Leeds -	Half 35	- -	315	108	- - -	- -	25,830
— Leamington -	12	- -	108	91	Coventry -	92	9,936
— Northampton -	14	- -	126	66	Blisworth, near Northampton -	60	7,560
— Pinner -	14	- -	126	15	Pinner -	12	1,512
— Rickmansworth -	14	- -	126	20	Watford -	15	1,890
— Rugby -	6	- -	54	82	Rugby -	82	4,428
— Tring -	12	- -	108	33	Near Tring -	32	3,456
— Two Waters -	4	- -	36	24	Near Two Waters	20	720
— Wellingboro' -	12	- -	108	68	Blisworth, near Northampton -	60	6,480
— Wendover -	12	- -	108	37	Near Tring -	32	3,456
						Total of Miles -	569,808

569,808 Miles per Week will be 29,630,016 Miles per Annum, which, at the Railway average Charge of 2*d.* per Head per Mile, will give £246,916. 16*s.* per Annum.

NOTES explanatory of the preceding TABLE.

Notes explanatory of Table.

1. That Coaches now run between London and the Places specified in the First Column.—Stamp Office Returns.

2. That they make the Number of Journies per Week mentioned in the Second Column.—Stamp Office Returns.

3. That the average Number of Passengers in each Journey is Nine for the Coaches and Five for the Mails, being the Numbers mentioned in the Third Column.

4. That consequently the Number of Passengers per Week between London and the Places mentioned in the First Column is the Number stated in the Fourth Column.

5. That the Number of Miles now travelled by each Coach each Journey (not including any Distance beyond Birmingham) is that which is stated in the Fifth Column.

6. That Reason, and the Experience of the Liverpool and Manchester Railway, show that all Passengers will travel the Whole or a Part of their Journey on the Railway, whenever they can save Time and Expense by doing so.

7. That in all the Journies mentioned in the First Column, Time and Expense will be saved to Passengers by their joining the Railway at some Part of its Course, instead of proceeding by the present Roads now in use.

8. That the Places at which it will be most convenient for Passengers performing the Journies mentioned in the First Column, to join or leave the Railway by existing Roads, will be the Places mentioned in the Sixth Column.

9. That upon this principle, the Number of Miles on the Railway that will be travelled by each Passenger in each of the Journies mentioned in the First Column, will be the Number mentioned in the Seventh Column.

10. That the aggregate Number of Miles on the Railway which will be travelled by the Passengers in each of the Journies mentioned in the First Column, will amount per Week (by multiplying the Number in the Fourth Column by the Number in the Seventh Column) to the Number mentioned in the Eighth Column.

11. That, consequently the aggregate Number of Miles on the Railway travelled by the Passengers in all the Journies mentioned in the First Column will amount per Week to 569,808 Miles, and for the whole Year to 29,630,016 Miles.

12. That if the Charge for travelling on the Railway be 2*d.* per Head per Mile, the gross annual Income of the Railway from Passengers only, who now go by public Conveyances, will be £246,916, 16*s.*

£246,916. 16*s.*

A Duty similar to Coaches would make £31,000. extra or 2*s.* 6*d.* or 3*s.* per Passenger.

The result of the Calculation in money is £246,916. 16*s.* A Duty on Railways, corresponding to the Duty paid on Coaches, would amount to one-farthing per head a mile, which would give a Gross Sum of £31,000. to be added to the above amount. In the event of the above taking place, I apprehend an additional Charge would be made to the Public; the Difference would be 2*s.* 6*d.* or 3*s.* each Passenger going to Birmingham.

Ex. Mr. PETER LECOUNT, *Naval Officer.*

Method of calculating the Returns.

There were 32 Persons employed on the Road to ascertain the Traffic, and their Returns were put into my hands. I have likewise ascertained personally the Number of Carriages that passed at the following places: Edgeware, Barnet, St. Alban's, Hockliffe, Daintry, Coventry, and two places at Birmingham, which exceeded the numbers stated in the Returns furnished me.——I consider the present Profit of the Grand Junction Canal is much more than we have stated.——I have made a Calculation of the Traffic, &c., as follows:

Calculation of Traffic.

General Results of the Traffic on the Line between London and Birmingham for One Year, and the Expences by the present Means and by the Railway.

Means of Transit.	Number of Journies of 110 Miles.	No. in each.	Total Carried.	Expence by the present Means.		Expence by the Railway, 12½ Miles.	Time. At present.	Time. By the Railway.
			Passengers.		£.	Passengers at 2d. per Mile each.	Hours.	Hours.
Four-horse Coaches, counted on the Road	21,641	9	194,769	316499,6				
Two-horse do. do. do.	4,221	6	25.326	44003,9		—		
Pairs of Post Horses do.	7,622	3	22,866	83842,0		£.		
Commercial Gigs do.	5,569	1	5,569	11138,0	455,483	227,819		
Contingent Coaches, from Stamp Office Returns	23,745	9	213,705	347270,6			12½	5½
Proportionate Number of Pairs of Post Horses	6,998	3	20,994	76978,0				
Ditto of Commercial Gigs	5,113	1	5,113	10226,0	434,474	219,827		
		Cwt.	Tons.					
Private and Stage Vans, counted on the Road	1,600	18 13	2,315½	18522			30	
Stage Waggons do. do.	3,665	70	12,827½	76965			60	
Errand Carts do. do.	11,543	10	5,771½	34629	130,116	58,821	40	
		Tons.						
Boats, counted on the Canals	$11,131\frac{7}{11}$	11	122,428		306,070	286,910	72	
	149 Miles.			£	1326,143	793,407		

(Signed) PETER LECOUNT.

Explanation of Table.

I was employed about 20 hours each day. Regarding the Gigs, I took those of Commercial Travellers only. I did not enquire whether they were going all the way to Birmingham; but when at Edgeware I calculated them as coming from London to Edgeware; when at Barnet, I calculated the number of miles between Barnet and St. Alban's, and so on, and the Canals were calculated upon the same principle; the Distance was taken for the separate thing, and reduced afterwards to one large sum; but I consider the question is one altogether relating to time, (it may not be necessary that the Cheese and Salt Butter should travel at the rate of 20 miles an hour) and not to money, the price of each being nearly the same. By Canal it is £306,070., and by the Railroad £286,940.

Traffic by Canal £306,070.

Expence of same by Railway £286,940.

Ex. Mr. W. MEADE WARNER.

Farmer of Thornly Hall, Oxfordshire.

I occupy about 200 acres of land near Leighton Buzzard, and about 400 in Oxfordshire, it is situated about one furlong from the proposed Line. I consider it will be an essential service to myself and the other farmers on the Line; had it passed 10 or 15 years ago it would have been a benefit to me of not less than £50. per annum. The farming of Grass Land in particular would be considerably improved; we should be able to send to London much better kind of produce. Lambs are sent to London from our neighbourhood by Waggons, which occupies a space of 24 hours; but they are usually sent on the road, which prevents us sending many, as they have not strength to bear the fatigue of the journey; if a Railroad was established no doubt the farmers would send more; it is important to send them early in the season, as they would feed off sooner. At Hemel Hempstead, and down in Buckinghamshire, below us, they send calves, to which the same observations will apply. I have been a Dairyman 20 years and have 40 or 50 cows, and I consider that a Railway would be still more important to the Dairy Farmers, as they would be able to send milk and butter to the London Market. I had offers made to me to supply a part of London with milk, but I could not undertake it for want of a conveyance; if we could get a Railroad we should increase the profits 400 or 500 per cent. on milk and butter, and I think we should be able to force more milk by artificial food, such as turnips, &c.——I now get about £10. per annum by a Cow, all casualty and loss being a drawback upon that amount; if the Railroad passes I expect to get upwards of £40. I have a Cow giving more than three gallons of milk at meals, which makes 7*s.* a week, and as they are selling milk in London at 1*s.* 4*d.* per gallon, six gallons a day or 42 gallons a week, is something like 50*s.* instead of 7*s.*; we reckon that the

Advantages of Proposed Railway to Farmers, &c.

& Partic. so in Reference to Grass Lands.

Present Conveyance of Lambs.

Ditto of Calves.

Impor. Railway to Dairy Farmers.

Railway would increase Profits of Milk 400 or 500 per cent.

Present Profit of a Cow £10. per ann. Anticipated £40.

A Cow gives six Gallons of Milk per day.

keep of a Cow in summer is at 4*s*.per week, and they will eat two and a half hundred weight of hay per week in the winter; in the butter system the keep of a Cow must come within £10. per annum to leave any Profit; in the previous case I calculate £30. clear Profit on every Cow, no doubt there would be a greater drawback in the expense of retailing milk, more compared with butter.——In the heavy seasons 1,500 head of Cattle pass through Hockliffe to London weekly on the line of road parallel to the proposed Railway. The Charge of the Road and Selling together is 10*s*., we estimate the Road Expenses, which is 40 miles, at 7*s*.; they are considerably injured by being driven up, it would be a great advantage to send them by a Railroad, even if it was dearer; sometimes the poor things are driven until their feet become sore, they are consequently sold on the road for what they will fetch, they are often driven until they have not a foot to stand on; if I paid double for transmitting my cattle I should be a very great gainer; besides the cruelty and exposure, they are a very great nuisance on the public roads, to carriages, &c. There are not less than 10,000 head of Sheep pass weekly, they are driven 40 miles at about 1*s*. a head. I should not imagine Sheep sent by the Railroad would be liable to be heated or hurt by rubbing against each other.——There is a considerable quantity of Poultry reared about us, much of it goes to Aylesbury. A rapid communication with the London Market would be an advantage to both the Suppliers and Consumers of Poultry. Our Lands are principally heavy, but potatoes might be grown and sent up, and at other parts of the Line garden produce might be cultivated.——The Manufacture of Staw Plait is carried on very largely in our neighbourhood; it is sent principally to Dunstable and the vicinity, some of it is manufactured into Hats and Bonnets at Liverpool, and some at Luton. I conceive a Railroad would be advantageous to the Buyers.——I am a Proprietor as well as an Occupier, my property lies in Land; I believe my estates along the Line would be increased in value 30 per cent.

1,500 Head of Cattle pass through Hockliffe weekly.

Charge for driving and selling 10*s*.

Objections to Driving.

10,000 Sheep ditto. Driving 1*s*. head.

Poultry.

Land about Leighton Buzzard heavy.

Straw Plait Manufactured at Dunstable.

Anticipation that the Railway will increase Value of Land 30 per cent.

Ex. Mr. CHARLES WHITWORTH,

Farmer and Landowner, Northampton.

I have seen both Beasts and Pigs conveyed on a Railroad, which were landed as fresh as if they had come out of the field.——A great quantity of Meat is sent from Northampton to London, although it occupies too

Beasts and Pigs sent by Railway advantageously.

Meat is sent from Northampton to London.

Also Asparagus.

Difficulty of sending Lambs & Calves to London.

much time. I have sent Meat from Northampton for which I have had no return, owing to its having become putrid, and consequently thrown away. There is a great quantity of Asparagus sent from Northampton; it generally goes by Coach, which is very expensive: the neighbourhood is particularly adapted for Garden Ground. We are almost precluded from sending Lambs and Calves to the London Market, as we are obliged to be at the expense of sending the Ewes part of the way with them, unless they go by Canal, which injures them very much. They are put into the Boats on Saturday, sometimes they get to Market on Monday morning, sometimes they do not; it would be a great advantage if we could send our Lambs up in three hours. When the Canal has been frozen I have sent goods by Coach, which was six times the price of the latter.———I have known Coals 5*s.* per Cwt., which is five times the regular price, in consequence of the Canal being frozen, and Grain has gone down 3*s.* 4*s.* and 5*s.* a Quarter. I remember a case, about three years back, in which we could not convey it either to Birmingham or Wolverhampton on the Market Day.———The Canal proved a great benefit, and a Railroad would be a greater.———There is a considerable Sale of Grain in Northampton, Barley in particular; it is sent in large quantities to Birmingham, at a considerable expence.———The nearest point of Northampton to the Railroad is about four Miles. I am an Owner of about 800 Acres of Land in Northamptonshire and Buckinghamshire; the Railway will go across one mile of it, (about 300 acres in a ring fence, which it goes right across) which I have no doubt will increase its Value, as it will give a Facility of Conveyance by which Produce may be exported from the Farm. We should also be able to get Manure from the London Market, which at the present time we cannot get within some miles. I do not hold any Shares in the undertaking.

Effect of Frost upon Canals and Roads.

The Canal was a great Benefit—the Railroad would be greater.

Considerable quantity of Grain sent to Birmingham.

Manner in which the Railway will increase the Value of the Land.

Ex. Mr. JOHN SHARP, *Butcher, in High Street, Marylebone.*

Supply of Lambs and Beasts for Smithfield.

Bad Effects of Driving Cattle.

Cattle sent by Steam.

I am in the habit of buying Beasts at Smithfield weekly, which Market is very well supplied, 25,000, or 26,000 is the general supply of Lambs on a Monday, and 2,000 or 3,000 Beasts. If carefully drove, and the weather is good, they do not suffer much; but if badly drove, which they always are more or less, from the negligence or ignorance of the driver, they suffer materially, they lose weight through the length and fatigue of the journey, and their value is lessened. I have seen Cattle sent by Steam from Scotland, although they arrive speedily, I consider the injury equal to about a weeks' careful driving; as a principle, the longer they are on the road, being away from comfortable pasture, the more they are deteriorated in quality.———The Line of Railway comes through a Grazing Country; few come that road from about Christmas to June, but from June to Christmas they are very numerous, after they have clipped the sheep they generally send them; these counties not being turnip counties (the sheep sent at this time are principally from Norfolk and Suffolk.)———I have known Lambs sent 60 or 70 miles to Market.———I think that the farthest from which Calves are sent to London does not exceed 30 miles; the principal part of the veal of the best quality is sent by the land carriage.———In winter the season admits of Dead Meat being sent a distance of 100 miles, but in the summer it can be sent only a short distance, owing to the warm weather.

The Line passes thorough a Grazing Country.

Lambs sent 70 Miles.

Calves are not sent more than 30 miles.

Dead Meat.

Ex. Mr. ROBERT ATTENBOROUGH,

Farmer and Grazier of Braybrook, which is 80 Miles from London.

From June to Christmas I attend Smithfield Market once a week; it takes Three whole Days and One Night, the Fare is £1. 12*s.* inside, and £1. 1*s.* outside, each way.———From 1,500 to 1,700 Head of Cattle come up the Stratford Road weekly, and I should think 6,000 or 8,000 Sheep are sent up every week; we send Sheep all the year round, but Beasts only six months. The regular Price for Driving Cattle up is 7*s.* per head in summer, some people do it for 6*s.* 6*d.*,—in winter it is 8*s.* From June to Christmas I would be willing to pay 10*s.* per head extra, rather than have the Cattle drove. The Charge for Driving Sheep is 1*s.* per head. I have always understood that a Sheep driven 80 miles loses 8 lbs.; we should be willing to pay more in proportion for Sheep than Beasts, as they are longer on the road.———I have heard Irish dealers say that Cattle are carried very conveniently by the Railway, and greatly to their advantage; a person holding a farm on the Railway would find it very convenient; for instance, he could put 20, 30, or 40 Beasts on the Railway on a Sunday morning and follow them, and having sold them could go home the same evening, which would do away with many of the Commission Salesmen; the Commission is 3*s.* 6*d.* a head on all, good or bad.——— ———The principal part of the Grain purchased at the Northampton Market, particularly the Barley, is sent by Canal to Birmingham. I have known depressions in the market from the frequent interruptions by the Canal.

Time and Expense of Travelling by Coach.

Price for Driving up Cattle 7*s.* a head

Ditto Sheep 1*s.*

A Sheep driven 80 miles loses 8 lbs.

Cattle are Carried very conveniently by Railway.

Advantages of Do.

Commission for Selling Cattle.

Much Grain is sent by Canal to Birmingham.

Objections to Can.

Ex. MR. OLIVER MASON, *of Great Malvern.*

I was a Merchant and an Inhabitant of Birmingham 23 years. I was also High Bailiff for that Town last year.———The Population according to the Parliamentary returns, including the suburbs, is 142,000, in the Town only 108,000; the increase of Manufacturing Population in the last 10 years is calculated at 25,500, yet the Poor Rates have decreased; the Population in 1821 was 85,416, in 1831 110,914, that is an increase of 25,498; the poor who received relief in the workhouse in 1821 was 561, and 1831 was only 469; the poor who received relief weekly in 1821 amounted to 4065, in 1831 to only 3701, of orphan children there has been a considerable decrease, the people have increased 25,500; the poor in the house have decreased one-sixth, those out of the house one-tenth, and the children in the Asylum one-fourth; this shows Birmingham to be one of the most flourishing towns in His Majesty's kingdom.———The Trade of Birmingham has likewise considerably increased, notwithstanding the imperfect means of communication.———The opening of Canals was a great benefit to the Town, and an increased communication would be a greater advantage to them, particularly as Merchants. We are often put to great loss, and cannot execute orders for want of time, if we lose the opportunity of shipping by a certain vessel which only sails once a year from a port, we lose the order; our correspondent says, "If you can ship my goods by such a ship send them, and if you cannot, do not send them." A Railway would also be a great convenience to Manufacturers, they would be able to calculate to a day the delivery of their goods, and be enabled to compete more successfully with the Great Opposition they experience from Foreign Merchants; at Altona and Liege they have made such extraordinary progress during the last few years, that we are beaten out of the Italian, Spanish, and Portuguese Markets, as they can get the raw material cheaper than we can; we excel them in dispatch and expedition of orders; they are establishing a Railway from the Hague to Antwerp; they frequently undersell us, and we require every possible advantage to enable us to meet the Foreign Manufacturers.———In manufacturing towns the population do not work the first days in the week, it is only at the latter end that the great strain of business is brought into the market.———I remember three instances in which we might have executed orders to the amount of £1,500, provided we could execute them in time for shipment; no doubt during a period of 13 or 14 years there has been repeated instances of my losing orders in consequence of not being able to execute them for want of

Population of Birmingham.

Comparative Amount of Population and Poor Rates for the last 10 Yrs.

Trade of Birmingham has likewise improved.

Canals were a great Benefit.

Railways will be greater.

Advantages of a Railwy to Merchts.

Do. to Manufactrs.

The great Opposition offered by Foreign Merchants, requires us to use every means of facilitating Trade, as we depend upon celerity of executing orders.

They can get the Raw Material cheaper than we can.

Instances of Losses arising through the want of a quick conveyance.

a quick conveyance.——As an instance that we owe much to the celerity with which we execute orders, I once received an order from Spain for a particular kind of brass work to be sent to Madrid, they stated they could buy it from five to 10 per cent. cheaper in Germany, but they could get the order quicker executed in England.——Foreign Manufacturers gain ground upon us daily; in the years 1808, 1809 and 1810 we sent immense quantities of Birmingham Hardware to Germany formerly, (although copper and iron was sometimes double its present price), we also sent immense quantities of Needles, now they can make them as cheap as we can; it is the same with Cutlery and Brass Foundery. (We certainly have many new branches of trade as Fancy Articles, also much Glass Work, which we did not make formerly.) Our Manufactures require every protection, as they copy our articles.——The Frankfort and Leipsic Fairs were attended by Manufacturers. I have known my father have an account of £10,000 with a single house. Leige is one of the principal places where Cutlery is made, but it is coarse: We excel them in finer Articles; they copy us in every description of Carpenter's Tools, and by sending an inferior article do us serious injury.——A great quantity of the Goods that go to Germany are made in Sheffield.——I remember in 1813 we were in complete possession of the German Market, until a rise of 25 per cent. took place in this country, the consequence was, that we lost possession of it for two years, and when the price was reduced we regained it.——The principal articles made at Birmingham are all kind of Brass Fancy Articles, Spades, Shovels, Axes, Tools, and the like; all kind of Hoes for the West Indies, which is chiefly supplied by Birmingham, many of these articles are manufactured at Sheffield, where, also all the Cutlery is made, therefore, this town will likewise be benefited by the Railway.——Birmingham Goods are inferior to those of Sheffield. ——Most of the Cutlery made abroad is merely Cast Iron, or Iron purified by refining.——The Birmingham Manufacturers get their Iron from mines in the neighbourhood, it is sometimes brought from Sheffield. The Copper Ore is brought by the Western Canal to Gloucester, so on to Worcester, and that way; the Railway would not assist us in this, in fact, very little raw material would go by the Railway. Regarding Brass, a large quantity of Spelter is brought over from Hamburgh and Yorkshire, and is very expensive, being sold for £70. per Ton; but in consequence of the immense quantity to be had from Germany, it may now be got for £12. per Ton; it is shipped at Hamburgh and comes to London, from whence it is taken to Birmingham by Canal.——One-third of the Goods Exported

Comparison between Eng. and Foreign Manufactrs.

Foreign Manufacturers are gaining ground on us.

Articles formerly supplied to the Continent.

Articles at present supplied.

They copy our Patterns.

Leige is one of their principal places for Cutlery.

We excel them in Fine Articles.

Sheffield send to Germany.

We formerly sent largely to Germany.

Articles made in Birmingham.

Supplies the West Indies.

Articles made at Sheffield.

The Goods of the former inferior to the latter.

Cutlery made Abroad.

Most of the Spelter is brought from Hamburgh.

Prices of ditto.

Mode of conveyance of ditto.

1-3rd Goods exp. from Birm. are from Sheffield.

from Birmingham come from Sheffield; the assortments required are a certain quantity of Cutlery with Hardware to make it saleable.———The Cases made up for South America consist of a hundred different things, there are many Sheffield Articles mixed with Birmingham, they are sometimes sent from Sheffield to be made up.———A Ship for the Baltic is generally shipped in London.———The Railway likewise offers an advantage, in reference to St. Petersburgh. I had an order for St. Petersburgh, and the goods arrived so late that the last vessel was on the point of sailing, we might have got them if we had only two days more.———The Line of Communication with the West Indies and America from Birmingham is principally through Liverpool, although, there is much through London. ———Canal Communication is inconvenient from the stoppages; I once had an action brought against me for the freight of some goods which I was engaged to ship, they were detained by a stoppage of the Canal, and I was called on to pay the freight, although the goods were returned upon my hands.

Cases for South America.

Trade with St. Petersburgh.

Birm. Trade with W. Indies and Am. prin. thro' Liverp.

Inst. of loss thro' the uncer. of Canals.

Ex. Mr. GEORGE BOWLES, *General Salesman.*

Butter sent to the London Market.

There is a large quantity of Dutch and French Butter brought into the London Market (the quantity ebbs and flows). I give the preference to English.———I sell among other things a considerable quantity of Fresh Butter, which comes 60 miles to Market, and it is my opinion that Farmers are prevented from sending much of it, in consequence of the risk of loss for want of a cheap and ready Conveyance.

Ex. Mr. E. TILSLEY MOORE, *Merchant at Birmingham.*

Instance of the want of a speedy Conveyance.

Losses through Stoppages by Can.

A Railroad would materially assist a Manufacturer in the execution of his Foreign Orders. I am unable to execute an Order for a Vessel that sails on the 14th of July, as I cannot get the Goods on the 10th, which is the last day a Boat sails from Birmingham; these goods are Copper manufactured Goods, Sadlery, and Leather Goods.———I have experienced Losses through the Stoppage of the Canal. I had some goods kept from the 24th of December, 1829, until the 20th of February in the following year, and when the goods arrived out, £ 1200. worth were rejected as being out of time, which compelled me to make a new arrangement, by which my returns were 18 months instead of nine. We pay 55*s*. per Ton by the Canal. We seldom send by Waggon, on account of the expence; when we are pressed for time, we are obliged to send by Coach, although the expense is 8*s*. 4*d*. per Cwt., or between £8. or £9. per Ton.

Freight by Canal.

Carriage by Coach.

Ex. Mr. FREDERICK BARRY, *Ship Broker, London.*

I have been in the habit of dispatching Ships to all parts of the globe, but more particularly to Spain and Portugal, for the last 20 years, and am Owner of several; they are loaded chiefly with manufactured Goods, also Woollen Goods from the North, but the principal part comes from Staffordshire and Birmingham. There are particular days for a vessel to sail, whether the goods arrive or not. There are 14 Days between the sailing of Ships to Spain, if to Lisbon nine Days, to the Havannah a Month. ———I am aware of disappointments frequently occurring from the freezing and cleansing of the Canal, also from other casualties; both Manufacturers and Merchants suffer materially by them. I have known it shut up by Frost for six or seven Weeks, and twice a year in cleansing. Many of the articles will not bear the Expense of Land Carriage, as Bar Iron, which must go by water, or we could not afford to export it. The Season likewise operates in some measure upon Land Carriage, but not to the same extent as upon the water.———Many of the Goods are bound to different Ports, some have to be transhipped from Manilla. The Spanish Ships sail for Cadiz at a certain time annually, and if we arrive too late for them, which frequently occurs, we are compelled to wait until the following year.———Some parts of Spain have imported a great quantity of Machinery, Steam Engines, and the like, from this country, and have improved very much.———In the course of the Season many Vessels take their departure for the Baltic, and if the Manufacturers are two or three Days too late they are thrown out all the Winter, as the Baltic is frozen up, and they are frozen in. Three or four days will make this alteration, and one night of Frost will stop the Canal.———Of late there has been a falling off in Merchandise Shipments from the Port of London, but it is caused more by the Quarantine Regulations than want of demand from Foreign Houses: The Quarantine Regulations are very rigorous; they will not let our Vessels into any of the Ports of Spain, but

Cargoes of Ships to Spain.

Time of Ships sailing for different Ports.

Cons. of Stoppages by Canal, &c.

Frost also affects Land Carriage.

Goods are transhd.

The Ships from Cadiz sail only annually.

Spain is improving.

Vessels for the Baltic.

Frost.

The falling off of the Port of London owing to the Quar. Regulations.

the Trade of London with the West Indies, and in all articles of consumption, has increased, and taking the average, with the exception of the unfortunate Quarantine Regulations, I think it has increased.——A Railroad between London and Birmingham must of necessity benefit the Commerce, as all manufactured goods sent to Germany, the North of Europe, and Russia, must travel eastward; therefore any improvement in the conveyance of same to London would be advantageous.——There has been a material alteration in the Port of London, arising from the competition in the Docks along shore; St. Katherine's and the London Docks rivalling each other. I recently got a Vessel of 400 Tons discharged in 24 Hours. Vessels that sail periodically sail more quickly after each other, and they are not detained so long in London as they were formerly; the great object of the Shipowner being to clear her out, and get her off. I think many Ships from the Baltic may go to Hull, but there is more tonnage out of the Port of London for St. Petersburgh (although Hull is nearest St. Petersburgh, and there may be greater facilities) than all England put together. I consider all Birmingham manufactured goods for the Continent will pass through London, with the exception of such Southern Parts that go from Liverpool.——Should the opposition to the Bill prove successful, and a Railway established between Birmingham and Liverpool, the Manufacturers in the Port of London would certainly consider that Liverpool has an unfair advantage.

Prop. Railway would impr. the Commerce & Port of London.

Improvement in the Docks.

St. Kath. and the London are Rivals.

Case of a Vessel being discharged in 24 Hours.

Trade with St. Petersburgh.

Trade of the Port of London.

Ex. Mr. JOHN TRAVERS,

Wholesale Grocer for Thirty Years in St. Swithin's Lane.

I have lately had dealings to the extent of £20., or £30,000. per annum with Birmingham; it was formerly inconsiderable: The Improvements are partly owing to the Increase of Population, and to the increased Facilities of Communication, which reduces the price of an article, as a great supply will always tend to increase the consumption of same. A Railway would give still greater facilities of communication, by which the Public would be benefited, it would lessen the Employment of Capital, thereby enabling Wholesale Dealers to turn their Capital oftener, also to reduce the number of their Agents and Travellers, and I think business better done by personal communication. I also consider the proposed Line

Trade of Birming. was formerly inconsiderable.

Cause of the Imp.

Advantages of a Railway.

will be an advantage to the Port of London generally.——We send principally Sugars, also Tea, Fruit, Spices, under which term a variety of goods are comprehended.——If the Canal was much cheaper than the Railway it would certainly command the preference; (although speed is a matter of great importance) but if it was only 3*d*. or 6*d*. cheaper, I consider the speed more than equivalent to so small a saving. The Canal must carry at least 30 per cent. cheaper than the Railroad to compete with it; with a difference of 10. per cent. only, the Railroad would decidedly have the advantage.——I have known a Winter when the Canal has not been frozen, but we always expect a considerable Stoppage.——As our Customers do not order the goods until they require them, they must consequently be sent quickly; certainly there may be exceptions. ——Perhaps there is more Sugar sent to Birmingham from Liverpool than from London; most of which is brought from the West Indies, and from the Mauritius. All the Tea is supplied from London.

Sugar, Tea, Fruit, and Spices are sent to B. from L.

Compar. between Canals and Railways.

A Canal must carry 30 per Cent. cheaper than a Railwy. to compete with it.

Stoppages on the Canal.

More Sugar sent to Birm. from Liv. than London.

Ex. Mr. HENRY HEMSLEY,

Director of the Union Flint Glass Company, which has been established Twenty Years in Birmingham and Two in London.

The House in London is supplied from Birmingham.——We have 18 Glass Houses in Birmingham. The Crown Glass comes from Newcastle.——Our communication with London is by Canal, and I believe 1050 are the Number of Tons sent to London annually, which I think comes to about £200,000.——A Railroad will be of considerable advantage to us, as it frequently happens that the Owners of Ships under engagements to sail on a certain day, will take less than the ordinary price, when they have not got their full freight. If we had a Railroad we could avail ourselves of the same.——There would likewise be a saving in Breakage, which at present averages 2½ per Cent.; by the Railway it would not exceed ½ per Cent. The Breakage arises from the carelessness of the men in putting the glass into the Boat, and trampling on it. A great loss also arises from pilfering.

Union Fl. Glass Co. have 18 Glass Ho. in Birm.

1050 Tons of Glass sent from Birming. annually, amounting to £200,000.

Advantage of Rail.

Breakage by Canal, By Railway.

Ex. Mr. FREDERICK BARNES,

Wholesale Ironmonger, of Fenchurch Street, also of Birmingham.

Advantage of proposed Line. Trade with Birm.

If a Railroad was established between London and Birmingham it would be a great advantage to our trade. I receive three or four packages by Coach daily from different Manufacturers at Birmingham, and always one or two from our own house; the Carriage of which we find exceedingly expensive.——We have had Instances of Packages being refused by the Coaches on account of their Weight, (as persons object to travel in a Coach with 800 or 900 lbs. on the top, or in the boot); the consequence of which has been that we have lost the order. This state of things is almost the ruin of our business. I have just received an order which I cannot fulfil in time for shipment.——I am sorry to say the Directors of the Grand Junction Canal Company have a very unjust and arbitrary way of ascertaining what kind of goods are in our packages, as they charge a freight according to the contents: They have an Instrument, I suppose a Hammer, with which they break a Hole in the Cask. I have a cask of nails in my house now, with the packages so broken, that the nails are subject to be damaged by the water, they are also liable to fall out. ——We have Cutlery and Edged Goods from Sheffield. They make Edged Tools at Birmingham, but very inferior; they cannot compete with Sheffield in light Goods, in heavy articles they do.——Metal Buttons were originally a Manufacture of Sheffield, but at present there is only one Maker of the same in the Town, the Birmingham people having taken it entirely from them.

Instances of Packages being refused by the Coach.

Method used to ascertain the Contents of Packages by Canal.

Sheff. exc. Birm. in Cutlery and light Goods.

Manufacture of Metal Buttons.

Ex. Mr. RICHARD PURKESS WESTALL.

Linen Draper, of Birmingham.

Advantage of proposed Line.

I consider the proposed Railway would be a decided advantage to us, both for rapidity, and cheapness of conveyance.——We make 24 journies in the course of the year, for the purpose of purchasing; if we had a Railroad we should go oftener, as it would not occupy so much time. I understand that the Carriages will go up in the Morning, and return at

A Train will go to Lond. & back same day.

Night, performing each Journey in six Hours.———Supposing our expense of carriage was £300. per annum, I understand that by the Railroad it would be reduced to £100.; and I find that many articles carried in the Winter, and charged six per cent, will be reduced to two, which will be a saving of four per cent. to the Public.———The bulk of our Silk Mercery, Haberdashery, and Drapery Goods come from London, also a considerable quantity of Woollen Goods; we receive the main part of them by Coach, (they are conveyed in the most expeditious manner, although very expensive) on account of the fluctuations and changes of fashion, as Silk Goods, which is 1*d.* per lb., or 8*s.* 4*d.* per cwt. We get them in about 15 or 16 hours from the time of the Coach starting from London. The Linen and heavy Goods are sent by Canal, which is 2*s.* 9*d.* for Mercery and Linen Goods, and 2*s.* 6*d.* for particularly heavy Goods, as Pins and the like. They occupy about five Days or sometimes a Week on the Journey. The Charge by Waggon is 5*s.*, occupying generally four Days.

The Journey will be perf. in 6 hours.

Savings made by the Railway.

Carriage per Coach 1*d.* per lb. or 8*s.* 4*d.* per cwt.

By Canal 2*s.* 9*d.* & 2*s.* 6*d.* occ. 5 da. or Week.

By Waggon 5*s.* occ. about 4 Days.

Ex. Mr. JOHN MOSS, *of Liverpool.*

If the London Mail could arrive in Liverpool the Morning after it leaves London (which by the proposed Line would be the case) it would be a great advantage. By arriving at Liverpool at seven o'clock in the morning, it would reach Dublin the same evening at eight or nine.———When the Liverpool and Manchester Railway was before Parliament, perhaps there were one-third Dissentients. The late Mr. Heywood, of Manchester, opposed the Bill, and afterwards complained of the Railroad not going through his land.———The Charge upon this Railway is about 2*d.* per Mile for each Passenger. The Charge for Posting from London to Liverpool is £21., with four Horses it would be Double that Sum. The Charge by the Railway would be £14. I consider that all the Posting, also all Carriages from Ireland and Scotland, would come by this Railroad, as it would be the shortest route. The Charge on the Liverpool and Manchester for putting a Carriage containing four Passengers on a Truck is £1. for the 30 miles, and if it contained an extra person, 5*s.* more would be charged as his fare; so that with double that number of persons in it, it would be double the price. According to the Bill in the House of Commons for taxing Railways, 4*d.* per Passenger is proposed for the Liverpool and Manchester; and if added to the Fare by the Company, which I do

The proposed Line would expedite the several Mails.

Charge on the L. and M. 2*d.* per Mile a Passenger.

Posting from Lond. to Liverp. £21., if 4 Horses double. By the Railroad it would be £14.

On the L. and M. charge for putt. a Carr. on Rail. £1.

Pro. Duty on Rails. 4*d.* L. and M.

not expect, would make it 5*s*. 4*d*. (5*s*. being the Fare by the best, and 3*s*. 6*d*. by the inferior Coaches). I therefore consider the Duty would not exceed 2*s*. 4*d*. per Passenger from London to Liverpool.——People come a considerable distance to avail themselves of the Liverpool and Manchester Railway; for instance, instead of going direct to Southport, (which is the most distant, and about 38 miles from Liverpool) they travel 30 miles on our Railway, and 25 on the Road, which effects a saving both of time and expense, and the same thing would probably occur on the proposed Line. During the late contested Election we sent a Steam Engine to Liverpool for some Voters, which completed the Journey there and back in two Hours. The Trains upon the Liverpool and Manchester travel at the same rate during the Night (I consider Night Travelling perfectly safe) as in the Day, viz. 20 miles an hour; I have gone at the rate of 30. They at first had a strong body of Police, which they have since discontinued, but they continue to carry Lights with them. I do not think a Stone lying on the Road would upset a Carriage.——I am acquainted with the Kenyon and Lee Railway, which is a short line, and connects the Liverpool and Manchester with the Bolton and Lee. The Plans were laid down under the direction of Mr. Stephenson, and Mr. Rastrick was appointed Engineer to do the work, Mr. Stephenson being very busy at the time with the Liverpool and Manchester. It was completed for a less Sum than the Estimate, which was likewise the case with the Wigan Line.

Proposed Duty, From London to Liverpool 2*s*. 4*d*.

Collateral Travelling.

An Engine on the L. & M. compl. the Journ. there and back in two Hours.

Night Trains.

L. and M.

Kenyon and Lee Railway.

Executed under the Estimate.

Wigan Line ditto.

Ex. Mr. HENRY CHEETHAM,

Cotton Manufacturer in the Neighbourhood of Manchester.

The principal Market Day at Manchester is Tuesday. The Mail arrives there at four o'clock in the afternoon, Foreign Letters do not arrive until the Market is over: If the Mail could be expedited six hours they would be delivered in time for the Country Manufacturers attending the Market, to make their purchases accordingly.——A saving of time between London and Birmingham, would likewise be a saving of time between Birmingham and Glasgow. The Mail for the latter place leaves Birmingham one hour after the London Mail arrives. ——The Fare by Coach from London to Manchester is £4. 4*s*. Inside, including Guards and Coachmen. The expense of Posting is from

The Railway would expedite the Mail six Hours.

Fare to Manchester

1*s*. 3*d*. to 1*s*. 6*d*. per mile.———Persons travelling from Manchester to Chester find it cheaper to go by Railway to Liverpool, although it is 12 or 13 miles round.———The Railway between Liverpool and Manchester has caused a saving to me of £150. per annum on the Conveyance of Cotton alone, and it has caused a Reduction of 2*s*. per ton in the price of Coals, or 20 per cent. to the Manchester Spinners.

Collateral Travelg.

Savings effected by the Railway.

Ex. Mr. THOMAS BADGER, *of Dudley*.

I am Magistrate for the Counties of Worcester and Stafford.——— I am engaged in the manufacture of Nails, Glass, and other hardwares, also Flint Glass.———The Population of Dudley at the last Census was 23,043; the greater part of these persons are engaged in the above trades. ———Two of the largest Manufacturers have established houses in London, where they keep large stocks. An increased communication would enable Glass Blowers to dispense with these large stocks and establishments (independent of other advantages).———The Glass and Nails are principally conveyed by Canal, which is objectionable on account of the Stoppages. There are also frequent instances of Breakage and Pilfering.———In time of Frost we are obliged to have recourse to Coach Conveyance, which is very insufficient; and also adds considerably to the expense: it is a total prohibition to Nails.———The proposed Railway would afford considerable benefit to the Poor in the neighbourhood of Dudley, which the Line runs next to. There are also many men in the neighbourhood who are accustomed to Canal work; and they often go 30 or 40 Miles to work upon Canal Cuttings, &c., similar to Irish labourers.

Manufactures and Population of Dudley.

The advantage of the Proposed Line.

Conveyance by Canal.

Insufficiency and Expense of Coach Conveyance.

Benefit of the Railway to the Poor.

Men go 30 or 40 Miles to work upon Canals.

Ex. Mr. JOHN CHEETHAM.

I consider that persons travelling from London to Leicester (which occupies 11 hours by Coach) would travel by the proposed Line as far as Rugby, where they would take the Coach to Leicester, completing the journey in 6½ hours, and I understand the tolls by the Railway would be about £1. 1*s*.; the expense by Coach is about £2. or £2. 2*s*.———It also occupies 15 or 16 hours in travelling by Coach to Nottingham, but by taking the proposed Railway to Rugby, as before described, the journey would occupy only 11 hours. The Fare by the Coach from Nottingham is

Saving of Time to Persons going to Leicester by Pro. Line.

The same to Nottingham.

35*s*., by Railway it would be 25*s*.———A corresponding saving would be made in the conveyance of Lace made at Nottingham; the greater part of which is at present forwarded by Van and Coach.

Ex. Mr. RICHARD WHITMORE.

The following paper is a correct Return, taken from the Custom House, and signed by Mr. Irving.

Provisions imported into London from the Netherlands.

An Account of the Quantities of Butter, Cheese, and Eggs imported into the Port of London from the Netherlands during each of the last Three Years.

	Butter.			Cheese.			Eggs.
	Cwts.	*qrs.*	*lbs.*	*Cwts.*	*qrs.*	*lbs.*	*Number.*
Year - - 1829 -	115,002	1	4	91,624	2	22	4,221,960
Year - - 1830 -	76,477	3	13	60,627	1	19	3,477,208
Year - - 1831 -	79,797	0	22	93,057	2	17	6,761,666

(Signed) WILLIAM IRVING,
Inspector General of Imports and Exports.

Inspector General's Office,
Custom House, London,
10th April, 1832.

Ex. Lieut.-Gen. Sir J. W. GORDON, Bart. K.C.B.,

Quarter Master General.

Barracks and Mil. Depôt at Weedon

Advantage of the Railway to same.

Troops should always march to their Stations.

I have been informed that the proposed Line of Railway will pass near Weedon, where there is a considerable Depôt for Troops, and Military Stores, and I consider that a more rapid conveyance for the latter would be of great utility. Regarding the conveyance of Troops, my opinion is that they should not travel by any conveyance whatever; but perform one of the most efficient parts of military duty, namely, march. In cases of emergency, where the presence of the Military may be necessary, (which frequently occurs in all countries) the Railway would be desirable.——A Railway certainly may be easily destroyed, but the same remark is also applicable to Canals; a dozen men could cut down the bank of a Canal, and interrupt its passage in a very short time; but this I consider an extreme case; means could also be taken by the civil Magistrates to prevent it, should such be attempted.——Regarding any imagined insecurity of the Military Stores, arising from the facilities which a Railway would offer to the multitude, (in the event of a riot) I am quite sure there is always a competent force to protect them.

A Railway could be easily destroyed if not guarded.

A Canal the same.

Ex. Mr. WILLIAM KAY, *of Tring.*

I am acquainted with the Silk Trade.——I was Chairman of a Committee of Silk Manufacturers in 1828 and 1829, in which Coventry, Macclesfield, Congleton, and London were represented.——The Value of a Commodity depends much upon the Speed with which it can be conveyed to the Market, and I consider it a material thing to get all Fancy Goods to Market as soon as possible.——The Railway passes through some of my property, but I cannot say to what extent; I have many fields, and should have no objection to the proposed Line passing through the whole of them, as I am sure it would increase their value.

The Value of an Article reg. by the Facilities of Con.

Fancy Goods require a speedy Con.

A Landowner satisfied the Railway will imp. his Land.

Ex. Mr. JAMES MARSHALL,

Secretary to the Provincial Bank of Ireland.

All the Bullion of this Bank is transmitted from Liverpool to Ireland (from the Bank of England).——The direct course of post from London to Dublin occupies 36 hours, provided the Packet can sail immediately; there would be a Saving of six hours in the conveyance of Bullion by means of the proposed Line.——I have known cases when the latter would have been of immense service. Some years back political circumstances occasioned a sudden demand for gold, and we were called upon to supply it *instanter* without any notice; the saving of an hour at that time would have been of great importance.

Bullion of the Bank of Ireland sent by Liverpool from the Bank of England.

Necessity of the above being conveyed quickly.

Ex. Mr. RICHARD CREED,

Secretary to the London and Birmingham Railway.

Taking the Traffic by Coach at Railway Prices gives 2*d.* per Mile for each Passenger.

I have examined the information and evidence supplied by the several Gentlemen, regarding the Traffic on the Road, by Coaches, Posting, Canal Boats, &c.; also the Parcels by Coaches, the Goods by Waggons and Fly Boats, and the Cattle.——Captain Moorsom's Calculation of the Coach Traffic, at Railway Prices, amounts to 2*d.* per Mile for each Passenger, as proposed to be levied by the Act.——I have made a Calculation of the expected Traffic on the proposed Line, (which is annexed). The Parcels from Birmingham to London, and from London to Birmingham, only are taken.

Summary of Traffic at Railway Prices.

Table of the expected Traffic on Proposed Line.

	£. s. d.
Statement A (Coach Traffic) - - - - - - - - - - -	246,916 16 0
Statement B (other Traffic) - - - - - - - - - - -	244,858 8 10
Mr. Henry Booth in his Evidence before the Lords Committees states, that the Number of Passengers on the Liverpool and Manchester Railway is nearly Three Times what it was by Coaches between Liverpool and Manchester before the Railway was opened, or in the Proportion of 1,200 to 450.	491,775 4 10
It may be presumed therefore that at least Twice the average Number of the Passengers by Coaches in Statement A, would travel by the Railway between London and Birmingham; say - - - - - - -	246,916 16 0
Total - £	738,692 0 10

(Signed) RICHARD CREED.

Note.—Statement B does not include:—

1st. Parcels by Coaches to intermediate Distances.

2d. Vans, Market Carts, and Gigs.

3d. Fly Boats conveying Liverpool, Manchester, Derbyshire, and Leicestershire Goods, which enter the Line at different Places between London and Birmingham, and only go Part of the Distance.

Nor is any Allowance made for the considerable Increase, which may certainly be calculated upon from the Carriage of Meat, Fish, Dairy Produce, and Eggs, and of other Articles, not now carried to any Extent by the existing Modes of Conveyance.

STATEMENT (B.)

Abstracted from the Minutes of Evidence taken before the Lords Committees on the London and Birmingham Railway Bill, in respect to Traffic.

Evidence.		£. s. d.	Rate of Expense as at present. £. s. d.		Rate of Expense at Railway Prices. £. s. d.
Coach Parcels: John Hart -	Coach Parcels by Fourteen Coaches and Two Mails which pass daily each way between London and Birmingham, Average £2. for each Journey, is for One Year		23,360 0 0	Same Price - - - - -	23,360 0 0
Posting: Matthew Holman* -	Twenty Pair of Post Horses on an Average pass Hockliffe daily. Taking the Average between London and Birmingham at* Fifteen Pair daily, and at 1s. 9d. each Mile, the Amount for 180 Miles for One Year is - - - - - -		51,738 15 0	Three Persons to each Pair of Post Horses is 16,425 for One Year; say at 2d. per Mile each Person for 112 Miles - - -	15,330 0 0
Goods by Waggon: John Shackell -	Goods carried on an Average by Waggon:				
	From Birmingham to London - 1,248 Tons at £5. per Ton	6,240 0 0			
	From London to Birmingham - 1,040 - £3. -	3,120 0 0			
	2,288	9,360 0 0	- - -	at £2. 6s. 8d. per Ton £5,338. 13s. 4d.	
	From intermediate Towns to London and back - 1,872 - - - - -	4,270 10 0	- - -	at £1. 3s. 4d. - £2,184. 0s. 0d.	
	4,160				
	Goods taken up, and delivered on the Road by Waggon, between London and Birmingham, direct, equal to Two Thirds of the Goods carried the whole Distance - - - - - - - - - -	6,240 0 0	- - -	Two Thirds of £5,388. 13s. 4d. £3,559. 2s. 2d.	
			19,870 10 0		11,084 15 6
Goods by Fly Boats: William Partridge†	Goods carried on Canal by 35 Fly Boats which pass up weekly, averaging yearly 27,300 Tons, at £2. 10s. per Ton	68,250 0 0	- - -	at £2. 6s. 8d. per Ton £63,700. 0s. 0d.	
	35 down 14,560 Tons, at £2. 2s. 6d. -	30,940 0 0	- - -	at £2. 0s. 0d. - £29,120. 0d. 0d.	
			99,190 0 0		92,820 0 0
Cattle: Robert Attenborough W. M. Warner -	1,600 Oxen drawn up weekly, on an Average for Six Months in the Year, between Coventry and Hockliffe. Estimates driving and Loss for 50 Miles at 17s. 6d. each Ox. Estimates driving and Loss for 40 Miles at 14s. Say 1,600 Oxen weekly for Six Months, or 41,600 Oxen for One Year, at £1. each for 80 Miles - - - - - - - - - -		41,600 0 0	at 3d. per Ox. per Mile for 80 Miles	41,600 0 0
Robert Attenborough	7,000 Sheep driven up weekly, on an Average throughout the Year; Charge for driving 1s. Estimates Loss in Weight 4s. 6d. Say 7,000 Sheep weekly is 364,000 for One Year, at 5s. 6d. each for 80 Miles		100,100 0 0	at ½d. per Sheep per Mile for 80 Miles	60,666 13 4
	Total - -		335,859 5 0		244,858 8 10

* Calculated in the Proportion of the Coaches between London and Birmingham, allowing Twenty Pair of Post Horses from Hockliffe to London, and Twelve from Birmingham to Hockliffe, the Mean is Fifteen Pair and a trifling Fraction for the whole Distance of 108 Miles.

† John Norton and Thomas Norton state that 366 Fly Boats passed Braunston in Fourteen Days up and down, equal for Seven Days to 183 Boats, being 113 Boats more than are included in the above Statement.

Ex. Mr. AUGUSTUS GODBY.

The proposed Line would expedite the Mail six Hours.

I am in the Department of the Secretary of the Post Office of Dublin.——I consider that expediting the Mail six Hours at Liverpool, Manchester, and Glasgow, would be a great advantage.

Ex. Mr. JAMES FORSTER, *Broker in London.*

The Leeds and Liverpool Canal has increased in Value since the opening of the L. and M.

I am well acquainted with the value of Railroad and Canal Property. Canal Property has increased in value since the opening of the Liverpool and Manchester Railway. The Leeds and Liverpool Canal, (of which the Duke of Bridgewater's is a part) which it was expected to interfere with most, has increased in value, and is improving half-yearly; the dividends are the same as they were in 1829, (viz. £20. per Annum) although a large sum of money has been appropriated to paying of their debt. I have always understood that they opposed that Bill under an apprehension that it would prove prejudicial to their property, and not on account of it interferring with some of the Warehouses belonging to the Canal, as advanced by the Counsel for the Opposition to the Bill.

Ex. Mr. JOHN DILLON, *Silk Manufacturer of London.*

People of Coventry mostly engaged in the Ribbon Trade.

Nine-tenths of which are sent to Lond. by Coach.

People at Coventry will be able to visit London and return in the same day.

I am in partnership with Mr. Morris, M.P. for Ipswich, and have been extensively engaged in the Coventry and Manchester trade for some years.——The great mass of the labouring Population of Coventry are engaged in the manufacture of Ribbons; about Nine-tenths of which are sent to London; they are always sent by Coach, (which is expensive), a correct and speedy communication being indispensable, as it is a fancy article, depending upon fashion.——We frequently have an order for a general Assortment of Goods, and two days and nights must elapse before we can procure our supplies from Coventry, as the intermediate towns do not manufacture them, which is a great loss to us.——The proposed Railway would be a great advantage to Tradesmen, it would enable them to visit London on the general Market Day, and return at Night; it would likewise be advantageous in a general point of view.

GREAT WESTERN RAILWAY.

ABSTRACT OF EVIDENCE,

Given before a Committee of the House of Lords, June, 1835.

COUNSEL AND WITNESSES FOR THE BILL.

COUNSEL.

Mr. St. George Burke, (Parliamentary Agent to the Bill).
Mr. Harrison.
Mr. Serjeant Ludlow.
Mr. Talbot.

WITNESSES.

Mr. Isambard Kindom Brunel.
Mr. Thomas Cabrey.
Mr. Joseph Locke.
Mr. George Stephenson.
Mr. Henry Robinson Palmer.
Mr. George Watson Buck.
Mr. James Copeland.
Mr. Henry Habberly Price.
Mr. Robert Stephenson.
Mr. Charles Vignoles.
Mr. Edward Driver.
Mr. Daniel Lonsley.
Mr. Francis Hawkes.
Mr. Henry Edmund Gooderidge.
Mr. Young Sturge.
Mr. James Hammond.
Mr. James Otto Heisse.
Mr. Richard Creed.
Mr. Henry Rowles.
Mr. William Warren Sutherland.
Mr. William Shearman.
Mr. John Scholes.
Mr. Thomas Jones Howell.
Mr. John Pugh.
Mr. Roland Jones Venables.
Mr. Robert Podmore Clark.
Lieut. Nicholas Chapman, R.N.
Mr. William Donne Bushell.
Mr. Charles Ludlow Walker.
Mr. John Harley.
Mr. Charles Wilkins.
Mr. Robert Cordwent.
Mr. Thomas Marling.
Mr. Thomas Reynolds.
Mr. William Montague.
Mr. Thomas Birch.
Mr. Philip Davis.
Mr. Thomas Morris.

*** The Bill was entituled, "An Act for making a Railway from Bristol to join the London "and Birmingham Railway, near London, to be called the Great Western Railway; with "Branches therefrom to the Towns of Bradford and Trowbridge, in the County of Wilts." (The Bill was passed.)

COUNSEL AND WITNESSES

FOR THE

OPPOSITION TO THE BILL.

COUNSEL.

Mr. Serjeant Merewether.
Mr. Joy.

SOLICITORS.

Messrs. Few, Hamilton and Few.

WITNESSES.

Col. George Henderson.
Dionysius Lardner, L.L.D.
Mr. John Urpith Raistrick.
Mr. George Leather.
Mr. Francis Giles.
Mr. Henry Smith.
Mr. Charles Fowler.
Daniel Mallet.
Mr. Zachariah Allnutt.
Mr. George Treacher.
Mr. John Cutler.
Mr. Charles Tull.
Charles Harding.
James Barnes.
William Hatch.
Mr. Benjamin W. Scott.
Mr. Stephen Leach.
Mr. Charles Baker.
Mr. Benjamin Wingrove.
Mr. George Henry Elliott.
Mr. William Ashman.
Mr. George Baillie.
Mr. Thomas Carter.
Mr. William Gifford Cookesley.
Mr. Walter Long.
Mr. John Bleeck.
Mr. William Gray.
Mr. William Stone.
Mr. James Hammond.
Mr. John Child.
Mr. William Pinnegar.
Mr. Joseph Hull.
Mr. Robert Hull.
Mr. Edward Sherwood.
Mr. William Shackell.
Mr. John Bulley.
Mr. Robert Palmer.
Mr. Albert Dangerfield.
Mr. William Seal Evans.
Mr. Pearson Thompson.
Lieut. Thomas Francillon.
Mr. Maurice Swabey.
Mr. Samuel Foss Dessian.
Capt. James Weeks.
James William Deale.
John Rouse.
Mr. Henry Scrivener.

Ex. ISAMBARD KINDOM BRUNEL, ESQ. C.E.

I was applied to about February 1833, by the Provisional Committee, on the practicability of making a Railroad from Bristol to London. The connection between myself and the Provisional Committee was sought for by them entirely; my Instructions were no further than asking my terms, stating the object was to Survey the country from Bristol to London, with a view to the best Line of Communication, without reference to any other district.——Shortly after commencing the Survey, I pointed out to the Committee two or three different Lines I had examined, one passed to the South of Marlborough Downs, and the other to the North. I stated the North line would embrace all the communication on the South and Gloster. The South line was by Reading and the Wiltshire Hills, and was very similar to the Basing line as far as Newbury; I then followed the valley to Newbury, and Newbury to London. When I had taken sufficient Levels I found the Southern line was inferior to the Northern on account of the general Levels, and comparative Population and importance of districts. A Survey having been made for a Turnpike Road, one of those took the North and the other the South of Marlborough Downs; these I got access to, and the Levels of the Canals. The Level of the Kennet and Avon Canal, which takes the Southern line, is about 130 feet higher than the Wiltshire and Berkshire Canal, which takes the Northern line; the former Canal passes that summit by a deep Cutting, while the latter is carried over it without any, so that the Summit Level on the one side is considerably greater than the height of the other. I first examined the Southern line, and, finding its difficulties, I tried the Northern; and the Directors likewise agreed with me, that embracing Gloucestershire and South Wales was important. When the prospectuses were issued, I made a Survey in a more regular and formal manner, and the Levels were taken to enable me to prepare the Estimates. ——The general course of the Line is governed to a certain degree by the position of the principal Towns. There is a great superiority in the Western line over the Basing in point of Population, which I have taken in various ways,—here is one statement where the tabular Population is taken of all towns lying within 10 miles of each Line, and no place unless it has 1,500 inhabitants, nor is there any Parish included that has not some central Town. (See page 73 for Table.)

Instructions from Prov. Committee.

The Southern line.

The Northern line.

Kennet and Avon Canal higher level than the Wilts and Berks.

Superiority of the Northern line.

A Line of Railway should be governed by the Population and importance of the adjacent Towns.

Superiority of the Northern in this respect.

Description of the proposed Line. Bristol Terminus.

The Harbour.

Description of the Line.—I will commence with the Terminus at Bristol, for which Plans were submitted by several parties; Temple Meads presented itself as the only eligible spot for a Depôt; the Harbour of Bristol is made by damming up a part of the river, and is in length about $2\frac{1}{2}$ miles.

Line between Bath and Bristol.

——Between Bristol and Bath the country is hilly, and very difficult; the river, which winds very much, being the only valley. The Levels are very favorable, the steepest being but 6 feet 10 inches per mile, which is nearly equal to a Level; the quantity of Cutting and Excavation is not great. The Line is carried across several ridges of old red Sandstone, around which the river winds; at these points I proposed making short Tunnels, but they might be made open Cuttings.

Description, Dimensions, &c. of the Tunnels.

——The first Tunnel is about $1\frac{1}{4}$ miles from Bristol, (under Langton Court farm) and through soft Sandstone rock; it is about 435 yards long, 25 feet wide and 30 feet high, (all the Tunnels are of similar dimensions) having a fall of $6\frac{1}{2}$ feet, which is at the rate of 1 foot 2 inches per mile; this Tunnel was made at the desire of the proprietor, the Estate being considerable, and having some good houses upon it. The next Tunnel is 132 yards long; the ground being higher a tunnel is more advisable. The next is 1,012 yards long, with an inclination of 1 foot 2 inches per mile; it is through hard Sandstone called Pennant stone, which will be serviceable in the works. The Line then skirts the river, and again enters Rock Cuttings,

Runs near Keynsham.

the extreme point being 76 feet deep; it then comes out near Keynsham, passing close to the village; it then passes under Saltisford in a Tunnel of 528 yards, through the Lias lime stone rock; it again skirts the river, crossing Keynsham meads, passing through the village of Tiverton.

Deviation from the line proposed last year.

(Last year the line crossed the river twice to avoid Tiverton, but we have since obtained the assent of one of the principal proprietors, also of the Commissioners of Turnpike Roads, as we divert the latter; the line then proceeds straight to the old bridge at Bath, crosses the river obliquely, and goes into Ham gardens, where the Depôt is situated, occupying about 8 or 10 acres of ground.

Description of the Depôt in Ham gardens.

This Depôt was settled by the Mayor and Corporation of Bath, after mature consideration; it will be close to the principal thoroughfares, and within 2 or 3 feet of the level of the streets. The Oblique Bridge by which we cross the river is on the Bristol side of the Depôt; it is estimated at £12,500.

Bridge at Bath.

Termination of the Basing line.

—— ——The Termination of the Basing Line is on the South side of the river, (80 feet above the level of same) and upon the side of a very steep hill. (Our line is about 34 feet above the level of the river.) I therefore consider the proposed Basing Line Depôt objectionable on account of its situation, and it will be necessary to make a zig zag approach to it, in order for the Bath flys or

Objections to the Depôt of the latter line.

carriages to get up; The Soil is formed of Clay and the Débris of Beechen cliff, which has sunk down; Springs run in all directions; the Water Works of Bath are supplied from springs in the direction of this Depôt; the ground is very bad, and you may see on the upper ground at Beechen cliff the cracks and slips made from the slipping of the Débris under it; the ground would not be accessible for carriages without the zig zag approach, in addition to which a portion of this Hill will have to be excavated for the Depôt; (from which the water is collected in Conduits, and led into Reservoirs for the supply of Bath water works, but still the slipping of the ground is most to be feared.)——Mr. Brunton prepared a Plan for a Railway from Bath to Bristol, which I have seen; it was totally different to this, (he has made four) and proposes to come from Beechen cliff straight to Bristol, being on a much higher Level; and terminating above the Level of Somerset Square, which is the very top of Redcliffe Hill; he avoids Tunnels by very deep Cuttings. In proceeding from Bristol, his Line follows the valley at the back of the Hills, through which my Line goes. (I avoided the valley on account of the Residences and Gardens being valuable, besides the Levels are inferior to the ground through which my Line passes.) At Keynsham the two Lines very nearly join, except that his is at a much greater elevation than mine. From Keynsham to Tiverton they are much the same, except at Newton St. Looe part of it passes very near to Mr. Gore Langton's park; the Line originally proposed crossed the river near to Kelston, through Bath on the North side.—— ——On leaving Bath we pass through Sidney Gardens in Cutting 230 yards long, and we have made an arrangement with the proprietor to cover the same part with wooden roof and skylights in an ornamental manner. At one part it is Level with the gardens, at another it is 3 feet below it. This covered way will be 16 feet from the Level of the Railway, which is 2 feet more than usual. I laid down two Lines in the direction of the Tunnel; one has an inclination of about 16 feet per mile, and is about 9 miles in length, the other immediately quits the ridge near Bath, takes a lower Level until at the portion of cutting near the Tunnel, and has an inclination of 9 feet per mile. My reason for laying down two lines on the Section in the direction of the Tunnel was as follows: In laying down a line of Railway for the whole distance, I found, with the exception of that 9 miles, I could reduce the Levels so that none should exceed 11 feet per mile, which would be an advantage, as the lower the maximum can be kept upon a Railway the better, because the power of the Engines is governed by the inclinations over which they have to take the load, except it is very

Description of Mr. Brunton's Lines from Bath to Bristol

Continuation of Description of the G. W. line.

Passes through Sidney Gardens.

14 ft. Headway for the Railway.

Reason of laying down two lines in the direction of the Box Tunnel.

A Rise of 16 feet a mile, or 1 in 330, affects the power of an engine.

short indeed; if there is an extension of inclination of 16 feet per mile, or 1 in 330, it affects the power of the Engine, consequently they must have an Assistant Engine, or be calculated to ascend it without one. The nature of the country along the Line is very favorable, and could be kept at 11 feet per maximum, except those 9 miles, which would be 16 feet per mile; and to pass this inclination of 9 miles either the Engines must be of sufficient power to overcome it, or the loads must be reduced. To have Assistant Engines upon the whole line able to get over 16 feet per mile, would be a serious loss; it therefore became important to have Assistant power of some kind, and it is best to have the Line steep at that part, in order to have the full assistance of such power, which will leave the rest as nearly level as possible: There is nothing new in this theory; the same principle is applied on the Liverpool and Manchester, as, in the middle of it, they have to get over a sharp ridge a mile and a half each way, one inclination is 1 in 96, the other 1 in 90, where they have an assistant engine. I preferred laying down two lines, in order that we might afterwards adopt either. ——I consulted Mr. Stephenson, Sen. and Mr. Palmer upon it, both of whom confirmed my opinion that the one containing a short and steep inclination was the best. The length of the Inclined Plane is $2\frac{1}{2}$ miles, including the Tunnel, which is nearly $1\frac{3}{4}$ miles, and the whole is straight; the last inclination at the foot of the Tunnel is 9 feet a mile; which is our steepest. We then go up the inclined plane at an inclination of 1 in 107, which is less steep than that on the Liverpool and Manchester.—— —— A considerable length of the Tunnel would be in Bath stone. (In the hill are some of the best Bath stone quarries in the neighbourhood.) It would then be in Cornbrash, then again in Bath stone, and the end next Bath in Clay. The descent of 1 in 107 is considerably less than the descent from the houses at the corner of Parliament Street to the Parliament Houses, and about half as steep as Burlington Arcade, and exactly as steep as the Lowther Arcade, therefore no effect could be produced upon the passengers arising from a feeling that they were descending something precipitous; the longest time occupied in passing through the tunnel would be about six minutes. There will be four working Shafts, and we shall probably make more for air: these will be left open to give light; should there be a desire on the part of the public, we could easily light it with gas. The deepest Shaft is about 220 feet; another would be 90 feet; the principal object of the shafts is for ventilation. The Bath stone continues down to about 120 feet, and is easily cut through (until some time exposed to the

The advantage of having assistant power at an inclined plane, in preference to making the locomotives able to go over it.

Instance of the above principle in the L. and M.

Description of the inclined plane and tunnel.

Descent of plane 1 in 107.

Shafts.

Description of the formation of Tunnel, &c. — strata passed through, &c.

air) and we should be very glad of it, as it would be useful. There is an establishment in the neighbourhood to cut it up into blocks and slabs: it would be harder than clay to work, but it would pay for the cutting. It would in a great degree make its own lining; in some parts we should blast it.—— ——In passing through the tunnels I do not apprehend any danger from the effluvia from the engines. There is a tunnel already formed upwards of a mile long, on the Birmingham line.———The inclination of our tunnel makes no difference; the smoke would certainly be greater with two engines than with one. Whether I should take the goods up by a moveable assistant engine, or by a stationary engine, I have not yet determined; but we should use that which answered best.———There is a tunnel at Leicester nearly a mile long, 12 feet high; they burn coal, which causes much smoke, yet persons pass through it.———If an assistant engine is used to carry the train up at full velocity, you have two engines at the same time, one of them behind the train. In passing down the inclined plane, the engines will not work, the steam must be shut off, and the damper put down; then, instead of smoke there is an escape of heated air; the sound and effect of which is very different to smoke; if they are the same engines that are used upon the level line, the fire must be kept up.—— ——Upon the Canterbury and Whitstable Railway a passenger train starts every hour; and a train for goods also starts every hour. I have been down the Canterbury Plane twice without a rope in a single carriage, and when at its full velocity, it was stopped by the break within a distance of 60 yards; there were five persons in the carriage; (therefore should the rope snap in descending the inclined planes, there could be no danger, as it is not absolutely required).—— ——The Break is a piece of wood that drops on the wheel, causing a slight friction; the resistance required to detain a carriage is very trifling, any thing applied to the wheel gradually stops it.———On the Manchester and Liverpool Railway the engines and carriages run down the incline (which is steeper than ours) every day, (and the darkness cannot increase the danger, ours being in a tunnel). Upon the same Railway, with all their experience, they are making a tunnel upwards of a mile and a quarter long, (with a steeper inclination than ours); this tunnel is intended for passengers to go into Liverpool, and they cannot have many shafts, as it goes under the town. The inclination is 1 in 100.———If coke is burnt, heated air issues from the engines instead of smoke. I believe the shafts would draw off the heated air, each engine being in the tunnel about 4 or 5 minutes; I do not

Working of the inclined plane.

Tunnel at Leicester.

Canterbury and Whitstable.

Inclined plane upon the latter.

Description of the break.

L. and M. Tunnel into Liverpool is 1¼ miles long, and an inclined plane of 1 in 100.

Necessity of shafts.

Tunnel on the Birmingham, $1\frac{1}{2}$ miles long, 25 ft. high.

Means of ventilating tunnels.

Assents and Dissents between Bath and Bristol.

Curve $\frac{3}{4}$ of a mile radius quite safe.

anticipate any inconvenience from it. There is a tunnel a mile and a half long upon the Birmingham line, and about 25 feet high, with no more than four shafts.——Should the ventilation be imperfect, it could be easily remedied by making a fire in one of the shafts, as they do in mines, or otherwise.—— ——Between Bath and Bristol the assents are equal to 9 miles and 26 chains, the neuters $50\frac{1}{2}$ chains, half-special $17\frac{1}{2}$ chains; total 10 miles 14 chains. The first Curve is about a quarter of a mile from the end of the plane, and is entirely in deep cutting, with a radius of three-quarters of a mile, which has been found perfectly safe for locomotive engines to work on; it produces but the slightest inconvenience possible in the distance, but no difficulty.—— ——From the tunnel the line proceeds in deep cutting for $2\frac{1}{2}$ miles, until it gets to the open ground near Chippenham and Wootten Basset; passes close to the town; runs northward, and crosses the Avon near Christian Malford: it then goes upon an embankment varying from 20 to 37 feet, through Grittenham great wood to Wootten Basset. Proceeding from Bath, the landowners almost entirely assent for a distance from near Corsham parish; and through Chippenham and the immediate neighbourhood they all assent; through Langley Bural, still assenting, until we come to the parish of Draycot, where we have not received an answer from Mr. Wellesley; we then enter the Marquis of Lansdowne's property, who assents, and then upon Lord Caernarvon's property at Christian Malford, which is an embankment, varying from 25 or 26 feet to nothing; it is wet land; the Railway will improve it, as we shall make drains, on each side of the embankment; we shall also make whatever bridges and culverts may be required, both now, and at any future time. The flooding of the land is owing to the stagnation of the water, but it does not reach us. I consider that the several culverts would be thoroughly sufficient, even in cases of great falls of snow and sudden thaws, as they will be much larger than the surrounding ditches; after Lord Caernarvon's there is a small portion of Lord Peterborough's; the next I am acquainted with is Lord Holland's, it is a large wood, which we go partly through. The next property of any extent is Lord Clarendon's, which is very extensive, between 3 and 4 miles in one piece; I believe he assents. The next is the Charterhouse property, which is in the parish of Lydiard Tregoz; they assent. There is no extensive property until we come to Stratton, St. Margaret's; but they are generally consenting parties. We then go through part of Lord Bolingbroke's; then they all assent until we come to another farm of Lord Caernarvon's, in the parish of Highworth; the line then passes

Continuation of Description of line

Assents and Dissents continued.

Provision for Floods.

through a farm, which is a small estate for sale; the line runs parallel with the turnpike road; we sever small portions of the fields, which we should purchase; we then come to Lord Barrington's property, he assents; we then come to Lord Craven's, he assents; we are now in the parish of Baulking, we go through a considerable extent of property belonging to Mr. Bastard and his family.——The greatest cutting would be on the hill near Grittenham wood; at one part it may be 70 feet, but by moving on one side I could make it less, and by moving it a little on the other, take it under by a tunnel. We then pass through some extensive property, belonging to one or two colleges at Oxford. We pass through much property, all assenting, until we pass to Coombe lodge, at a distance of ½ or ¾ a mile from the house, and we have agreed to plant the Railway within a mile of direct distance from same, either way, to keep it out of sight. The line is still in cutting, and passes through a farm belonging to Sir Francis Sykes, who assents; we pass through several estates, who assent, until we come to Mr. Willder's, Purly House; he assents: it is a large house, having a considerable park. We have carried a Tunnel through Chalk hill, at the back of the house, to avoid it; this tunnel is 1914 yards long. I should work it by shafts; it is a very fine chalk, and could be easily made, and ventilated. The line then goes into deep cutting, and comes out in embankment: it soon gets into the property of Mr. Knowles; we make this tunnel to avoid going through the house; the expence of the house and grounds would be worth incurring to avoid it. We pass through several estates, on embankment, who assent; we then come to Crown Property, where we cross the river Kennet, (which is the Kennet and Avon Navigation) close below Reading, on an embankment, 20 feet high. We pass the meadows, (which are low, but well drained) sufficiently high for the roads to pass under.——We pass through several estates until we come to Mr. Palmer's, who strongly objects; we pass within ¾ of a mile of his house, (at the back) which is in a park. The tunnel we now propose is about ⅝ of a mile long; last year it was 1 mile long. The alteration has not been made with a view of annoying Mr. Palmer in any way; if it would satisfy him, we should be happy to return to the longer tunnel, although we object to it on account of the expense. The expence of tunnel would be about £40. per yard; the soil removed it would go to the embankment over the Loddon, and in front of Reading. We then cross the valley of Loddon, on a 25 feet embankment. There will be two or three Bridges on the Loddon, the arches of which will be in 2½ bricks, and the estimated cost

Fields that have small portions severed will be pur.

The greatest depth of cutting 70 feet.

Expence of a tunnel, about £40. per yard.

Bridges over the Loddon.

£4462. We then come to (where we cross the Colne there will be twelve arches, of 20 feet each, the estimated cost of which is £3965.) the property of Mr. Leverson Gower, at Twyford, who assents; we continue on embankment, 15 feet high, until we go into deep cutting in chalk, through a small portion of Lord Baybroke's, who dissents, and a considerable length of Mr. Vansittart's, who assents, and proposes making a depôt at one of his lodges. After passing through some estates that assent, we come to the valley of Maidenhead, where we cross the Thames, about $\frac{1}{4}$ of a mile below the present bridge. Ours will consist of three arches, of 80 feet span each, over the river, and 30 feet above high water mark, and two on each side on the land, of 40 feet; the piers will be 12 feet wide, the width of the bridge is 31 feet from out to out, and the estimated cost is £21,000. We then get into Taplow Parish, and cut through two or three fields of Lord Orkney's, near Clifton; after passing through several estates, we cross the road at Slough entirely in cutting, we approach within $\frac{1}{8}$ of a mile of Bayley's school, and pass $\frac{1}{4}$ of a mile the other side of the turnpike road, which is $1\frac{1}{2}$ mile from Eton College, to which a Branch was proposed, but at the request of the College, we gave it up; they likewise had an objection to the Railway, on account of the facilities which it would offer to the Scholars; we are willing to remedy the same, by building a fence, or wall, for $1\frac{1}{2}$ or 2 miles of either side of the College, thereby excluding them. I think it should be 8 or 9 feet high; a single wall, 8 feet high, and 4 miles long, would cost £8000. or £9000., and it is intended to be on both sides the Railway; but where it is on an embankment, (which is but for a short distance) it would be easy to line it with other materials.——We pass on until we come to the parish of Hillingdon, passing through the property of Mr. Charles Towas, who assents; I met an Engineer on the subject, who apprehended danger from the floods, but we have arranged as to the arches. We pass on to Hanwell, and then to Ealing, passing through the property of Messrs. Wood, with whom we had some difficulty, on account of our cutting up their farms, to which they attach great value; we have however fixed a price, and arranged with them as to going across the fields with a belt of trees, &c.; (there will be a covered way under the road at Mr. Wood's) this brings us upon the Birmingham line, close to the canal.——The total amount of assents upon the whole line is equal to 66 miles, $6\frac{3}{4}$ chains; neuters, 17 miles, 26 chains; special answers, 8 miles, $39\frac{1}{2}$ chains; making a total 91 miles, $72\frac{1}{4}$ chains: and the dissents, 23 miles, $36\frac{1}{4}$ chains: Total upon the whole line, 115 miles, $28\frac{1}{2}$ chains.——By Special Answers I mean such as "Declines to give

Bridge over the Colne.

Bridge over the Thames, at Maidenhead.

A Wall, 8 ft. high, 4 miles long, would cost about £8,000. or £9,000.

Total amount of Assents & Dissents.

Special Answers.

an answer," and "abroad." If a party says, " I will give no answer," we class it Neuter;* we generally consider Neuters equivalent to Assents, if they do not Dissent before the bill passes they are Assents.——The Proprietors and Occupiers are returned separately. It is customary to pay for the Land at a Valuation, and give a sum over and above for the injury or imagined injury done.—— ——The Great Western Railway will leave the London and Birmingham Railway 4 miles from the depôt, just beyond the New Cemetry; thence pass through Southall, close to the market, within 2½ miles of Uxbridge, just North of Slough, and North of the Great Western Turnpike Road, touching Maidenhead, through Reading, within 3 miles of Wallingford, within 4 of Abingdon, through which the communication with Oxford would be effected, which is within 10 miles of Oxford, close to Wantage, within 6 miles of Cricklade, within 4 miles of Cirencester, 19½ miles of Stroud, close to Wootton Bassett, within 6 of Malmsbury, within 11 of Tedbury, within 6 of Calne, through Chippenham, between which place and Bath there is a Branch, included in the Bill, and another Branch to Melksham, Bradford, and Trowbridge, within 7 miles of Devizes.—— ——

Neuters.

Syst. of Valuation.

Route & Distances of the line, from the adjacent Towns, &c.

POPULATION.

Comparative Table of Population of the different Towns lying within Ten Miles of the Great Western Railway and the Basing and Bath Railway.

Parishes which are spread over a considerable Surface, and have no central Towns, are not included.

N.B.—No Town is taken into this Calculation lying nearer to London than Fifteen Miles, or containing fewer than 1,500 Inhabitants.

Great Western Line.

Uxbridge and Hillingdon	6,885	Highworth	3,127
Windsor and Eton	10,335	Swindon	1,742
Maidenhead	6,817	Wootten Bassett	1,896
Marlow	4,237	Calne	4,876
Henley	3,618	Cricklade	1,642
Reading	15,595	Malmsbury	2,169
Wallingford	2,563	Chippenham	4,333
Abingdon	5,259	Corsham	2,952
Oxford	20,649	Laycock	1,640
Farringdon	3,033	Box	1,550
Wantage	3,282	Carried forward	108,200

* There appears to be some little ambiguity here, as "Declines to give an Answer" is classed Special Answer; and "I will give no Answer" is classed Neuter.—*Editor.*

Table of Population.

Brought forward		108,200

Proposed Branch to Gloucester.

Cirencester - -	5,420	
Tetbury - - -	2,939	
Stroud - - -	8,607	
Minchinhampton -	7,255	
Painswick - -	4,099	
Avening - - -	2,396	
Horsley - - -	3,690	
Kingstanley - -	2,438	
Stonhouse - -	2,469	
Uley - - -	2,641	
Bisley - - -	5,896	
Dursley - - -	3,226	
Carn - - -	2,071	
Gloucester † - -	11,933	
Cheltenham - -	22,942	
		88,022
		196,222

Kingston Parish - -	7,257

† Gloucester and Cheltenham both contain a greater Population than is here stated, in consequence of the new buildings.

Suburbs of Gloucester -	8,494
Do. Cheltenham -	3,632
Making a Total of -	12,126

Basing and Bath Line.

Oakingham - - -	3,139	
Newbury - - -	5,959	
Speenhamland - -	3,044	
Hungerford - -	2,715	
Burbage - - -	1,448	
Marlborough - -	3,426	
Puzey - - -	1,588	
Devizes - - -	4,562	
Carried forward		25,881

Brought forward		25,881

Southampton Line from London to Basingstoke.

Ewell and Epsom - -	5,082	
Chertsey - - -	4,795	
Guildford - - -	3,924	
Odiham - - -	2,647	
Basingstoke - -	3,581	
Farnham - - -	8,228	
Kingston - - -	3,151	
Windlesham - -	1,912	
Woking - - -	1,975	
		35,295
		61,176

Towns common to both Lines.

Trowbridge - - - -	10,863
Bradford - - - -	10,102
Warminster - - - -	6,115
Westbury - - - -	7,324
Froom - - - -	12,240
Melksham - - - -	5,866
	52,510

Great Western Line -	108,200		Basing and Bath, including the Southampton to Basing Line	61,176
Branch to Gloucester	88,022	196,222		
Suburbs of Gloucester and Cheltenham - - - - -		12,126		
Towns common to both Lines -		52,510	Towns common to both Lines -	52,510
Total - -		260,858	Total - -	113,686

Without Suburbs of Gloucester and Cheltenham.

Great Western Line - -	248,732	Basing Line - - - -	113,686

We have made all the requisite Borings to ascertain the nature of the Soil, which I should say upon the whole is rather favorable.———The only London Clay we have is for a short distance after leaving the Birmingham line; it is in Cutting and the greatest depth is but 22 feet, which is not sufficient to cause any difficulty. From London to Reading it consists almost entirely of excellent Gravel, at Reading there is some Chalk; in the first hill through Reading there is a small quantity of Clay with the Gravel, upon leaving Reading it is very hard Chalk, which in the neighbouring quarry stands upright. The line thence through the North of Berkshire, being upon the surface, the quality of the soil is of no comparative consequence. The Oxford clay, which is as bad as the London clay, ranges in that basin until beyond Swindon, where it is almost entirely Clay, and continues so as far as the Avon at Chippenham; upon crossing the Avon it is a Stratafied Stone, rising in beds: it lies above Oolite of Bath. We wall the sides of our clay banks, as we shall have Stone at hand. Passing Chippenham the cuts are still in Stratafied Stone, which is very easily worked; they form dry walls of it, in place of hedges, in the neighbourhood. It continues the same until we arrive at Box Hill, which is in Bath Stone; the entrance of the Tunnel is in Cornbrash, which is Stratified Stone. We then pass through the bed of Bath Stone, and enter the Clay which lies below it; the first hill after that is in the same Stone, and the next in Clay. The Cuttings are not of any extent until we come to Bath and Bristol; it is principally Lias Lime Stone and Red Sandstone. Most of these soils will be useful in constructing the Railway: the Cornbrash will be useful, the Bath Stone, of course, and the Lias is used for walling and building purposes; the Red Sandstone will also be useful; the Gravel would be used for ballasting, it likewise makes good embankments; Chalk is used for the same purposes.———The Soil of the Northern is much superior to the Basing line, as the valley of the Avon up to Bradford is

The Borings are favorable.

Description of the Soils passed thro'.

The sides of the Clay banks to be Walled.

All the Soils will be useful, particularly the Bath stone.

The Soil of the G. W. superior to that of the Basing, & less difficult to work.

The Basing passes thro' much Waste Land, but gains no advantage by it.

formed of slippery Débris from the hills; they are frequently moving, and a deep cutting through them would be both difficult and dangerous. We have no cuttings in so difficult a soil. A great portion of the Land on the Southern line, at least from Newbury to London, is of inferior value, being principally Heath and Waste Land, which would not influence it much; as there is no surrounding population, it was an object with me to select a direction that would admit of Branches to neighbouring towns and districts.

Table of the Gradients.

Gradients of the Great Western Railway.

Distance from Bristol.	Total Rise from Bristol.	Length of Gradient.	Difference of Level.		Gradients.	
			Rise.	Fall.	Per Mile.	Proportion to Base.
Mls. Chs.	Ft. In.		Ft. In.	Ft. In.	Ft. In.	
4 53	5 7	4 53	5 7	—	1 2	1/4409
8 59	21 6	4 6	15 11	—	3 11	1/1845
11 0	37 0	2 21	15 6	—	6 10	1/771
11 24	38 6	0 24	1 6	—	5 0	1/1056
11 45	34 0	0 21	—	4 6	17 2	1/308
11 75	34 6	0 30	0 6	—	1 4	1/3960
16 57	77 6	4 62	43 0	—	9 0	1/586
19 13	199 0	2 36	121 6	—	49 7	1/106
21 18	176 9	2 5	—	22 3	10 9	1/489
24 38	140 6	3 20	—	36 3	11 2	1/473
26 73	140 3	2 35	—	0 3	0 1	1/31680
27 41	141 6	0 48	1 3	—	2 1	1/2534
41 17	289 0	13 56	147 6	—	10 9	1/490
50 4	240 0	8 67	—	49 0	5 6	1/952
59 11	172 0	9 7	—	68 0	7 6	1/706
59 53	171 0	0 42	—	1 0	1 11	1/2772
62 0	157 9	2 27	—	13 3	5 8	1/931
66 4	126 0	4 4	—	31 9	7 10	1/673
76 76	105 0	10 72	—	21 0	1 11	1/2741
77 78	100 3	1 2	—	4 9	4 8	1/1139
83 31	89 6	5 33	—	10 9	2 0	1/2656
85 20	96 3	1 69	6 9	—	3 7	1/1457
88 69	92 0	3 49	—	4 3	1 2	1/4488
96 40	50 0	7 51	—	42 0	5 6	1/960
104 77	46 6	8 37	—	3 6	0 5	1/12766
107 40	61 6	2 43	15 0	—	5 11	1/893
108 1	61 6	0 41	—	—	Level.	Level.
110 0	50 6	1 79	—	11 0	5 6	1/934
110 75	50 0	0 75	—	0 6	0 6	1/9900
115 27	67 0	4 32	17 0	—	3 10	1/1357

—— ——The Levels, starting from London to Bristol, are as follows: for 4 miles and 32 chains, it rises from the Birmingham Railway at the rate of about 3 feet 10 inches per mile, or 1 in 1367, which in reference to Locomotive power is nearly level; then, for nearly a mile, it rises 6 inches per mile, or 1 in 9900; then, for two miles, it rises 5 feet 6 inches per mile; or in other words,—For the first 50 miles from London, there is nothing above 6 feet 7 inches per mile: it is generally running from a Level to 4 or 5 feet per mile, then there are some inclinations of 7 feet 10 inches per mile; but in the first 74 miles there is nothing above 7 feet 10 inches per mile, or 1 in 673; to the 96th mile there is nothing above 11 feet and a fraction per mile; so that the highest inclination we have to encounter with the Locomotive power is 11½ feet per mile, which is 1 in 473. At the Box Tunnel comes an inclination of 2 miles, 36 chains, at 49 feet and a fraction per mile, or 1 in 107.—— ——The Levels of the Basing Line are as follows: about 6 continuous miles and 54 chains, or 1 in 202, which is about 26 feet per mile, near Devizes; there is also a Plane near Hungerford, 7½ continuous miles, 1 in 250, which is about 22 feet per mile; there is another Plane near Burbidge, 3¼ miles, 1 in 264, which is about 20 feet per mile; there are several shorter ones, 1 in 330, or 16 feet per mile.—— There are several upon the Southampton line 16 feet per mile.———The Proportion of Power required at those different inclinations, supposing the power of the Engines to be the same, would be as follows: taking the Friction at 280, or 8 lb. per ton, which is about what it is in practice, supposing the Weight of the Engine to be 10 tons, and the gross Load drawn upon the Level to be 110 tons, it is found to be 1 in 473; the gross Load drawn would be 59 tons and a fraction: This is the greatest Inclination on the Great Western, except the Plane near Box, where it is intended to have an Assistant Power of some description to overcome it. At 1 in 202, it would be 1 in 36, and the proportion between those two would be as 100 is to 163½: so that supposing the Engines to go along the whole of the Basing line, and to carry their load up the long Inclination, the proportion of the Load would be diminished in the proportion of 100 to 163, a Loss of Power of 63 per cent.; but supposing them to have an Assistant Engine upon that Plane of 6 miles, as we have near Box, 2½ miles, they having to attain rather a greater elevation, but doing it more gradually, and as the other Planes that remain upon the line would govern the power of the Engine, the proportion of the power would be for the Plane of 264, as 100 to 136, or 36 per cent. over the power required upon the steepest Plane upon the Great Western; and supposing them still to have an Assistant

The levels of the G. W.

3 ft. 10 per mile, or 1 in 1367, nearly equal to a Level.

The highest inclination encountered by locomotive power will be 11½ ft. per mile, or 1 in 473.

Description of Box Tunnel inc. Plane.

Levels of the Basing and Southampton.

Friction equal to about 8 lb. per ton.

Comparison of Power required on the opposing Lines.

Superiority of the G. W. in point of Gradients.

engine upon those two Planes, making three long Planes altogether, still the general run of the Inclinations of 16 feet per mile would require an increase of power of 20 per cent. more than would be required on the whole length of the Great Western, with the exception of the plane; that is upon a supposition of three Assistant Engines being upon different parts of the Line: one at the Plane of 6½ miles, another at the Plane of 7 miles and 28 chains, another of 3¼ miles; all these Planes are between Basing and Bath. By going slower at these Inclined Planes, Engines of less power may be used, but I have supposed Engines of equal power, carrying equal weights, with equal speed; this induces me to say, that there can be no doubt as to the Northern Line being the best. On the other some of the Levels are very Steep, whereas we are able to keep them low; even if the communication with great Towns was equal, I should prefer the Northern.

A short steep Inc. pref. to a long one, altho' not so steep.

——I stated, that in their long Inclination of 1 in 202, they have to rise a greater total Height than we have in our short one of 1 in 107. We preferred a short steep Inclination to a long one, and 1 in 202 would be very steep, and to carry Engines all along the Line, of sufficient power to get up, would be very expensive. Our Plane would not be so steep as absolutely to require additional power; as upon the Plane on the Liverpool and Manchester Railway, which is steeper, they do not always use an Assistant Engine.——

Terminus of the Southampton R.

The Terminus of the Southampton Railway, as laid down, is about 35 chains along the Turnpike Road, above the foot of Vauxhall Bridge.——

Amount of Cuttings upon the G. W. with averages 78,000 per mile.

The Amount of Cuttings upon our Line are as follows: between London and Reading, 2,402,173 cubic yards; from Reading and Bath, 6,386,042, making a total of 9,750,156, which includes the Branches: this gives upon an average about 78,000 to the mile.——

Spoil Banks.

Spoil Banks are made where there is more earth than is required for the embankments, and land is obliged to be purchased, upon which it is laid in heaps, to save its being carried away. We have no Spoil Banks, they are generally considered objectionable.——

Side Cuttings.

Side Cuttings are made to get earth to make embankments, when there is not sufficient; we have none upon the Great Western.

General run of deep Cutting, 30 to 40 ft. Do. Embankments 25 to 30 ft.

There is but one place where we have any Cutting above 40 or 50 feet deep, and that is 70; the general run of our deep cutting is 30 to 40 feet; our Embankments are from 25 to 30 feet; there is a short one near Box, nearly 40 feet; it is 37 feet near Chippenham; over the Brent, near London, it is more than 40 feet.——

The Cutt. and Emban. should be about equal, which is the case with the G.W.

We regulate the height of an Embankment according to the quantity of Cutting we have to form it, although it must be regulated by the general level of the Railway, yet that level is regulated by the line of the country.

——Our Cuttings and Embankments are nearly equal.——At the Brent there is a Viaduct of nine arches, from 60 to 63 feet high, and 60 feet span, the estimated cost of which is £22,000.; and at one end of the Viaduct, for a short distance, the embankment is at that height, which would be principally in gravel.——The Cuttings upon the Basing line are as follows—from London to Basing about 10,000,000 cubic yards, as stated by Mr. Giles in the House of Lords last year; from Basing to Bath I make 11,500,000, (some Engineers state it may be reduced to 10,596,000); and from Bath to Bristol 2,500,000,—that would make altogether 24,000,000 of excavation. ——There are several deep cuttings on the Southern line from London to Southampton; there is a Cutting near London of 116 feet, I believe they have reduced it 10 or 12 feet; there is one of 116 feet through St. George's Hill, near Oatlands; there is a Cutting at Frimley from 60 to 90 feet; from Basing to Bath there is much heavy Cutting; there is a short Tunnel of half a mile near Bradford, where the Cutting is from 70 to 75 feet at one end, and 6 feet at the other; then they come to a hill with a Cutting of 114 feet; then there is a Cutting 100 feet deep, and another 90.——The Tunnel through the hill at Claverton is just upwards of one mile in length, and in 110 feet Cutting at one end, and 68 feet at the other; it is laid down Level, but the Engineer stated he should give it an Inclination probably of 16 feet per mile; it is intended to be worked without shafts, and the soil taken out at the ends; he proposed carrying a small Driftway, and then enlarge it, in order to have a number of places to work at once; but still the materials must be taken in, and the earth brought out at the ends. The extreme Height of earth above the Tunnel is 375 feet; there is a height of 355 feet for half a mile, it then drops; it is almost out of the question having Shafts, although they are necessary for the Ventilation. This tunnel would occupy about three minutes to pass through. (The Box tunnel would occupy about 5 to 6 minutes at the same speed.)——We might have had all our Tunnels Open Cuttings, except the Box Tunnel, without exceeding the amount of Cuttings upon the Basing Line.——I consider a deep Cutting through open pleasure Grounds more objectionable than a Tunnel beneath them, and an Embankment still more objectionable.——If I had expunged all the Tunnels, with the exception of the Box Tunnel, and a small one near Bristol, it would have raised the Line a little, and made an addition of 2,000,000 cubic yards of Cutting.——A Cutting, 14 to 20 feet, is the most advantageous Cutting, both in reference to the expence (if it was 30 feet, it would make a difference in the cost of a Bridge) and to the convenience of the Landowner; being just the height of the Bridges, and gives a free com-

Brent Viaduct.

Cuttings upon the Basing Line.

Their great Depth.

Tunnel at Claverton.

To be worked without Shafts.

A deep Cutting is worse than a Tunnel, and a high Embankment is much more objectionable.

A 14 ft. to 20 ft. Cutting the most advantageous.

munication upon a level; where Cuttings are deeper, it causes greater gashes in the land, and makes a greater number of communications necessary. The same thing applies to Embankments, the higher they are the longer the Arches must be, and it causes greater expence to proprietors should they wish to construct an arch for their private use. The severance of Land is also more expensive where in deep Cutting or high Embankment; therefore the Basing Line, independent of expence, interferes more with the country, the amount of Cutting between Basing and Bristol being 14,000,000.; the expence of Land alone between these two points is 40 per cent. greater than ours, which is under 10,000,000; and the Bridges also would be more expensive.——The London and Southampton Railway up to Basingstoke, and from Basingstoke to Bristol, averages about 200,000 yards a mile of Cutting. The London and Birmingham averages about 110,000. The Liverpool and Manchester 100,000, and the Great Western 78 or 80,000.——From London to Basing there are no Tunnels; from Basing to Bath one a mile in length, and another ½ a mile; our total length of Tunneling is 4 miles, 54 chains; we have a few chains less Tunneling than the London and Birmingham, which is 111 miles long. The total amount of Tunneling upon our Line, including removing of soil, lining, &c., amounts to £280,000.——The Tunneling upon the Basingstoke amounts to about 2,600 yards, of much worse Soil than ours; the Engineer of the Line admits it to be Fuller's earth, with springs in it, and take it at £40. per yard, which is less than I have put it at, it amounts to £104,000., which is much below what it will cost.——The aggregate Amount of Cutting and Tunneling upon our Line amounts to £767,500. The quantities upon the other Line from Basing to London, at the same prices, although their lead is a mile longer than ours, amounts to £810,000.—— ——The Estimate includes all expences attending the Works, as well as the expence of the Railway, allowing a large amount for Contingencies and unforeseen difficulties. I was not limited to any precise Amount to be expended; my original statement was about £2,700,000., £2,800,000. or £3,000,000., but I found it would not amount to that sum; my Estimate has been approved of by several Engineers. The Details are as follows:—

Objections to Deep Cuttings and Embankments.

Expence of Cutting 40 per cent. more upon the Basing than the G. W.

Average Cutting per Mile upon several Railways.

Amount of Tunelg. — Basing line, — G. W. — L. and B.

Estimate.

Original Estimate.

Details of Estimate G. W.

Excavations and Embankments, 9,750,000 yards, at 1s. per cubic yard	£487,500
Masonry, including Bridges, Viaducts, drains, and walling	459,725
Tunneling	279,195

Forming the road, fencing, &c. at about £5,000. per mile	630,400
Depôts	57,000
Locomotive power, carriages, &c.	57,000
Land and compensation, allowing a large overplus for contingencies upon the land	280,000
Making a Total of......	£2,250,820
And leaving something more than 10 per cent. for contingencies	249,180
Making a Total of......	£2,500,000

Estimate.

The prices of the above are about the same as are now being paid upon the London and Birmingham Railway. To excavate Gravel, which is a tolerable easy Soil, and carry it a distance of three miles, would be very cheap at 1s. per cubic yard, which is the price I have allowed for Cuttings and Embankments on our line; the average lead being $2\frac{1}{4}$ miles, (we have leads of 6 miles, but the average is $2\frac{1}{4}$) the leads on the Basing line are longer. One shilling is the price paid on the London and Birmingham Railway. We have allowed about four Bridges per mile for the accommodation of the Public and Owners of Land; and the annual expence of keeping them in repair is not much.—— ——In the event of there being an Enclosure Act, the expence of Bridges for any new Roads would fall upon the Company; but the cost of Bridges is greater in an unenclosed country than in an enclosed, because there are a greater number of small narrow Roads, than where it is better laid out ——I have examined the country between our Railway and Gloucester, with a view to a Communication, and find to the eastward of Gloucester there is a range called Cottswold Hills, which to a certain extent cut off Gloucester from any communication; these hills are at that part of the Line between Cheltenham and Oxford: and a Line between Tring and Gloucester would be across the highest part of these hills, where all the Streams run that supply the numerous Rivers both East and West of this part of England; the Thames is close by, and all the Streams that run down into the Severn, so that I do not consider a Line between the latter places altogether practicable. The Branch to Gloucester leaves our Line at Swindon, and there being no break whatever in the before mentioned range of Hills, except one which was taken many years ago by the Stroud Canal Company, we should pass through the same gap, the summit Level of that Canal being 90 ft. lower than the Kennet and Avon

Prices same as the L. and B.

1*s*. per cubic yd. for Cutt. and Embank. —lead $2\frac{1}{4}$ miles.

Four Bridges to the mile.

Enclosures.

Branch to Gloucester.

Canal. The water communication between Stroud and Oxford branches off at Swindon, and gets to Oxford North East by East; it then branches into two, and joins the Wilts and Berks Canal. It is in this same valley that the Railway passes, and on the North of the Marlborough Downs. If I was employed to point out the best line between Gloucester and London, I should bring it in this same direction, as I do not think I could carry any Line from Oxford to Gloucester without coming as far as Byberry Colne and St. Aldwins, which would bring it within five or six miles of the Canal I am speaking of.—— ——The Great Western is the only Line between Bristol and London which affords the means of Communication with other Towns in that district. Between London and Bristol there is the large Chalk range, that extends from the Bristol Channel a considerable way up to Wantage, and the centre of the County of Berks, and drawing a meridian at Reading, and another a little East of Bath, there is a complete table land, very elevated, between the two places; and there are no large commercial Towns between the two; so that a railway would be almost impracticable. But there is a valley between Basingstoke and Bath; by following this valley, Salisbury Plain and the hills about Highclere are to the South, and Marlborough Downs to the North. By following the gap, although you are cut off from branches to the South, you get into a valley 150 feet lower, and have the North open to you. The large Towns West of Bradford and Trowbridge are open equally to the one and to the other.—— ——A Communication may be made from Bristol to Southampton, by a Branch from our Line at Reading or Twyford; but the communication between the two places is very inconsiderable, as a proportion of two Coaches and a half only leave Bath and Bristol for that direction daily, so that it would never pay for a Railway.———I have not yet constructed a Railway, but I have been engaged upon works where Railways have been used.—— ——I was Engineer to this same projected line of Railway, during the last session of Parliament. It was then from London to Reading, and from Bath to Bristol only, omitting the part between Reading and Bath.———I calculated it would take about two years to complete that part of the line between Bristol and Bath.—— ——The Expences incurred in Parliament last year were about £35,000. to £40,000. which will be included in the cost of the work, and allowed for out of the item of £200,000. for Contingencies.—— —— I examined three spots with a view to a Terminus in London: one near Waterloo Bridge, one near Millbank, and one near Vauxhall Bridge. (I stated before the Committee last year, that the Termination next the River

Facilities of Communication afforded by the G. W.

Communication between Bristol and Southampton.

The Line last Session was from London to Reading and Bath to Bristol only

Expences last Session £35,000. or £40,000.

London Terminus.

from the Hoop and Toy at Vauxhall, by a Viaduct, would cost £180,000. not £800,000. as erroneously reported.) There was also another near Paddington.———A Terminus on the banks of the Thames certainly would be desirable, if other advantages were combined with it. I thought, and still think, that the Terminus we had at Vauxhall Bridge was better than joining the Birmingham Railway. I also thought that two large Railroads bringing their Traffic to the same part of the Town an objection, besides the difficulty of making arrangements between the two Companies.——— The Terminus of the London and Birmingham Railway, according to the Extension, is at Euston Grove, (I think at an Inclination of 1 in 86 or 90) which situation is convenient. Light goods could be distributed over London in Carts and Wagons, and Heavy goods would go down the Regent's Canal, to be carried by the River; a great increase of Trade is expected upon the Canal. I therefore think it is perfectly capable of carrying our Trade. ———I stated last year, when the Terminus was intended to be at Vauxhall, that I considered a Line like the present, by the Paddington Canal and the Regent's Canal to the Pool on the whole objectionable, as there are twelve Locks, occupying seven or eight hours in going down; and is therefore expensive, and I am still of opinion that it is objectionable compared with a better. ———The main Depôt upon the Liverpool and Manchester Railway is in the street, above 200 yards from the nearest wharf. There are no means of lifting goods from the ships to the wharf by a Crane, therefore anything going by the Railway heavier than a man can carry must be put into a Cart and taken to the Depôt; in fact, Vessels that are engaged in the Manchester and Liverpool Trade do not come to that Dock, but stop half a mile off, where the Goods are generally unshipped, and taken to the Warehouse of the Consignees, and afterwards sent to the Railway. The distance from the Terminus of the Railway to the Wharf is not more than 50 yards.——— (This same Railway crosses one Turnpike Road upon a Level.)———At Bristol the trade is carried on much in the same manner, except to a greater extent, the Warehouses not being at the Wharf.———To sum up, the general advantage of this Line are as follows: the Country is naturally Level, affording an opportunity of making a Level Railway; there are great facilities for making Branches to many parts of the West of England, and the Soil also offers very great facilities of Construction, as the greater part of it is in Gravel or Stone, which is valuable and easily worked. (Of course Devonshire and Cornwall are upon both to the Great Western and Basing lines.)

A Terminus at Vauxhall would be better than joining the L. and B., but the latter is convenient.

Advantages of the Terminus.

Objections to the Terminus.

Depôt L. and M.

General advantage of the G. W. line.

Ex. Mr. THOMAS CABREY, C. E.

I have had experience in Locomotive and Steam Engines. I was employed on the Stockton and Darlington Railway; upon which Line there are both Locomotive and Stationary Engines. I have likewise been Engineer upon the Bolton Railway.——— I am at present engaged on the Canterbury and Whitstable Railway, which I became acquainted with in 1829; it opened in 1830, and is a Single Line of Railway Six Miles long.———I erected the Steam and Stationary Engines upon that Line; there is a great variety of Power used upon it.———There are three Inclined Planes. The first is within a Quarter of a Mile of Whitstable, and is between 1100 and 1200 yards long; the Rise is about 1 in 54; it is worked by a Stationary Engine of Fifteen Horse Power, and a Rope the same length as the Plane, the weight of a load being from 25 to 30 Tons; we travel up it at the average rate of Eight Miles per Hour, but there would be no difficulty in increasing the speed; it is then propelled by Locomotives. This Plane ends in a Curve of a Quarter of a Mile radius, which they go down at the rate of 24 miles an hour; when near the termination it is checked, making an average of 15 miles an hour: It is principally on an Embankment.———(About 10 Trains pass this Line daily from each termination.)———The bottom of the next Inclined Plane is about Two Miles from Whitstable, and is nearly a mile long, with a Rise of about 1 in 34; it is worked similar to the others. I have seen the Carriages stopped within a distance of 100 yards, when going at the rate of 24 miles an hour, with 48 Passengers. (Sometimes it amounts to the speed of 30 miles per hour.) This Plane is partly on a 15 feet Embankment and partly in a Cutting.———The Inclined Plane that goes down into Canterbury is near Two Miles long; the average Rise is 1 in 54: The first half-mile from Canterbury is in a Tunnel, 12 feet high and the same in width, (it has no shafts), and a Rise of 1 in 47: It is worked by a Stationary Engine of 25 horse Power, at an average speed of about nine miles per hour, and an average load of 15 tons; (it sometimes goes 12 miles an hour, with a light load) with a rope $4\frac{1}{2}$ inches circumference, or $1\frac{1}{2}$ inch in diameter, and of the same length as the Plane, and is used for Goods as well as Passengers. The Passengers' Carriages are detached from the Train and descend of themselves, each Carriage containing 16 Passengers: sometimes Three Carriages descend connected together, without being attached to the rope, a Break (which is a lever acting upon the circumference of the wheel) being fixed, to check them. In the case of the Merchandise Train, the rope is attached, and the other end passed to the foot of the Plane, and connected to the other Train, which it brings up, and at the same time checks its own speed; as they go down there is a slight tension upon the rope, (they have sometimes to wait a short time for the return Train.) We ascend the Plane at an average rate of nine miles, and descend at twenty miles an hour. On their moving down the Plane, they commence at the rate of 10 miles an hour, increasing to 30, (which averages 24 miles), when it is gradually checked, without any difficulty.——— We had an accident upon the middle Plane, at the termination, owing to the Carriage passing too quick into the Siding, the impetus caused it to pass again into the Straight

C. and W. opened in 1830, is a single Line six miles long.

Descrip. of Plane. 1 in 54.

Working of ditto.

Weight of a Load 25 or 30 tons.

Rate of Transit, &c.

Plane 1 in 34.

Train may be stopped within 100 yds.

Descrip. of Plane. 1 in 54, which is partly in a Tunnel.

Dimensions, &c. of same.

Description of Working same, Speed, &c.

An Accident.

Line, where it ran over the Embankment, (it was intended to stop in the siding). There were four Passengers in the Carriage; one had an arm dislocated, another a leg broken, and the others were injured.——Upon the Canterbury inclined Plane, the rope once became undone, the Train, which was descending at the rate of 15 to 20 miles an hour, was stopped by the Break, and the rope re-attached.——I have known two instances of the rope snapping, which merely detained the Carriages until it was spliced; they were ascending a Plane at the time at the rate of nine miles an hour.——I see no difficulty in working a Plane 2½ miles long, even through a Tunnel, having a Rise of 1 in 107; with a Stationary Engine and rope, there would be no difficulty in moving up the Inclined Plane at the rate of 20 miles an hour. An Incline with a load of 50 tons, at 20 miles an hour, would require a rope 5¼ inches circumference, or 1½ in diameter. ——Suppose the rope was to break on an incline of 1 in 107, it would be unattended with danger.——We change our ropes once in two years, sometimes they will go three years. The danger of the rope breaking might be obviated by frequently changing it, which is rather expensive with us.——Whitstable is much frequented as a watering place, and as many as 130 Passengers have been carried in one Train. ——The number of Passengers between April 1833, and April 1834, was 22,909; from April 1834, to April 1835, there was an increase of 4000, but there was no increase of Goods.——Coals and Merchandise are the principal traffic on the Line, as Canterbury is supplied entirely by it, the Goods being sent from London and landed at Whitstable Harbour.——Forty Minutes is the average time of performing the Journey, but it frequently takes but Half an Hour.——The Railway was constructed under the direction of Mr. Stephenson, and I consider it pretty well executed; it may be improved, which remark will apply to every Railway that I have seen.——There are parts of the Line where the Cutting was made too deep, (in some parts two feet) which was filled up again, and some of the walls have bulged.——I believe the Shares are not in a very flourishing condition.

Rope escaping.

Rope breaking.

No difficulty in the G. W. Incline.

Description of Working and Speed of such a Plane, &c.

Expence, &c. of the Ropes.

Amount of Passengers.

Traffic.

40 Minutes occ. in the Journey.

Regarding the Execution of the Works.

Ex. Mr. JOSEPH LOCKE, C.E.

I was connected with the execution of the Liverpool and Manchester Railway. ——I have made a Railway in the neighbourhood of Newcastle.——The Whitstable and Canterbury Railway was surveyed and the Act got by Mr. James; Mr. Stephenson was applied to to set out the Line, and I was sent down by him, where I remained two weeks, but we had no control over the Line.——I am Engineer to the Grand Junction Railway. I am not paid by a per centage, neither for the measuring during the progress of the works; but by an annual fixed salary, (which is the usual mode of payment) not in proportion to the work done.——The Contracts are drawn up thus: we make a Drawing of every Bridge, we fix the Levels, the Number of Cubic Yards, and the Chairs and Sleepers for any length, say six, eight, or ten miles, and the manner in which they are to be executed is described in the Specification. The Directors advertise for Tenders, and call upon me for an Esti-

C. and W. Railway.

Description of Grand Junction.

Mode of Letting the Work.

Security required.

mate, the most eligible Tender is then accepted, and Security required for not less than £500. I am not aware of any advantage derived by the Engineer out of the Contracts; it is entirely in the hands of the Directors.——I have known Contracts let to Persons without any Security or Agreement; in such cases it is divided into small Lots, and let to the Class of Men employed by a Contractor, that is to the Head of a small Gang.

Method of letting to Sub-Contractors.

When a Gang is set to work the Contractor finds all Materials and Waggons, Wheelbarrows, Rails, Sleepers, &c.; he has to keep the Cutting free from Water, and various things of that kind.

Excavate for 4½d. or 5d. a Yd. finding Labour only.

The under Contractors merely excavate for 4½d. or 5d. a Yard, and have nothing to do with other items and contingencies. A Portion of our Work has been done for that Price, the original Contractor taking all risk of slips, &c., so that tools and tackle would amount to very nearly half the expense of the Cutting.

Objections to the above mode.

It would of course vary, according to the Lead and Depth of Cutting; the Depth of Cutting and Embankment materially increases the Danger of Slips, as does the Length of the same. In the event of the Works costing more than the Engineers's Estimate, he may accommodate them to the latter better with this sort of Contract. I have known the petty Contractor execute the profitable, and throw up the unprofitable part of the work, which produces an advance of price.

The Commencement of a Cutting the cheapest part of it.

The early part of a deep Cutting is of course the most profitable, the Lead being then but short; and the tools and tackle being new they require no repairing, when at the latter part they require a great deal, also many more hands.——

Eight Contracts are let on the G. J.

We have let Eight Contracts on the Grand Junction, to the amount of £600,000., which is within £4000. or £5000. of my Estimate.

At an Average of 9d. a Yard for Cutt. with 1 Mile Lead.

Do. 11d. 1¾ Mile Lead.

1s. 1d. a fair Price for a 3 Mile Lead.

We have let about 4,400,000 cubic yards of Cutting for about £160,000. or £170,000., being very nearly 9d. a Yard. The average Length of the Leads is under a Mile, the longest being 1¼ of a Mile, which is nearly 11d. a Yard. The extent of the Lead regulates the Price; for a Lead of Three Miles I should think 13d. per cubic yard a fair Price, but it would depend on the Inclination of the Road it was to be led upon.——

Descrip. of Work upon the S. R.

Prices of the above, &c.

I was examined last year for the Opposition to the Southampton Railway; the average extent of the Lead there was four miles; I am sure it could not be executed for 4½d. or 5d. (St. George's Hill was stated at 5d. by Mr. Giles) per cubic yard; it would be at least 1s. 2d., but it would depend on the Levels. I have no doubt Mr. Giles lets the work upon the Southampton Line at less than 5d. per yard, but the Company find Waggons, Barrows, Rails, and all other articles, and they will not know what it costs them until the completion.——

The Economy of large Contracts.

I think works may be executed much cheaper in large Contracts. When a Contractor has a number of things to do he makes one part fit in with another; but where one man has to build a Bridge, another to make an Embankment, they will not study each others convenience; the man building the Bridge requires materials brought down to the spot, and the Contractor for the Embankment will not let him have his Rails, &c.

The System of Employing Sub-Cons. was tried upon L. and M. and failed.

Particulars of same.

Upon the Liverpool and Manchester Railway the system of letting out the works, and finding the materials was tried; that is, the Company found Rails, Chairs, Sleepers, ballasted the Road, and took upon themselves all Risks, and kept all the Cuttings free and clear of Water; but it was found not to answer.——

Cost of Waggons.

As an instance of the Value of Tackle when done with, (a Waggon costs from £25. or £27. to £35.) our Waggons, which cost upwards of £30., were sold for £5. 10s. and we paid all expenses for keeping them in order.——The Estimate for Fencing upon the

Southampton Line per running yard was 1*s*. 6*d*. We are paying upwards of 4*s*. per running yard for our Fencing. I consider Mr. Brunel's Estimate a veryfairone; (the Contingency Fund of 11 per cent., is one per cent. more than is generally allowed) he has some at 4*s*., and some at 5*s*. in his Estimate; 1*s*. 6*d*. is out of the question; 4*s*. is good, 4*s*. 6*d*. is better, and will do the work.——The Estimate for the Railway upon the Southampton Line was 4*s*. per running yard, we pay 10*s*. for the same upon the Grand Junction Railway; I cannot tell how it can be done for 4*s*., as it includes the Ballasting, which is the Road Materials, such as broken Stone or Gravel, put under and between the Blocks forming the Roadway.——For an average Lead of 1½ Miles they pay 4½*d*.; we pay 9*d*. for an average Lead of One Mile.——Deep Cuttings and high Embankments involve expensive Bridges, Drainage, and unforeseen contingencies; it may likewise be considered injurious as defacing the Land.——St. George's Hill is I believe a Cutting of 116 feet deep. I stated that it would take seven years to complete it by the best methods I have seen used; calculating that 800 yards may be put over the Embankment (that is teamed) in one day, and allowing 250 daysto o ne year, (deducting for wet days) will produce 200,000 cubic yards per annum. In the Cutting there was 3,500,000 cubic yards, and only 1,500,000 of it is required for the Embankment; therefore, there are 2,000,000 more than is wanted, which may go to spoil while the Embankment is being made, but it being thrown out at the same time will not facilitate the progress of the Embankment.——The Levels have been altered since the Act was passed, the Embankments raised, and the Cuttings diminished. By lowering the Levels you increase the amount of Earthwork upon the Embankment, The Embankment is the proper measure of the time that will be required to complete the works, because it must be worked from one end; a Cutting may be worked at both, and much of it may be thrown into spoil.——Regarding my calculation that 800 Yards would be the amount of a Day's Cutting, I judge from my experience upon the Liverpool and Manchester Railway; we could not exceed it, even working 18 or 20 hours a day, and we had a great number of faces, so that the men were all concentrated. Now, the width of an Embankment limits the teaming, as we could only bring the Waggon up to the face of the Embankment, when of necessity it must be teamed before we could load another.——10 or 12 Men is the maximum that can be employed to team.——In a Cutting 1000 or 1,500 Men may be employed, but they cannot team more than before stated; certainly, upon extraordinary occasions, 1000 may be teamed by working the 24 hours, but Night-work is very expensive, and it requires some time to replace the Rails.——In forming an Embankment there are generally two Roads, one to bring the loaded Waggons up, the other to take them back when teemed; after arriving at a certain point, they branch out into two, three, or more different Roads, in order to get as many teaming places as possible. The teaming is always taken to the working end, never to the sides, except to fill up the Slopes, and Contractors get as many as the width of the Embankment will admit of.——I have heard of Mr. Grahamsley's Method of Excavating, but have never seen it: I believe he proposed carting and unloading 2000 or 3000 cubic yards a day; if this method is found practicable it must be to the interest of the Contractors of the Grand Junction Railway to adopt it.——One year has elapsed since I gave the above evidence, and two days

Comp. above with the G. J. in Prices. 4*s*. per Yard for Fencing G. J.

Railway, including Ballasting, &c. 10*d*. per Yd. runng. G. J.

Average Lead of 1 Mile 9*d*. per Yard cube.

Objec. to deep Cutt. and high Embanks. S. George's Hill.

Calculation as to Time of Completion

800 Yards of Embank. is all that can be teamed per Day. The above proved upon the L. and M.

Night-work expens.

Method of working an Embankment.

Mr. Grahamsley's Method of working.

Description of the Progress of the S.R.

ago I inspected the works, and found they had removed 20,000 instead of 200,000 cubic yards: At that rate of proceeding it will take 35 years to complete. There were some 4000 or 5000 yards taken to spoil, which would not affect my calculation. (If you have a deep Cutting, as St. George's Hill, the best way to get rid of it is to throw it into

Material of Saint George's Hill.

spoil; forming an Embankment to receive it is certainly no economy.) The Material was stated last year to be Gravel and Sand, which is favourable; since I have examined the spot I find the portion of Cutting that is worked to be Clay, which is much worse than the former. I therefore consider it will take seven years to complete the Cutting and form the Embankment, even from this time, and using extraordinary means, and I do not think you can reduce the Cuttings and Embankments on this Line, and keep it as good as it is now.——I did not know how long they had been at work, but I judged from what was done; neither do I know the Number of Men employed. At Shapley Heath the Depth of Cutting was about 20 feet for about 70 feet into the

Method of working a Cutting.

entrance of the Cutting. The Cutting is taken off to the required Level for a certain distance, and then stopped, and an Inclined Plane is formed to take off the top lift of Stratum, above which I think it was 15 or 20 feet. Between Basing and London they are at work at 10 or 11 places: the first place from London is on the road to Wandsworth, where they have done a good month's or six weeks' work; the next place is at St. George's Hill, I before stated that they have only commenced working at one end

Continuation of Description of the Pro. of the Works Southampton Line.

of the Cutting; the next place is Goldsworth, which has been commenced 10 days, it is a Cutting, there are seven men at work, and they have done about 220 yards, which is scarcely a day's work for an ordinary number of men; the next place is at Frimley, where they have commenced one end of the Cutting, the Culvert having been put on Moss has tumbled in; the next is a very small Cutting this side of Shapley Heath, I suppose a week's work for six men has been done; the next is at Shapley Heath, which I have spoken of, at a guess I should say they have worked from 15,000 to 20,000 yards; the next is at Hook Common, where there is not as much done as at the last place by a considerable quantity. I do not recollect having seen more than three Bridges in operation. I understand Mr. Giles intended to have but one Chair in 15 feet at the

Description of Rails & Sleepers S. Line. Objections to.

joint of the Rail, and the Sleepers are to be Scotch Fir, Kyanized, split, and laid across; I consider the method bad, because the wood would be likely to decay, and the weight would force the Rail into it, as it is not sufficiently hard to resist the pressure; neither would Kyan's preparation render them fit for the purpose, although they should

Larch or Oak Sleepers will last 15 to 20 Years.

almost resist the operation of the saw. I prefer Larch and Oak until the Embankment is settled, as they will last from 15 to 20 years, which is quite sufficient. Sleepers are used upon Embankments, as they do not settle so much as Stone Blocks, which

Remarks upon Rails, Chairs, and Sleepers.

are also very difficult to raise. I believe that one Chair in 15 feet is not sufficient, if there is no intermediate substance between the Rails and the Sleepers, the Rails will twist sideways, and there is nothing to prevent the Engines getting off; certainly a Chair may be constructed upon the Rail, i. e. made in one piece, but sufficient bearing cannot be got by a Rail of 3½ inches base, a Chair being 10 by 5 inches;

Replacing Chairs & Sleepers is very expensive.

upon the Liverpool and Manchester Railway they are 9 by 4½ inches. Taking out defective Sleepers, and replacing them is attended with much Expense, as the Men are obliged to loosen every thing, and watch the progress of the Engine, which sometimes comes, and they have to put it in again before they are ready, which renders the work both tedious and expensive; it is best to put them down right in the first instance. I have known Contractors put Rails down in that manner in a Cutting to save themselves expense, as an Engine getting off is nothing to them. If an Embankment is made of good materials, and of tolerable height, it will consolidate in Two or Three Years; if it is made of slippery Clay, very high, it will not consolidate in 10 years, and until the Embankment is settled Wood Sleepers must be used.—— ——The Southampton Line proposes crossing some Roads upon a Level, which I do not find the case upon Mr. Brunel's Line; although the most expensive, a Bridge is the cheapest in the end, on account of a Policeman being required at every crossing, to keep the Gates shut. —— ——A short time back a new Line was projected to Manchester, and was stated to be very superior, there being but few Inclined Planes upon it; to judge of which I made enquiry in two different ways, by which formulæ I have tested these competing Lines. The first was ascertained thus: it is apparent that the resistance on a Railway is generally composed of two elements, Friction and Gravity, and the Power required to overcome this resistance will in ascending be equal to the sum of these elements, and in descending to their difference. The aggregate amount of mechanical force expended on a Railway is equal to the sum of the products of the resistances, and the respective lengths of the Planes. This is the first way by which I tested them.———From London to Bristol, by the Basing and Bath Line, the Inclinations are such that a Locomotive Engine, that could drag on a Level 110 tons, would not be able to take more than 45 tons upon the Basing and Bath Line. Upon the Great Western Line, going from London to Bristol, on the greatest Inclination, it will drag 60 tons; therefore, in that direction the proportion is different as 45 is to 60. Upon the Basing Line, at the greatest Inclination, a similar Engine will take 40 tons; and upon the Great Western Line, by having an assistant Engine upon the Inclined Plane of 1 in 107, the Engine would be able to take 63 tons, which would be rather more than 50 per cent., deducting the cost of the assistant Locomotive Engine.———Upon the Basing Line assistant Locomotive Engines would be found expensive: for instance, there is one Plane which rises 1 in 102, where it would carry 40 tons in 6¼ miles in length; therefore, every Engine which assisted up this Inclined Plane would have to travel 13 miles for every trip, and taking 100 miles, a day's work for an Engine going backwards and forwards, then one Engine could assist but eight Trains per day; whereas, on the Inclined Plane upon the Great Western Line one Engine would assist 20 Trains per day, and an Engine travelling from London to Bristol will by the present Western Line take 60 tons, whereas on the other it will take but 45. Upon returning from Bristol by the Basing Line it would take 40 tons, and by the Western 63, deducting the expense of the assistant Engine upon the Great Western Line, which deduction would probably amount to 5 per cent. The above test refers to the Wear and Tear of the Locomotive Engine, Fuel, and every thing connected with those Engines. I now give a Return as to the Cost of Fuel, according to the second mode of formula. Upon the Basing Line, from London to Bristol, it is 88,284 lbs., and back again from Bristol to London, 84,342 lbs., the average being 86,313 lbs.; the Great Western, from London to Bristol 87,007 lbs., from Bristol to London 87,760 lbs., the average being

Embankments will consolidate in Two or Three Years.

Observations on paved Crossings.

The new Line between L. and M.

Formulæ to test the Gradients.

A Railway in reference to the Power required.

Long Planes bad for assistant Engines.

Test of the Wear & Tear of Engines on both Lines.

Ditto of the Power requi. (or Fuel) on both Lines.

87,888: therefore, there is a slight advantage in favor of the Basing Line, in the proportion of 86,313 to 86,888. I give a total 88,391, to allow for the distance from Camden Town to Euston Square. Now, taking the aggregate amount of power used upon the Basing Line, and on the other, there is a slight difference in favor of the Basing Line.

Fuel is $\frac{1}{4}$th or $\frac{1}{5}$th of the Expense of working an Engine.

By power I here allude to that portion of the expense of working a Locomotive that consists of Fuel, which is but $\frac{1}{4}$th or $\frac{1}{5}$th of the whole Expense. The other items are the Repairs, which are not affected by the result of the calculation I have given.——

Wear & Tear greater in desc. a Plane by Gravity alone, i. e. without Steam Pow.

I consider the Wear and Tear more upon a Descent than upon an Ascent, and in descending a Plane merely by Gravity it would be greater still, as the Wheels must take the same number of evolutions, and as the Pistons are not at work, they are more liable to injury than when the Steam is acting upon them, and the measure of Wear and Tear is in proportion to the Line;——therefore,

Result of the above Calculation.

there is a saving of between 20 and 30 per cent. upon four-fifths of the expense of working Locomotive Engines upon the Great Western Line, and there is only a saving of 1 or 2 per cent. in the remaining one-fifth upon the Basing Line.

Method of working the L. and M. In. Planes.

They use assistant Locomotive Engines upon the Liverpool and Manchester Inclined Plane. There are two Tunnels on the same Railway: one is a Passenger Tunnel, 300 yards long, and is worked by a Stationary Engine and a Rope: the other is $1\frac{1}{4}$ miles long,

Tunnel for Goods L. and M. 1 in 48.

through which the Goods pass, having an inclination of 1 in 48; the Rope used for this Tunnel is $5\frac{1}{2}$ inches circumference, (ropes are measured by the circumference) and nearly $2\frac{1}{2}$ miles long: and a third is just now completed to carry the Passengers more into the centre of the Town; (they were previously conveyed by Omnibuses, which was

The new Passgers'. Tunl. $1\frac{1}{4}$ mile long, 1 in 106.

considered a nuisance) it is $1\frac{1}{4}$ Miles long, and has an inclination of about 1 in 106; it is 25 feet wide, and 18 high; there were six Shafts, two or three of which are still open, the others are closed; the Machinery of this Tunnel is not yet completed, nor is the

Description of do.

Rope ordered. Regarding the necessary size for a Rope, much depends on the number

Ave. Load of Goods is 20 or 25 tons.

of Passengers or Goods going at the same time, and how the Line is worked. An average Load with us is from 20 to 25 Tons.——Nine Trains travel this Railway per

Pass. Train about 30 or 40 tons.

day each way.——A Passenger Train is generally 30 to 40 Tons, with the Engine and Attendant it amounts to about 50 Tons.——(A Passenger Train sometimes

The Eng. & Tender make it about 50 tons

amounts to nearly 50 Tons.)——I might prefer under certain circumstances an ascent of 1 in 200, with 1 in 100 for a quarter of a mile, to 1 in 107 for $2\frac{1}{2}$ miles.——

G. W. In. Plane.

I see no objection to the Tunnel or the Inclined Plane on the Great Western, even if it

Inclined Plane L. and M. continued.

had a Rope five miles long, but I do not know that the latter is at all necessary. The Rope upon the Liverpool and Manchester Railway (which is the longest I know of)

Tunnels do.

is $2\frac{1}{2}$ miles, but the inclination is double. The Tunnel last mentioned is at the end of the Line, so that the Passengers could avoid it if they pleased.——In the Tunnel which carries Goods down to the Sea, upon the same Railway, I know of one instance

The Rocket Eng.

of the Goods being overthrown and crushed.——The Rocket Steam Engine, before its late improvement, has carried 30 Passengers up the Inclined Plane at the rate of

Power of Engines.

24 miles an hour.——An Engine, made to move at the rate of 10 miles an hour, would not go up an inclination of 1 in 200, unless travelling with a moderate load; with half a load she may do it. It may accomplish it with a Passenger's Train, but the Engine would be too powerful for the other parts of the Line.——The great evil of

employing Steam Engines of too great power is the extra expense arising from the wear and tear of the Machinery, and in case of accidents, both of which are heightened by increasing the weight upon the Rails.——Upon the Grand Junction Railway we had two ascents; one 1 in 200, and the other 1 in 100, but they were altered, the latter distance of 400 yards to 1 in 100, the other of about a mile to 1 in 180. The reason for our making the alteration was that there was another of 177 at another part, and the Directors desired to have the inclinations as near Level as possible; by these means we could arrange our Trains better, and get our Engines of the same power. ——I consider an accident not so likely to occur to a Train running up a Plane with a Stationary Engine, as it would to a Train running on a Level with a Locomotive. I have got into a Carriage at the top of the Tunnel, falling 1 in 48, and moved down to the bottom, without touching the Break, so that there cannot be much danger.—— We propose lighting the Tunnel at Liverpool, by attaching Lamps to the side of the Carriages; several experiments having been made in the old Tunnel. The moment the Trains arrive the Lamps (which give a great light) will be attached. The old Tunnel is lighted with Gas, and is attended with no danger.——I do not conceive any difficulty in ventilating the Tunnel, which being inclined, is similar to a chimney; there would be a constant current, except when the wind was blowing in a contrary direction, or a fire lighted in one of the Shafts would ventilate it. I do not consider the Shafts to a Tunnel requisite for any other purpose than that of working it, and quite unnecessary for ventilation.

Evil of empl. Engines of too great Power.

G. J. In. Planes.

Ascending an In. Plane safer than upon a Level.

Lighting the Tunl.

Ventilating do.

Ex. MR. GEORGE STEPHENSON, C.E.

I have had my attention directed to Railways for the last 20 Years.——I was Engineer to the Stockton and Darlington Railway, which was the first applied to the conveyance of Passengers as well as Goods. Before that time Railways had merely carried Coals and the like to the water side.——I am likewise the Engineer of the Grand Junction Railway, of which Mr. Locke is the Acting Engineer. I generally spend two weeks out of two months upon the line.——I was principal Engineer upon the Liverpool and Manchester Railway. Mr. Locke was my assistant, and superintended the heaviest part of the Line; the other part was divided between John Dixon and William Holcart. The Company continue to pay me a salary for advice, but Mr. Dixon is their Resident Engineer. (The Directors of this Railway have lately attempted to get leave from the Dock Company to lay down a Railway along the Dock, which is desirable.)——I laid out the first Line between Liverpool and Manchester to the North of Noadley; it was almost a dead Level from end to end. It was defeated by the Landholders upon Mr. Giles's evidence; he stated it would cost more money to execute the part across Chat Moss than was intended for the whole Line. His statement was that the Moss ought to be scooped out, and filled up with solid materials; but it has since been executed according to my scheme, namely, merely laying a Platform down upon the Moss. There are seven miles of it floating, and it is the best and

Has had 20 Years' Experience, &c.

His Works.

The first L. and M. Line was defeated by Mr. Giles.

Des. of Chat Moss.

Considers G. W. a good Line.

Termini.

Principle of laying out a Line in refer. to Branches.

Branch to Stroud.

Reasons for having the Inclined Planes L. and M.

Method of working the same.

Tunnel preferable for an In. Plane.

Delays thro' Snow.

Descrip. & Dimen. of the Pass. Tunnel L. and M.

Incliation 1 in 100.

Plan of ventilating a Tunnel

Size of Tunnels L. and M. &c.

Another Plan of vent. a Tunnel.

cheapest part of the Line, taking the value of the Land into consideration.—— ——I consider the Great Western Line a good and eligible Line.———A Terminus at Euston Grove is better than that at Vauxhall, and I think there will be plenty of room at the latter place for both Railway Depôts.———In laying ont an extensive Line, where a number of Branches are expected, it is very important to keep the Levels low, so that the Branches may descend from each side of the Valley; if you take one side of the Valley it is almost impossible to get the Branches on the other side to join it.———I think Mr. Brunel has taken the best Levels, and that his Estimate is a very fair one.—— —— Regarding the practicability of connecting of the Great Western and Stroud, as there is a Line for a Canal there must be one for a Railway.———My reasons for having the Inclined Planes upon the Liverpool and Manchester Railway is to allow the Engines to bring the heaviest loads possible to the bottom of the Inclined Planes, and having an assistant Engine to get it up. If I had distributed the inclination to a longer length, the Locomotive Engine could not have taken them up, and it would have been too long for an assistant Engine. ———It is curious that my Report to the Liverpool people is almost the same as Mr. Brunel's in this case. I stated that either a Stationary or Locomotive Engine might be used. The Trains frequently go up without an assistant Engine. If the Engine was out of order, even if it was half that ascent, it would not ascend, therefore the assistant Engine is always ready. Had it been in a Tunnel it would be much easier to work with a Locomotive Engine, as the morning dew often settles upon the Rails, and inclines them to slip, (the Engines hold entirely by the Rails,) which would not be the case in a Tunnel, as there would be no dew. In the event of a slight fall of Snow or Sleet the objection would be much stronger. The Engines are often stopped by the Snow till an assistant Engine is brought to assist them.———The Passenger Tunnel now constructing is 18 feet high, and 25 wide; descends about 1 in 100, and as it is at the end of the Line, it is more convenient to work it with a Stationary Engine; but it is so arranged, that if the Stationary Engine is out of order, the Locomotive may go down.———I do not consider there is any occasion to attend to the Ventilation of the Tunnel through which the Goods pass; the current is sometimes moving one way and sometimes another; if it was for Passengers we must have Shafts, but even then there is a difficulty in keeping the current passing up.———The plan I have prepared for ventilating a Tunnel is to have a Shield formed in it, to take off part of the segment of a circle, to prevent its coming immediately under the Shafts; if it was to do so, it might make the current come down as it does in a chimney. but the Shield prevents that. I have also been desired to contrive some means of keeping the temperature the same in the interior of the Tunnel as in the exterior, and think I shall accomplish it; I have no objection to a Tunnel.———The Size of the old Passenger Tunnel is only for a single Line; it is 14 feet high, and 10 or 12 feet wide; that for Goods is 16 feet high, and 22 feet wide. Some of the Shafts of the Tunnel going into Liverpool are to be stopped up, (of which there are five or six.)———I have a plan for ventilating the same by making another Shaft nearer the mouth of the Tunnel than those we have yet made, (at the highest end) which is to be the principal ventilator; but if we should find much difficulty, we could but resort to the common means used in Collieries, that of having a fire in the Shaft ———The Box Tunnel, being longer, would be much easier ventilated. They would go up the Plane with a fixed Engine, but so arranged, that if any thing was the matter with the fixed Engine the Locomotive could take it up, and in descending the Plane, should the Breaksman neglect to put on the Break, there would be no danger, as the friction of the Engine and Tender would be sufficient to control the Carriages; upon the Liverpool and Manchester the Breaksman has frequently neglected to put on the Break, without an accident occuring—— ——The Line from Basing to Bath is in a Valley, Salisbury Downs being on the left and the Marlborough Downs on the other side. It appears there are 23,000,000 of cubic yards of Cutting upon it, which makes the undertaking, in my opinion, quite impracticable.———The Summit Level of the Basing Line is about 153 feet higher than the Great Western, which is a great disadvantage. The Great Western Line is about the same Level as the Stroud Canal at the end.——In selecting a Line from London to Basingstoke, I should go by the Great Western as far as Twyford, instead of taking the Basing Line, in order to avoid "St. George's Hill," which will never be finished according to the plan which they laid down last year. At the rate they are proceeding, it would take near 40 years, as they are not going on very fast. I got my data by ascertaining what a waggon would contain, and calculating accordingly, I found it barely held one cubic yard; I found the quantity teamed was under 200 yards a day; I found the Men had 14s. a week, they worked one and a half days over, which made it 17s. I then got the number of Horses upon the Line, and considered what the Waggons and Horses might cost per yard, by this calculation I found a cubic yard came very short of 1s., last year I estimated it at *7d.*, but it cost more for distance of 600 yards that it is led, I consider that *3d.* was sufficient for filling it; it is not sufficient now, and *3d.* for Waggons, Rails, and Sleepers; the average lead is four miles, at *9d.* makes it come to *1s. 5d.*, and I am sure they cannot do it for that sum now.———The Cutting appears at 15, 16, or 17 feet of London Clay, (I cannot say whether it is London Clay at the bottom,) or Plastic Clay, above the Clay is eight or nine feet of Sand.———I likewise stated last year, that in a heavy storm the Railway would be turned into a Canal. I spoke from my experience upon the Liverpool, having seen Water rise six inches in a length of 300 yards. St. George's Hill being two miles long and level, if it was a Tunnel, it would be impossible to get the Water out, and a thunder-storm would bring down such a quantity of Sand as to stop the Railway altogether. We had a Cutting of Sand and Clay with a slight fall, in which the above occurred; certainly giving it a slope of one in 330 would enable you to carry off the Water better.—— ——The Wear and Tear is greater going down an inclined Plane than on a Level, as the gravity of the Waggons over runs the Engines; it is true the Breaks are put upon the Waggons, and a man applies them, but still the Waggons will over run the Engines and drive the Pistons the wrong way, producing an action upon the Engine that is injurious, it is a long lever working a short one, the wheel is the long lever and the crank the short one, if it was the short one working the long one it would be better. The Waggons begin to follow the Engines at 20 feet a mile, which is one in 260. We went to a considerable expense upon the Birmingham Line, in keeping the Tunnels under 20 feet, no part of that Line rises more than 16 feet, with the exception of the short distance from

Box Tunnel.

Working the Plane, &c.

Objections to the Basing Line.

Bas. summit 153 ft. higher than G. W.

His view of a Line to Basingstoke.

St George's Hill.

Detailed Prices for Cutting.

Prices for Waggons, Rails, &c.

Des. of the Soil of Cutting.

Difficulties with deep Cuttings.

Wear and Tear greater on an Incline than a Level.

Waggons Descend by their own Grav. at about 20 feet per mile.

The Ex. Wear and Tear upon the Basing Line far overbalances the Saving of Fuel spoken of by Mr. Locke.

Euston Grove, so that the extra Wear and Tear far counterbalances any advantage that would arise upon the Basing Line as spoken of by Mr. Locke, that " in going down a long Line there would be a saving in the expense of Coals."——Regarding the inclined Planes upon the Great Western Railway I see no objection to them, and in reference to the danger to be apprehended from them, I may instance that upon the inclined Plane in the Tunnel upon the Liverpool and Manchester Railway, before it was opened, I have seen as many as 20 or 30 people pass down in a Carriage, it was started by its own gravity, and regularly increased to 30 or 40 miles an hour, and the Break was not applied till near the bottom. 40 or 50 Waggons loaded with goods from Manchester frequently go down together, one Man managing the whole, there is a Break to each, which by the chain to it he can tighten as he likes, they go by their own impetus at any rate the Man pleases. Regarding the working of the Planes, an endless Rope of five miles long for the incline on the Great Western Railway was never thought of; it can be performed three ways, I should prefer an assistant Locomotive Engine, or it can be worked by having a small Locomotive to take hold of the Rope, and run back with it as soon as the Train is drawn to the top; but I should prefer a Locomotive Engine behind to push it up, as upon the Liverpool.——A common Engine upon an ascent of one in 200 could with difficulty proceed at the rate of 20 miles an hour, particularly in bad weather. I could make an Engine to do it, but it would be at a very great expense. And it would be a great loss of power in order to encounter one in six or seven miles, at 1 in 200 you must lessen the load, which produces 50 per cent. disadvantage.——One in 240, or about 18 feet per mile, is generally called the Angle of Repose.——In the Grand Junction there is an inclination seven miles long of one in 210, which is very objectionable, but we could not get over it any other way, (it is now reduced and altered so as to bring in a steep inclination at one point, the same as the Liverpool and Manchester), it being too long for an assistant Engine; we must have a load adapted to our Line, and of course the same system might be pursued on an incline of one in 330 upon the Basing Line, it is not impossible to get over it.——Regarding the Canterbury and Whitstable Railway, I merely gave the general Levels from the map presented to me, and sent Mr. Locke to lay out the Line in the best way. They had not enough money to execute the Line properly, and could not pay more than one Resident Engineer. I did not see it more than once or twice, and observing that the Walls of one of the Bridges were bulging I remarked that it would come down, however, the Bridge is still standing; the Engineer did the best he could, but he was a young beginner.——The Sheffield and Rotherham Railway put my name down as Engineer without my sanction, they having laid out the Line. I told them I would do all I could, but it was not the right Line, and by my persuasion they did not go to Parliament.——For the Hartlepool Railway I made out an estimate which they returned on account of its being too high, but I declined going to Parliament with the lower Estimate.——I was consulted about a Line from Edinborough to Glasgow. The Directors chalked out a Line and sent it to me, and I told them I could not support it in that case, all that I had to do was to make the estimate and give my opinion upon the same, and the practibility of the Line.——The amount of a contract generally comes to about £100,000.——Upon part of the Liverpool Line, also upon

The Great Western Inclined Planes.

Safety of the Liverpool and Manchester Inclines.

An Engine on an inclination of 1 in 200 made to do 20 miles an hour is expensive, as it would be a waste of power in other parts of the Line.

The Ang. of Repose.

The inclination of 1 in 210 on the Grand Junction very objectionable.

Canterbury and Whitstable.

Sheffield and Rotherham Railway.

The Hartlepool Railway.

Edinborough and Glasgow Railway.

Stockton and Darlington Railway, we divided the work into small lots, and measured every fortnight, and paid accordingly. The price depended upon the material we had to cut, and varied from 5*d.* to 9*d.*, except at Chatmoss, where it is very soft; the Company found Planks, Tools, and Rails of every kind, but it did not answer, as we could never depend upon the Contractors completing the work, if they found it did not pay them, they gave it up, and they could not find security; this method likewise required additional Overlookers, and it was found that it cost the Company at the rate of 1*s.* 2*d.* per cubic yard, if not more; there was some Cutting through stone at 3*s.* per yard, which was allowed for extra.——Upon the latter Railway considerable sums, in the shape of a Bonus, were given to facilitate the completion of the Line.——I have not seen Mr. Grahamsley's method of forming an Embankment. I understand that it has been used on the Hartlepool Line, and every one that used it *broke*, therefore it has been abandoned. Upon the Newcastle and Carlisle Railway, Grahamsley and Tredgold contracted for the Cutting of Cowran Hill at 5*d.* per yard, they finding Barrows, and Tools, but the material taken out was almost like snow, so easy to remove, therefore the price was quite as good as 1*s.* 3*d.* was for St. George's Hill; it was a pointed Hill that went up suddenly and came down suddenly, and there was plenty of room at each side to make Spoil Banks a very few yards from the Cutting of those Lines connected in a variety of places. And he let plenty of it to the Workmen at 2*d.* or 3*d.* per cubic yard; there was only a small portion of it conveyed 1½ miles by the side of the Railway, which cost him more than 5*d.*, but the great bulk was thrown to spoil; allowing 3*d.* per cubic yard per mile for the leading and teaming, and 3*d.* for filling, would bring it to near 9d. per yard, and, in point of fact, I do not think it cost less than that. In a contract 1*d.* or 1½*d.* per cubic yard is considered a great difference.——You cannot get a good Navigator under 3*s.* per day, but the Contractors employ the Agricultural people, as they get them for less wages, yet they find an advantage in having good men, although they pay them more.——I stated last year it was impossible to use Locomotive Engines upon a temporary Road with advantage, and I now repeat it would be much dearer than horses; if you can get a mile or two of permanent Road I think they may be used, but it is madness to attempt it upon an Embankment.——I and my son were the Engineers to the proposed Manchester and Sheffield Line, the summit Level of which was very high, 300 feet, on purpose to get the Limestone; there might have been a Plane of a mile, one in 18, another three-quarters of a mile, one in 32, another 1¼ mile, one in 18, and a Tunnel about 5½ miles long, one in 98. Out of the 43 miles there is about six miles of Tunneling; it was not laid out for Goods or Passengers, but merely to get at the Lime, yet it was thought they might pass (with care) by the Break; it amounted to £12,000 per mile, which is not very expensive considering the Tunneling, we expected the gravity of the Vans would be sufficient to bring the Goods to or from Sheffield without any other power. The Plane of one in 18 would be very steep for Passengers if they were to go down by a Rope, or any other means than a Break; it is not so steep as the Line I constructed for the Hetton Colliery, which is principally for Coals, and is the most difficult ever constructed in England, perhaps in the World. There are three 60 Horse Engines to get coals to the Summit of the country, and 4 self-acting Planes, and after passing the Summit the Rope is taken off and put on without the Carriages stopping.

The system of letting the work to Gaugers, and finding all Tools, &c. was tried on the L. & M. Line, & failed.

Bonus given on the comp. of a Contract.

Mr. Grahamsley's method of Embankg

Cutting of Cowran Hill.

Wages of Navigators 3*s.* per day.

Fallacy of using Locomotives upon temporary Rails.

Man. and Shef. Rail. made to get at the Lime.

Has a Plane 1 in 18, &c.

Plane on the Hetton Colliery steeper than the above.

Ex. H. R. PALMER, ESQ. C. E.

His Experience.

I have been engaged in Works of a Civil Engineering description for the last twenty years, seven years of which I was with the late Mr. Telford. I have never executed or had the conducting of a Railways, but I have been frequently consulted to a very considerable extent upon Lines of Railroads, and have made Surveys for my Estimates of same. I have also given a great number of opinions on the subject, and devoted much time and attention to it.

His Work upon Railways, explaining a Plan for a single-line Railway.

Speed was a short time back considered of but little moment, although it is now the most important consideration. In a very early stage of Railway Practice, (about fourteen years back) I directed my attention to the subject of Speed, and published a Pamphlet, which I believe to be the first work that contained any comparative measures of resistance; I wrote it with a view of ascertaining the difference between the various kinds of Railways then in use, and one that I had invented for a particular purpose. The Project alluded to was for Goods only, and consisted of a single Rail moving upon a Railroad, with the Goods suspended on each side. The Book merely described the Plan, without particularly recommending it, since the period of the Publication, Locomotive Engines have been much improved, to which I attribute the great prosperity of Railways. When the Liverpool and Manchester Railway was under discussion, the greatest velocity anticipated was ten miles an hour, therefore nothing like the velocity which has since been attained was contemplated.——I was Engineer and Superintendant of the London Docks, (which is the only work I have personally superintended) but have ceased to be so for the last two or three months. I constructed the Pier Head there, which, like other buildings of the kind, has become undermined by the operation of the water, and the imprudent management of the sluices. I therefore allow that it requires some support to make it permanently substantial.——There have also been some alterations and repairs made to some of the Dock Gates; some defects the Contractor had left, which I understand have been remedied, for which the Engineer is certainly responsible, but as he cannot be there at all times, he is obliged to trust to others, and although he is responsible, he should be provided with means to hold that responsibility. I consider myself the least responsible for the defects that may have arisen there.——During the period I was with Mr. Telford, he was engaged in the execution of works of considerable magnitude, many of which I saw the progress and execution of, and I was sufficiently acquainted with his concerns to know the prices of the same (which is the kind of knowledge I allude to when I say I know so and so, but not of my own knowledge. Much of my information as an Engineer is derived from similar sources.)——Certain persons having thought proper to project a Plan for the supply of the Metropolis with Water from wells, solicited me to become their Engineer, to which I certainly acceeded, and took some pains to endeavour to get them information upon the subject, but I really know very little of their proceedings. I consider it as much the duty of an Engineer to give his opinion as a Lawyer, when called upon to do so; and it was in that capacity and character that I acted in reference to the Project. ——I made an Estimate of the London and Birmingham Railway when the Bill was

The Improve. in Locomotives is the reason of Railways flourishing.

10 Miles an Hour, the Velocity originally intended on the L. and M.

Was the Engineer of the Lond. Docks.

Explanation of the state of the Works upon the latter.

Difference between Personal Practice & Personal Observation.

The Company for supplying London with Water from wells.

His Estimate of L. & B. Railroad.

before Parliament, and the work is being executed at very near the prices of my Estimate.———My Instructions were to make an Estimate of the Line, in order to corroborate the Estimate made by the Engineer; and I formed it upon my own actual observations of the Line. I certainly made no measurement of the ground, neither borings; the dimensions of the several Bridges and Arches were also supplied me by the Engineer; but I surveyed the Line, and very carefully examined the specimens of the borings, from which I made a conjecture of the probable cost, and upon which I founded my Estimate also from the Data furnished me by Mr. Stephenson, who acted upon the precise same data. Had I been the Engineer of the London and Birmingham Railway I might have been anxious to obtain more information. I knew the amount of the assumed Capital, but I made my calculations perfectly independent of their Engineer.——

Explanation of same.

——I have examined the proposed Great Western Railway throughout, and the proposed London and Basing Line as far as Bath, and I think the Great Western Railway will be more serviceable to a greater amount of Population, while at the same time it is more accessible to that quantity of Population, within the same distance, than the Southern or Basing Line would be. I think a line of Railway serviceable, useful, or profitable in proportion to the amount of Population that could take advantage of it. There is no difficulty that could not be overcome in making an Extension of the Great Western to Oxford and Gloucester, by Swindon, to the North. The Stroud Canal passes between this Line and Gloucester, and I do not think there would be any difficulty in carrying the Line on to Exeter and the West.———There are only two passages that lead at all convenient from the Basing Line Southward; the one is by passing the Chalk Range by Basingstoke, and the other by the western extremity of it, near Frome. I should never think of crossing at any point between those ranges, unless it was for a purpôse greater than any I know to exist. The Great Chalk Hills called Marlboro' Downs being on one side, Salisbury Plain on the other. The Great Western Railway, extended to Basing by Twyford, would be the line for any places on the Southern Coast. ———The chief profit derived by a Railway is from the Passengers; Goods travel more by the road and canal, but still the profit by Goods would be great.

His reasons for preferring the G.W. Railway to the Basing Line.

Capabilities of extending the G. W. to Oxford, also to Exeter.

Bad access from the Basing Line southwards.

Gt. Westr. may be extd. to Basing by Twyford.

Passengers most adv. to a Railway.

It is essential to adapt the Railway to the Carriages employed upon it; the desideratum is, "to produce the greatest effect at the least possible cost," which is done by bringing the Gradients of the different planes as low as possible: if a Line varies considerably in its Gradients, a power must be provided sufficient for the greatest, and these variations may be such as to require this same power to be continued throughout; when the latter is the case, you must provide in all respects for this "maximum of power" thus:—if you have an "extra powerful Engine" you thereby increase the weight, and must provide for the action of the extra weight by a proportionate increase of power.———It therefore appears preferable (at least to me) to continue the lowest Gradients to the greatest possible extent, and to concentrate the steeper rises; providing an especial force to overcome them. You are thus enabled to work the Railway generally with the best effect, and as you have less weight of Engine, you require less solidity of Railway: you are likewise less liable to Repairs, which is a great advantage; a Plane of 10 miles, at 16 or 17 feet per mile in a line 112 miles long, would be improved by altering the Plane to 11 feet per mile, and concentrating the inclinations in about 2½ miles; in the

Theory of Gradients.

Explanation of the Method of laying out the Great West. Gradients.

former case there would be a continual loss of power and expence upon 10 miles; —— —— I therefore concur with Mr. Brunel in his opinions that a short inclination to which an assistant Engine can be applied, is preferable to a long Plane; although not so steep, as the Gradients generally are rendered more favourable. I am not aware of any Railway of the same extent as the Great Western which has so favorable Gradients. —— ——The country between Bath and Bradford presents very considerable difficulties, thereby involving very great expense; when I speak of Engineering difficulties, I merely refer to the comparative expense; in the above case, greater expense would be incurred in a few miles than is usual elsewhere.———The laying down of a Line of Railway, or Canal, depends upon the comparative altitude of the surface of the country. When I surveyed the Line I did not take the several Levels. (Ranges of hills are never measured, except in particular cases; for instance, the Marlboro, and other Downs are of sufficient magnitude to enable a person to judge of their height without taking their Level.) I supposed Mr. Giles's section to be correct, and therefore I did not test it; I held the section in my hand as I passed through the valley, and Mr. Hennett (Mr Brunel's surveyor) pointed out the competing Lines to me.—— ——The Tunnels (on the Basing Line) will involve a more than ordinary expense; a considerable Side Cutting is also requisite in the Cliff, and on the side of the Hill (at Bath) they will also incur expense; a very considerable Viaduct is also proposed over the River at Claverton Hill, and a Tunnel near it, in the formation of which very considerable difficulties are expected, but my knowledge of it is simply drawn from information afforded me by Geologists, whose investigations have been particularly directed to the hill, and who have described the Strata to me; I have not examined the Hill sufficient to form an opinion as to the propriety of cutting a Tunnel through it. I have observed openings in the Hill, which merely exhibit Rock, it does not discover to the eye such great difficulties; but the openings were not numerous, nor in the right places for me to judge, (an individual has published a description of the strata, which he describes as very bad.)———The Tunnel passes through the Hill at so great a depth, that it would be exceedingly expensive to sink Shafts, although they are necessary, in order to get out the material excavated, also for the transmission of the working apparatus, and for ventilation; and the difficulty and expense would be greatly increased by working it at each end instead of by Shafts, (particularly in a hill of this description).———I do not know an instance of a Tunnel of equal length, 1¼ miles, through a hill having been attempted without Shafts, excepting the proposed Highgate Tunnel, which fell in, and was made into an open Cutting; a Driftway had been made through it, to ascertain the soil, which was found to be bad: I know the description of soil well, and have seen it run much more than it did in that case. I do not mean to say that any analogy exists between this Tunnel and the Claverton Tunnel; but if the proposed Tunnel should pass through a soil consisting of clay and water, it would probably fall in.———When ascertaining the cost of a Tunnel, we take into consideration the facility of executing it by Shafts; I have not constructed a Tunnel of any extent without shafts, therefore cannot say what increase of expense and time there would be; but I should think it would be very considerable, in fact, it causes a diminution of working space; now with Shafts, operations can be going on in different places at the same time, and the materials are coming in as others are going

He approves of the Box Plane G. W.

Country bet. Bath and Bradford very difficult.

An Engineering difficulty merely a matter of Expense.

An Enginr. judges the height of Hills without levlg. same

Difficulties of the Basing Line.

Claverton Tunnel.

Difficulty of same.

Advantage of Shafts

The Cla. T. to be worked without Shafts.

Account of the failure of the Highgate Tunnel.

Comparison of Expense of Working a Tunnel with and without Shafts.

out, without impeding each other.—— ——I am of opinion that a Line could not be conveniently made in the vicinity of Bath without passing through some Hill; if possible, the Line would be very circuitous, (whereby the principal advantage of a Railway would be sacrificed), in consequence of the situation of Bath being in a basin. A Line has been projected from Bath to Trowbridge, without a Tunnel, (the Bath and Basing goes from one town to the other, but it has two Tunnels in that distance).—— ——I have examined the Depôts of the proposed Lines at Bath, and I consider that of the Great Western the most convenient, both to the inhabitants as well as for the general purposes of a Railway; there are many objections to the Basing, it is placed on the side of a Hill, the materials of which indicate a very great inclination to slip, (considerable slips have taken place); and from the very great steepness of the hill a convenient depôt cannot be made, and its position is also inconvenient, as it must be ascended by a very steep road; I believe it is intended to go about a furlong up this steep hill, which is about 1 in 9, (this is steeper than Holborn Hill; there is a road at Lewes, in Kent, about 1 in 7, and it frequently occasions accidents to the different vehicles, if not to the coaches; additional power is also required to get up it).—— ——I have ascertained the mean cost per mile of the several great Railways throughout the Kingdom, taking one Line with another, to be about £18,000. per mile, which sum comprehends "the whole expense" Parliamentary and Law included. I have assumed the capital formed as being sufficient for all purposes, (*i. e.* the Parliamentary estimates). I divided the same by the number of miles, and added 500 miles of my own Estimates, taking the mean of the whole of that sum. (The London and Birmingham, and the Southampton are the only Railways I have absolutely calculated myself, it being a process usually performed by assistants). ——The average cost of the London and Birmingham is £22,300. per mile, but refering to that Line generally, it ought to have been put at a higher average rate than the Great Western. I do not remember whether I included the Great Western and the Southampton, the former of which (*i. e.* the Parliamentary estimates,) averages £20,800, the latter £13,000; although I believe the Basing Line will not be executed for that sum (£13,000.), yet I presume it to be correct; I do so for the purpose of shewing how inferior it is to all other Lines.——Upon Railways of the above description, Land forms a small item of expense, compared with the remaining portion; a Line may go through a country where the Land is of inferior value, without making a considerable difference in the sum total.——Should the Newcastle and Carlisle Railway cost but £8,000. per mile, and the Liverpool £25,000. (speaking hypothetically), I should not consider it a comparative price, nor do I think that the mileage would afford any means of calculating the comparative merits of these Railways. I know nothing of the Land upon the two Lines.——The number of cubic yards of Excavation on the Great Western is under 10 millions, and on the London, Basing, and Bath Line, 22 millions, *i. e.* from London to Basing there are 10 millions, from Basing to Bath 12 millions, therefore the Basing Line has 12 millions more cubic yards of Excavation than the Great Western. —— ——According to the estimate I delivered in, which, from actual experience, I am sure is rather below the mark; I believe the Line from London to Southampton*

Bath is situated in a Basin.

Comparison of the Depôts at Bath.

Explanation of his Table.

The mean exp. of the sev. Rail. thro. the Kingdom gives £18,000. per mile.

Mean of L. & B. £22,300.

Mean of G. W. £20,800.

Mean of Basing £13,000.

Land but a small item in a great Rail.

Memo. N. & C. Ry. L. & M. Rail.

Amount of Excav. G. W. 10 millions.

Lon. to Basing 12 millions.

Bas. to Bath 10 mil. making 22 millions.

* It is not quite clear, according to the evidence, whether Mr. Palmer's estimate refers to "between London and Basing" or "between London and Southampton," both places being stated.—Ed.

Mr. P.'s Estimate L. and S.

will cost £600,000. more than a million, which is the amount of the estimate put in; this amount does not include any extra sum that may arise between Basing and Bath.——— A short time back I examined part of St. George's Hill; and instead of finding Sand as I was led to expect from the evidence given last year, I found they were working in a considerable depth of Plastic Clay; and I am quite confident that not only the estimate given in by Mr. Giles was fallacious, but that the estimate I then gave in (which was three times the amount of Mr. Giles's) was not sufficient.———The Plastic Clay lays under the London Clay, at this place it crops out the Plastic Clay, and Bagshot Sand; it extends horizontally at a certain distance; the strata was rising up southwards from the basin of the London Clay, so that I expect it will increase in thickness, rather than diminish.———I found 43 Men, and 3 Horses, at work; 2 Teams were at work, each containing 3 Waggons; and from experiments and observations which I made, I discovered that the work was being executed for about 11¼*d.* per cubic yard, *i. e.* the actual cost of labour only. The Lead, which necessarily increases the expense, was only a quarter of a mile. I allowed but 16*s.* per week for the Men, and a moderate charge for the keep of the Horses. I believe 18*s.* to be the medium price paid at this time to Navigators; the last time I employed any was at the London Docks (about three years ago). I paid from 18*s.* to 21*s.* per week. I employed from 200 to 500 at a time; 18*s.* was the lowest I paid, (even for Irish Labourers). I employed some boys at 6*d.* per day.———There was no new Machine at work, calculated to facilitate the progress of the work, and I did not perceive that spirit among the men which I have often noticed when a good Contractor is on the ground, superintending his own Workmen. If they worked at the same rate throughout the day, their work would not exceed 125 cubic yards.———There is a large Embankment in the same direction, about 1,500,000 cubic yards, which was estimated at 5*d.* per cubic yard. I last year put it at 1*s.* 3*d.*, but now I know the nature of the soil, I am sure it could not be formed for less than 1*s.* 6*d.* per yard, (the difference between the soil they are excavating and sand being considerable), which would bring the Estimate to £112,500. instead of £31,250., the amount of Mr. Giles's Estimate, which is priced at 5*d.* per yard.——— I have no conception how one Engineer can get that done for 5*d.* which costs another 1*s.*, as all Engineers are anxious to get the work done as low as possible. If there was any possible means of executing the work at a lower rate than usual, it must immediately become known, and universally adopted; it is absurd to suppose the subject admits of being kept a secret; Contractors would adopt it in order to make the most of their contracts.—— ——I should say, upon a rough guess, it would take 48 years to complete it, working at the same rate they did when I was there; but if it was found practical to put 10 times that strength upon it, (which I do not think possible), it might be done in one-tenth of the time.—— —— I inspected St. George's Hill a second time, and found that the distance was 700 yards from the cutting to the end of the embankment, instead of a ¼ of a mile as I before stated. I arrived there very early, and found 35 men at work; 28 were employed in digging and wheeling, 4 in filling the carriages at the end of the embankment, and 3 in teaming, that is driving the horses. In order, therefore, to ascertain the cost of the embankment, I deducted 3 men out of the 35, who were employed in digging, and conveying the earth to Spoil, near the face of the cutting. I ascertained that 32 were

Acc. of his 1st visit to St. George's Hill.

Description of the Soil, &c.

He found the Exp. of the Cutting was about 11¼*d.* cu. yd.

The way it was ascertained.

Navigators' wages 18*s.* to 21*s.* per wk.

Account of the Works at St. George's Hill continued.

The Embnkt. will cost 1*s.* 6*d.* per yd. instead of 5*d.*

At the rate the Work is now prog. it will take 48 years to compl. the Cutt.

Acc. of his 2nd visit to St. Geo. Hill.

Des. of the Works.

Ex. to asc. the price the Work was being executed for.

employed in forming the embankments; I observed that a Team of 3 Waggons was loaded and discharged over the Embankment every 10 minutes, the waggons did not contain ¾ of a yard, therefore 135 cubic yards would be discharged in a day (10 hours.) The Waggons did not hold ¾ c. yd. ——I have taken the men's wages, at 16*s.* per week, and 3*s.* per day for the expence of each horse, of which there were 4; thus I found the whole expence for these 135 cubic yards amounted to 1,168*d.* of $8\frac{6}{10}d.$ per cubic yard, and at the time I made the experiment every thing was in their favour: by adding the 3 men to the 32, it amounts to 9¼*d.* per cubic yard, independant of any allowance for capital or the expences of management, &c. Result of Exp. was 9¼*d.* cubic yard for labour only. ——I observed on the sides of the cutting that the recent heavy falls of rain had produced an effect upon them, and that a quantity of Sand was washed down: this is by no means unusual, particularly in this description of soil, therefore a certain quantity of labour would be required to remove the same, also for keeping open the Drains, and the Temporary Rails, and repairing of the tools and carriages. I think 3 men a day for every ¼ mile throughout very moderate for the above. Memo. Estimate. ——By ascertaining the length and depth of the Embankment, I found that about 17,000 cubic yards had been excavated, exclusive of a small quantity that was taken to Spoil; and the Embankment is the proper measure of the time the work will take. It would take 6 years to complete this Embankment at a rate of 800 cubic yards per day, allowing 313 days to the year, *i. e.* merely excluding Sundays, (some state the average number of working days in the year to be only 250.) At 800 c. yds. diem, it will take 6 years to finish, allow. 313 days ann. some allow. but 250. ——The limit to the number of men is not determined by the number that can excavate, but the quantity of earth that can be discharged by the waggons at the end of the Embankment; the number above stated is all that can be discharged per day, neither would it be prudent to advance the Embankment more than 800 cubic yards per day. Memo. Formation of an Embankment. There have been cases of embankments slipping in consequence of being carried on too rapidly. ——The Embankment requires some time in order for the earth to consolidate effectually; the occasional rains produce a gradual deposition of finer particles among the interstices, which the matter thrown down by the gradual accumulation of the Embankment fills up, thereby rendering it solid; whereas if carried on too rapidly, the probability is that interstices may be left below, although the upper surface may be hard; thus spaces are left not only for water to have access to at a future time, but by the action of the weights passing over the Embankment, the consolidation is constantly proceeding, thus disturbing the Level of the Rails. The reason why an Embank. should not be carried up too rapidly. ——I have heard of tracks running sideways, being used to facilitate embankments, but as they discharge at the sides, the area for carrying on the headway is diminished; I have had an embankment made in this way, but it being the subject of Contract, I did not pay much attention to the means which the Contractor resorted to in the execution of it. Method of Side Tracks. ——I have not seen a drawing or a model of Mr. Grahamsley's method; but was there any new method of expediting embankments, Contractors would adopt it. Mr. Grahamsley's Method. ——I have no practical knowledge of Locomotive Engines, but merely general information upon them, and it is my opinion that they cannot be usefully employed in forming embankments, unless they travelled upon a perfectly firm Railroad, instead of a mere temporary road. Loco. unfit to run upon tempy. Rail. ——I can readily conceive that the earth I described as being taken to spoil at St. George's Hill was intended to be used for bricks. Spoil is a term applied generally to earth put aside, and not used in forming Explanation of his Estimate.

the Embankment; certainly a term more appropriate might be found: but it does not necessarily follow that there should be a considerable difference in the expence whether the earth was thrown to Spoil or used for Bricks; in the latter case you make use of that earth which otherwise would only occupy the land, which merely decreases the expence of the Bridges, the Excavations remaining as they were.——I expect the Plastic Clay extends a considerable distance, and that more will come into the cutting as the works proceed: I judge from the course of the strata, and from the distance that I believe the surface is from the Chalk-bed, upon which this Plastic Clay rests. There is a stratum of Sand immediately below the Plastic Clay, between the Clay and the Chalk; this stratum varies in thickness perhaps from 6 to 20 feet. I judge from the borings, but in that particular district there appears to be an exception, for the whole of this Sand for a considerable extent is brought to the surface, forming the Bagshort Sand. The Plastic Clay is the Soil above the London Clay; none is yet exhibited, nor should I expect they will find any if the regular course is preserved. The Soil really found above this Plastic Clay at St. George's Hill is a loose sand with patches of gravel interspersed, on the lower extremity of the face of the cutting; at the commencement it is all Sand, and at about half the height of the cutting the Plastic Clay gradually rises up until 14 or 15 feet in thickness, and the last 7 or 8 feet is in Sand.——I have allowed $1\frac{1}{2}$ millions of cubic yards for the embankment on this side of St. George's Hill, (the Mole Embankment).—— ——I have likewise seen the Embankment that is being formed upon the Southampton Railway at Wandsworth, and although it is not a very large embankment, there is but a small portion made.——The Basing Line crosses some roads upon a level, which the Great Western does not; crossings upon a level increase the annual expence; they also interfere with the Traffic, and are frequently the cause of serious accidents. I am aware that there are some Railways already formed which occasionally cross roads upon a level; and I am sure the Proprietors would remedy the same if they had the means (even at a considerable expence.) ——I was employed in opposing the Liverpool and Manchester Railway when before Parliament, and I have since walked over the line three or four times, occupying about two days each visit.

Description of the Soil of St George's Hill.

Mole Emb. $1\frac{1}{2}$ mill. c. yards.

Embankment at Wandsworth.

Crossing of Roads upon a Level by Railways.

He opposed the L. & M. Railway in Parliament.

Ex. Mr. G. W. BUCK, C.E.

I have been engaged in the execution of various Engineering Works for the last 20 years, but they were not of a very extensive character; I had not superintended any Railway until I was appointed one of the four Assistant Engineers, under Mr. Stephenson, upon the London and Birmingham Railway; my district is between London and Tring, and embraces a distance of 30 miles, and the work upon it is heavy compared with the rest of the Line, *i. e.* there is more Tunneling, and the Embankments are higher and the Excavations deeper.——My office is to see that the Works are properly executed, and the amount of work done is measured monthly by my Assistants, and tested by myself, as I am responsible for its correctness.——The Soil in the neigh-

Asst. Engr. upon the L. & B. Raily.

His District is bet. London and Tring.

The Work upon it is heavy.

His Assis. measure the Work monthly.

bourhood of Watford is Gravel and Chalk, also Gravel, Sand, and some Clay, and it is very favourable for the execution of the Cuttings, which in the immediate neighbourhood are of considerable extent. The extreme depth of the Watford Cutting is 42 feet; the height of the Embankment is 45 feet, and is upwards of a mile in length; it is let in two Contracts, one half to one man, whose Contract is 5 miles long; and the other half is let to another Contractor, and is 9 miles long: and they are bound to complete the Work within 2 years.———There are 2 Gangs, or Shifts, employed upon the Watford Cutting, and 80 men in each; they work 20 hours out of the 24, each Shift working 10 hours.———I think that about 1000 cubic yards of Soil is the utmost I have seen moved in one day; the average for the last month has been about 700 yards; 1000 yards cannot be moved, except under favourable circumstances; for instance, when the Soil will make a permanent foundation for a Road, or when it is Gravel or Chalk. Clay hangs to the waggons, and requires shoveling out; but Gravel, Chalk, and Sand, tilts well, or slips out at once. The Watford Embankment is at present about 32 feet high; the Permanent Angle of the Slopes is 2 to 1; but the Natural Angle at which the Soil will stand, immediately after teaming, (the Angle of Repose), is about 1½ to 1, which enables us to make the Embankment temporarily wider at the head (as it is called by the workmen,) and get in 6 Teaming Lines of Road: after the Line has passed that particular point, the work is trimmed down to 2 to 1, as it is no longer required. When our Teaming Road is the natural width of the Embankment, viz. 33 feet, we cannot team more than 700 or 800 yards, although working the same number of hours and presuming our arrangements perfect. During last month we moved 15,800 yards, and there was 22 working days, merely deducting 2 Sundays and 2 wet days; we cannot work as long in the winter nights; the expense would also be increased, as the Waggons are apt to get off the road, torches are also required. Taking the whole year round, you cannot reckon upon more than 5 working days in the week.———I have seen a method of Teaming tried different to the usual method; 8 square Baulks, about 60 feet long, were placed at the end of the Abutment of Watford Bridge, with their ends projecting over the Abutment, for the purpose of placing the Rails on, in order to run the Waggons on same and to team them; each trip was drawn by one Horse, consisting of two Waggons; the first Waggon was pushed on to the beam to make way for the second, which was then teamed, and both Waggons returned together. This method was continued until the Embankment was raised to the full height of the Abutment, when the Baulks were removed from the Abutment and placed upon the new Embankment, where they became quite unmanageable; the method was consequently abandoned. The Contractor's object was to expedite the Embankment; after the first Waggon was tilted, instead of drawing it back into the road it was pushed on a little farther, in order to make room for the other, so that when it was tilted both Waggons returned together. I do not think more Soil was teamed by that method than we are at present teaming, although we have only 4 teaming places at the present time. We have been using 6, and shall again, (6 cannot be worked unless the Slope is equal to 2 to 1), we have teamed 718 per day with 4 roads, and 1,000 per day with 6 roads.———A flat Slope can be executed quicker, in a certain proportion, than a steep Slope, as the greater Slope does not increase in the same ratio as the width of the Road.—— ——The

The Soil at Watford.

The Watford Cutting 42 feet deep and 1 mile long.

Worked by 2 Shifts of 80 men each.

1000 cu. yards the most he has known teamed in a day.

Gravel, Chalk, and Sand tilt quickly, Clay does not.

Method of working an Embankment with 6 team. places.

Particulars of the Watford Cutting and Embankment.

5 working days per week is the aver. number taking the year round.

Method of Teaming tried on the Works.

Description of same.

A flat Slope can be exd. pro. quicker than a Steep Slope.

Detailed Prices upon the Watford Cutting.

Navigator's Wages, &c.

Materials 2*d.* per Cub. Yard.

A Lead of 1 Mile costs 2½*d.*

It costs the Cont. 9*d.* per cub. yard, indep. of Drains, &c.

The Cont. of the Watd. Cuttg. is paid 1*s.* 2*d.* per cub. yd.

Method of letting the Works in large Conts. better than the Co. exec. it.

Paupers cannot be got to work upon the Railway.

Tring Cutting cont. about 1,200,000.yds

St. Geo. Hill about 3,725,93.

Watford Cutting costs the Contractors about 6¼*d.* per cubic yard, without the materials, and the present Lead is but 30 chains long. I make the Cutting and Filling 5*d.*, the Teaming ¼*d.*, the keeping the Roads in repair ½*d.*, the dressing the Slopes and stripping off the Sod ¼*d.*, the Lead about ¼*d.*, and oiling the Waggons ¼*d.*, which makes 6¼*d.*——Very few of the men are paid less than 3*s.* per day, which is the price I have allowed. Agricultural labourers get 4*d.* or 6*d.* a day less than the others; (they all find their picks and shovels), there are very few Irish on our part of the Line. Employing more Agricultural labourers might reduce the price ¼*d.*, but not more. I think the Waggons, Barrows, Planks, Rails, Chairs, Sleepers and the like, cost about 2*d.* for the present Lead, (⅜ths of a mile, which costs 1¼*d.*)——Upon a temporary Railroad, I consider that 1*d.* should be added for Haulage or the extra expense of Horses and Drivers, for every additional mile after the first; (this price comprehends Haulage only, not the expense of the temporary way, taking both together makes 2½*d.*,) perhaps ¼*d.* should be added for the repairing the Roads, making 1¼*d.*——Two and sometimes three horses are used in leading the stuff from the Cutting to the Embankment.——Another item must be added, viz., the re-sodding the Slopes, as I have allowed ¼*d.* for taking it off, I must allow the same for putting it on the Slopes again, making it 9*d.* per cubic yard, which is the price it actually costs the Contractor, who has to take the riskof all Slips, contingencies, &c., also Drainage, (Slips are more likely in deep Cuttings and Embankments.)——I do not think an Embankment, with a 3 miles Lead in Gravel or Sand, can be executed for much less than 1*s.* per cubic yard, if in Clay the price would be greater. The Contractor of the Watford Contract is paid 1*s.* 2*d.* for Cuttings, (*i. e.* the whole of the cuttings in the contract), he has to find security to the amount of 10 per cent., and is to complete the Works in two years, and keep it in repair for one year afterwards.——I consider the method of letting the Work in large Contracts to respectable Contractors is better than the Company executing it themselves, both for durability and cheapness.——The Select Vestry at Hemel Hempstead applied to the Directors of the London and Birmingham Railway to employ the Paupers upon the Works, and a portion of the Line was let to them at 9*d.*, they finding materials; although the length of the Lead was only 200 yards, and the Excavation in Gravel of very little depth, they worked but a few days and left it, as the men could not be got to the work.——I believe the Tring Cutting contains about 1,200,000 cubic yards, perhaps ⅓rd of that quantity goes to Spoil, all the rest goes one way, (St. George's Hill is stated to contain 3,725,93 cubic yards).

Ex. Mr. JAS. COPELAND.

Execug. a Contract on the L. and B.

I am a Contractor, and was engaged upon the Liverpool and Manchester, the Leicester, the Swanington, and the Whitby and Pickering Railways. I am now executing a Contract upon the London and Birmingham Railway, which I obtained by open tender;

my Contract amounts to £117,000., and extends from Watford to Kings Langley, and is about 5¼ miles in length. I found two Sureties, (two Directors of the Leicester and Swanington Railway, satisfied I had done my duty to them, came forward as friends and become my sureties. I certainly have a stock of materials of £14,000. or £15,000. value, which the sureties would look to in the event of any thing transpiring), to the amount £11,700., (or 10 per cent. upon the amount), each in half that amount. I do not allow my sureties any thing, but we take into account the difficulty of obtaining security.——— There is about 700,000 cubic yards of Cutting, and 600,000 cubic yards of Embankment, also a Tunnel of 1,716 yards in length, (the Cuttings and Embankments are nearly equal). Ten or twelve Tenders were submitted for the Contract, and I believe my Tender was about the lowest; it consisted of a gross sum, and a "Schedule of Prices" attached, in which the prices for open Cutting was 1*s*. 2*d*. per cubic yard, (the price for the Cutting also includes the Embankment), the average Lead of which is about a mile, the Tunneling was £28. per lineal yard, and the Fencing about 2*s*. 6*d*. or 2*s*. 9*d*. per yard for each side of the railway.———The price of the Fencing upon the Leicester and Swannington Railway was 3*s*. per lineal yard, for each side of the Railway, which I consider a good price, (it was not my contract, neither were Tenders delivered for same). I would have done it for 5*s*. but not 3*s*. 6*d*.———I have done about 120,000 yards of one Cutting, and nearly 100,000 yards of another, (in my Contract on the London and Birmingham). We let the Earthwork in small Contracts at about 8½*d*. per cubic yard, including the heavy cuttings we are now making; which are through Chalk, Gravel, and Sand, and the Lead is rather better than a mile. I find Waggons, Barrows, Planks, Sleepers, Chains, Keys, and Pins, which amounts to about 2*d*. per yard more; *i. e.* for the actual cost of Materials, exclusive of the Wear and Tear; as I have expended £15,000. for materials, there is the interest of that sum, and the Wear and Tear to be allowed for extra, there is a continual expence in the repairing of Waggons, &c. The value of the materials, after the conclusion of the Contract, may be about ½*d*. per yard, which, deducted from the original outlay, would make the price of the materials 1½*d*. per cubic yard, the cutting therefore costs 10½*d*. per cubic yard; and we take the risk of Slips and Contingencies, and prepare proper Drains, which in Clay cuttings are very considerable; also the Sodding of the Banks, as some of the soil is removed twice, which I have allowed for in the average of 2*d*. per yard. I therefore consider that a Contractor must pay great attention to his business, and practice considerable economy, to make a profit out of the 1*s*. 2*d*., as it is barely sufficient; (the Engineer derives no advantage whatever from the Contract).———I employ about 400 men at present, and the Sub-Contractor pays them; I find by experience that Labourers who receive the least amount of wages, are the most expensive; a mixture of Agricultural Labourers and Navigators is advantageous, (the wages would then average 3*s*. per day), but Agricultural Labourers only are not. I let some work at 9*d*. to the Overseers of Hemel Hempstead for the Paupers to execute, but they only worked two days, and the Overseers did not make any demand for the same, as it was so trifling; not more than 10 cubic yards, (about 4*d*. each man per day).———I have not seen Mr. Grahamsley's Machine, but if it is a machine that would merely enable us to do more work in a given time, it would not be advantageous, except in particular cases; for instance, at this

Amount of same, £17,000. it is 5¼ miles long.

Particul. of the Security required.

Particulars of the Contract.

His price for Cutt. which also includes the Embt. is 1*s*. 2*d*. per cub. yard, and 1 mile Lead.

Tunnel £28 per yd. Fencing 2*s*. 6*d*. or 2*s*. 9*d*.

Price of Fencing Leic. and Swanntn.

He under lets the Cuttings, which are in Gravel, Chalk, & Sand, and about 1 mile Lead, for 8½*d*. *i. e.* Labour only.

Materials about 2*d*. more.

Particulars of the same, &c.

It therefore costs him 10½*d*. independent of Slips, Contingencies, &c.

Agri. Labour. although cheap are not advantageous.

Instance of Paupers being emp.

Account of their failure, they did not earn above 4*d*. each man per day.

Mr. Grahamsley's Machine.

time of the year the days are very long, therefore any thing that could expedite the emptying of the Waggons at the end of the Bank would be an advantage; in wet seasons

Remarks upon same

the stuff cannot be teemed as fast as it can be excavated, more time being required to keep the roads in order, as they become muddy and sodden, which occasions the Bank to sink: the road, at this time of the year, stands very well.

Machine upon the Hartlepool Raily.

I have seen a Machine upon the Hartlepool Railway, but I do not know who invented it, or whether it belonged to the Contractor or to the Company; it consisted of a wooden frame placed at the end of

Remarks upon same

the Embankment, (which was about 30 feet high), upon which the Waggons ran and were emptied, it did not appear to work very well; I heard that several men were killed in moving it, as it was a very awkward operation; the Soil they were working was Clay and Sand, and their difficulties would increase as the Embankment was raised. I do not know how long it was used, but a short time afterwards I heard that it was taken down, and that they had resumed the old plan, therefore I did not attempt using it; if there was a machine that would save us 20 per cent. we should most likely use it.

Observations upon using Locomotives upon the Works.

I have used Locomotives upon a Permanent railway for carrying the earth, but they cannot be used with advantage upon a new made Embankment; 12 months at least should elapse, as it takes some time to settle. We are constantly altering the Rails upon the Embankment, although we work with Horses.——

Method of Workg.

The Sub-Contractors work the 24 hours, having 2 Shifts containing about 100 men, each Shift working 12 hours, they rest about 2 hours, therefore we get 10 hours actual work out of them; ⅔rds of the men are in the day Shift, and ⅓rd in the night; it would not be possible to put any more men upon it with advantage, the fewer men a Contractor employs at a time, the more profitable it

The fewer Men emp. at a time the more advantage.

is to him, provided they do the work. If the Company had it in their own hands, they could not employ 200 men upon it profitably; perhaps, at the present moment, about 40

700 or 800 cub. yds. a good day's work for any number of Men.

or 50 might be added, but even then they would not be able to cut more than 700 or 800 cubic yards per day; on particular occasions we might cut 900 yards, but we could not get "half as much more work done by employing half as many more men."

By employing twice the number of men, you cannot get twice the quantity of work done.

The greatest amount of cubic yards I have known them remove in a day, was under 600, or about 7000 in a fortnight, of 12 working days; even in dry weather, and although the Soil was Chalk, Gravel, and Sand, and the lead about a mile long; it was more at the Watford

Observations upon Working, and cost of Cuttings, &c.

Cutting.——260 is about the number of working days in the year, or 5 days in the week. If the work could be done for 5*d.* per yard, I should not pay 9*d.* for it. If the Lead was 3 or 4 miles, I should pay the Sub-Contractor 1*s.* 4*d.* per yard. I am going to pay 8½*d.* for a portion of 140,000 cubic yards of the whole work, (690,000 cubic yards), the Lead will be about 2 miles, and when they have finished, the price will be advanced.

The Sub-Contractor does not let the Work, but pays the Men by the day.

The Sub-Contractor does not let the work out again, but pays the men by the day, and my belief is that he will not make a reasonably good thing of it. If the Company were to undertake the work themselves, with a small number of men working by the day, it would cost them half as much more, it would cost 1*s.* 4*d.* instead of 8*d.*

Remarks upon a Comp. undertaking the Work.

I have tried it often, having measured a piece of work and put a man upon it that I was satisfied with by the day, and it has often cost double the sum it did when let by Contract. If the Company made sub-contracts with inferior men, they would be unable to obtain Security, and therefore could not depend upon them, as their practice generally is to execute the easy part of the Work, and leave the difficult.——I have expended about

£14,000. or £15,000. in Materials, &c. including expensive Engines for making Tunnels, also Steam Engines; and I receive 80 per cent. a month for the work done (20 per cent. being left in the hands of the Company), which comprehends the work done and completed, and the expence of materials; the outlay for materials and other things is comprised in the price of the work. I have never said that I expected to make £30,000. by the Contract; I might possibly have said that I should make £15,000. by it, but I have no recollection of it, as the Contract will barely clear itself. I would not refuse £10,000. for the profits derived from it, much more £20,000. (as enquired by the Counsel for the opposition.)

Mode of Payment.

Prospects of a Contractor, &c.

Ex. Mr. HENRY HABBERLY PRICE, C.E.

I have practised as an Engineer for the last 20 years, and have been much connected with the several Railways. I was for a short time engaged as principal Engineer upon the Clarence Railway, which is situated in the North of England. Mr. Thomas Rhodes is their present consulting Engineer, and resides at the works. (I am not aware whether he was appointed by the Exchequer Bill Commissioners, or by the Company.) I have not been consulted by the Company for the two last years until last month.——— I am also engaged in the manufacture of Locomotive and other Engines.———I have not been engaged as consulting Engineer to the Great Western Railway, although I should be very happy to give advice if I was called upon. I was employed last year by Lord Jersey (who has very large estates in South Wales) to oppose the Great Western Railway, on account of the Railway stopping at Reading.———I have since been engaged to examine the Line laid down by Mr. Brunel, with the contemplated Branch to Gloucester.———I attended the Meetings held at Stroud, Cheltenham, Penzance, Cornwall, Plymouth, Devonport, Exeter, Taunton, and Bridgewater, to discuss the subject of this Railway; the plans and sections of which were always on the table. At Exeter, Taunton, and Bridgewater, the comparative merits of the opposing Lines were discussed, and I think the Resolutions were carried unanimously in favour of the Great Western.———At Exeter it was considered of great importance to open a communication with the Midland Districts, particularly with Birmingham; which cannot be conveniently effected by the Southern Line, without following the Great Western from Trowbridge and Bradford, and up the valley of the Avon to Gloucester, by way of Stroud, as proposed by the latter. If the Southern Line was made, it would require a new Cut in point of distance to Gloucester.——— ———I have heard that *5d.* is the price allowed for the Cutting of St. George's Hill, with a Lead 3 or 4 miles long; I consider it will cost more than double that sum.———At the commencement of my connection with the Clarence Railway, they were in the habit of employing Gangers, or men without capital, who often undertook the work for less than its value; the Company finding materials, and men to keep the Line in order: I advised them to advertise for Contractors to undertake the whole, including the keeping of the roads in repair, &c., and

His Experience.

He was Engineer of the Clarence Railway.

He Opposed the Great Western last year.

At the several Provin. Meetings con. a Line of Railway the Gt. Western was always preferred.

Inferiority of the Basing Line in ref. to a Com. with the Midland Counties.

5d. per yd. not half enough for Saint George's Hill.

Superiority of letting the Works by Contract compared to Day Work.

It cost 1*s.* per yd. by Con. and 1*s.* 4*d.* or 1*s.* 3*d.* by Day Work.

Mr. Brunel's Prices quite sufficient.

Disadvantages of Day Work.

700 or 800 cub. yds. the most that can be teamed in 1 Day.

Case of a Premium of £1000. being given to a Con. to expedite the Works.

Account of his visit to St. George's Hill.

Description of the Works.

Remarks upon Mr. Buck's method of extending the Teaming places.

Theory of Cutting and Embankments.

who should find security: they adopted my recommendation in two cases, and upon analyzing the accounts, I found that according to the latter system, the Cost of Excavation, including the keeping of the Roads in order, &c., was under 1*s.* per cubic yard, and by the old System of Day Work 1*s.* 3*d.* or 1*s.* 4*d.* or more. The Leads were generally short; in one Contract the average Lead was under 2 miles.——The Price at which those Contracts were let fully corroborates Mr. Brunel's Estimate, and I think his Prices are quite sufficient. I have often known the Contractors who had executed the profitable part of a Cutting, leave the remaining portion, and run away. The work was consequently at a stand for 2 or 3 weeks, when it it was re-let at an additional price, which was constantly increasing with the distance.——Under the old system the men were frequently found under a hedge, smoking their pipes, instead of working; there certainly was not a sufficient superintendance, (the Railway was then under the management of the late Mr. Steele) but an efficient superintendance would have put the Company to considerable expence. No men will work by the day as they do by the job, neither do they regard putting a Company to expence, but they do a Contractor. It is the rule in large works to let them entirely by Contract, if otherwise there must be a good superintendance;—viz: a person in every Cutting; the expence of which, added to the small quantity of work men do by the day, makes it very disadvantageous. Such a Cutting as St. George's Hill may require but 1 Superintendant; but where roads extend over several miles it is not sufficient.——I never knew more than from 700 to 800 Yards teamed over an Embankment in 1 Day; in which Case a Premium of £1000. was promised the Contractor provided he executed a particular Cutting by a certain time; the men worked longer hours, and made double shifts; therefore, it was not all profit to him; and for a short time during the summer months he averaged 700 Yards per Day, certainly not more.—— ——I visited St. George's Hill about a fortnight since, and examined the Works, and I am of opinion that the same quantity of work that is excavated may be done in 6 weeks, with good men, and by the usual method. (The commencement of such a Cutting is not a fair criterion whereby to judge of the progress of the works, as the men are obliged to work with barrows.) The face of the work was plastic Clay, 15 feet thick, and some gravelly substance on the top; there was no Clay at the beginning of the Cut, although it is getting thicker every day; the men appeared to be working different to the usual manner, they were using smaller waggons, but larger waggons were being prepared; the Road was also of a different gauge, having small bars of iron, with holes punched through them for Rails,—I saw none longer. There was only a single Road upon the Embankment with turn-outs; the width of the Embankment would not admit of more than 2 teaming places upon the top—*i. e.* a double Line of Rails; if 4 Waggons were required to be discharged at the same time, they must shift them, or run them forward, which is inconvenient.——I have not seen the plan alluded to by Mr. Buck in his Evidence for extending the teaming places, but from his description it appears advantageous, in cases where it is applicable; it could not be used on an embankment of 1½ to 1, as it requires greater slopes.——Engineers generally endeavour to get the Cuttings and Embankments nearly equal, (which is accomplished upon the Great Western): if much earth is carried to spoil, ground must be purchased to place it on, which consequently adds to the expence.—— ——I am not aware of any Railway that has so favourable Gradients as the Great Western; when there is Traffic both ways a level line is cheaper than an undulating line; it would be different with a line entirely descending, because the summit level being low, it requires but a small quantity of power to work it, which diminishes the expences; an Engine would carry a greater load upon a level, than upon an acclivity; if an undulating Railway could be made even at a less cost than one upon a level, I think it would ultimately be the most expensive.———The first time I met Mr. Brunel at Stroud, (after the plans were deposited) I mentioned that I approved of his method of Concentrating the Inclinations at one spot, and employing Assistant Power to overcome the same; if Assistant Power is not used, the load must be lighter and the speed less, or it may not go up.———If the same load is to be carried with the same velocity up the Plane of 1 in 202 upon the Basing line, an Engine of greater power must be used, which will occasion a great loss of power upon the rest of the line; and will no doubt entail additional expence upon the future working of the Railway; if the Rocket of the Liverpool and Manchester Railway was worked upon such a plane, it would require to be put to its utmost power, and would therefore be more likely to get out of repair; consequently it is not an economical way of working a Railway, an Assistant Power is far more preferable. If Mr. Stevenson had no objection to a Plane of 10 miles 1 in 210, I should infer there was no other alternative, as he has had more experience in Railways than any Engineer of the present day, but if he proposed to overcome that Plane without either an Additional or a Stationary Engine, I cannot consider that he is right, unless the rest of the line approximated to it. A good Engine may take an average load up an inclined Plane of 1 in 202 at the rate of 20 miles an hour for a certain distance, but the same Engine would carry a much greater load upon a level Railway, such as the Great Western.———There is no increase in the consumption of Fuel by having more Power upon a line, but Fuel, although a permanent expence, forms a very small portion of the expence of an Engine, being only $\frac{1}{5}$th, the Wear and Tear in working of the Engine being the remaining $\frac{4}{5}$ths. If heavier Engines are obliged to be used, there must likewise be heavier rails and larger blocks, also a stronger foundation, as the same Engines will be used upon all parts of the Line, (the weight of an Engine is very much in proportion to the size of the Cylinder,) thus if you require upon the Basing Line an Engine that will mount 1 to 207, the rails must be sufficiently strong throughout the whole Line to carry such an Engine; as Goods may be carried at a lesser velocity, their weight will not be so destructive to the Rails, but heavy Engines travelling at a greater speed are very destructive, therefore in determining the strength of the Rails it becomes a consideration as to whether the Railway is used for Goods or Passengers; a Rail of 60 lbs. per yard is very strong, being sufficient for any traffic we have at present seen, and perfectly safe for the ascent of an Engine at a rate of 20 miles an hour. I think the original weight of the Rails upon the Liverpool and Manchester Railway was 35 lbs. per yard, and I have seen Passengers carried at the rate of 30 miles an hour upon it; great injury has been done to the Rails by allowing Engines of 10 and 12 Tons weight to run upon them, whereas they were only intended to carry Engines of 5 Tons weight.———In making a calculation of the Gradients of a Railway, whatever height you may have to reascend must be added to the height of

Theory of Gradients

A Lev. Line better than an Undu. Line.

His rea. for approv. ing of Mr. Brunel's method of concen. the Inclina.

Observ. upon the work. of Inclined Planes by Locomo. and Station. Eng.

The Rocket would not work advan. upon a Plane of 1 in 202, it should be worked by a Stationary Engine.

A Plane of 1 in 210 cannot be worked without Add. Eng.

Fuel is $\frac{1}{5}$th, & Wear and Tear $\frac{4}{5}$th of the Exp. of Work. an Eng.

Observations upon the work of Incli. Planes contin.

A 60 lb. Rail sufficient for any Traffic.

35 lbs. orig. Rails on L. and M. Rly. intended for Engs. of 5 Tons only, and greatly injured by Engs. of 10 or 12 T.

Mems. regarding a comp. of the Gradients of the 2 Lines.

the summit; if the level of one Line is lower than the other, and the undulations upon the lower line added to its summit level amounts to more than the other, the Gradients of the former Line are considered the best, it would be grossly fallacious to take the height of the summit levels only.——Upon comparing the Gradients of the two Lines it is greatly in favour of the Great Western, which has but 2 or 3 slight undulations upon it. I do not consider the Plane of 1 in 107 objectionable, on account of being situated in a Tunnel, as it must of necessity be placed somewhere, nor do I apprehend any more danger than in an open cutting.——I have often had occasion to adopt the plan of driving to higher levels in Mines to assist Ventilation; there will always be a current of air ascending, which prevents the Choke Damp taking effect. A Tunnel upon the Incline, looked at in reference to Ventilation, is better than a dead level. If Locomotives are used in a long Tunnel, it should be ventilated, which could be done without any difficulty; if a Stationary Engine is used it is not so essential; for instance, if the Tunnel through Claverton Hill, which is 1 mile long, was traversed by Locomotive Engines, it would be necessary to ventilate it. If a Line could be found from London to Bristol equally advantageous as the Great Western, and without a Tunnel, I should certainly consider it the best; there are 7 Tunnels upon the Great Western, (exclusive of those upon the Birmingham,) 4 of which are between Bath and Bristol, but 2 of them are of no consequence, being very short, the others are long but could not be avoided without great expence.——I think a Tunnel less objectionable to a Land-owner than an open cutting; if a cutting was made at Mr. Palmer's instead of a Tunnel, it would be 30 or 40 feet deep; and a Cutting of 116 feet would of course be more objectionable.—— ——I have not minutely examined the collateral country of the Great Western, but I have gone over the Line, and do not know a better.——The Bath Depôt of the Basing Line is intended to be formed upon the side of a hill which is shelving ground; I do not think there is sufficient space for a Depôt, and it is considerably above the principal parts of Bath; the Valley of the Avon being between them, the section shews a rise of 70 feet from the valley to the depôt at a rate of 1 in 9, which is a very objectionable inclination for carriages to pass up and down from the Railway with goods. The Great Western Bath Depôt is in Ham Gardens, which will afford a very convenient Depôt; it is at a considerable elevation above the level of the Floods, which I am informed are very considerable; the Railway will likewise be constructed in such a manner as to guard against them.—— ——My attention has been particularly called to the country between Swindon and Gloucester, and I think it would be practicable to make a Branch from the Great Western towards Gloucester at about the summit level, near the North Wilts Canal, not far from Swindon, and thence passing over the higher ground to the summit level of the Thames and Severn Canal, in which direction the Line would run past the Sapperton Tunnel, which is 2¼ miles long, but as the Railway would be upon a higher level than the Canal, our Tunnel would be less than a mile in length; there is also a pass through which it might be easily taken. The highest summit would be next the Stroud Valley; I have calculated the Line as not exceeding a rise of 6 feet in a mile, which I think particularly favourable: from thence between Stroud and Gloucester there would be no difficulty, as we get into the Valley of the Severn.——I certainly think the country between Gloucester and

The Gradients of the G. W. superior to the Basing Line.

The Inclination of 1 in 107 not disadvan. by being placed in a Tunnel.

Observations on Ventila. Tunnels.

A long Tun. work. by Locomo. should be ventilated.

Of the Tunnels on the Great Western.

A Tunnel is less objectionable to a Land-owner than a Deep Cutting.

Objections to the Basing Depôt at Bath.

Great Western Depôt at Bath.

Branch from the Great Western to Gloucester.

Bristol following the River Severn, is better adapted for a Railway than that between Gloucester and Swindon; but I apprehend there is not sufficient Traffic to pay for a Railway, and the distance would be increased 30 or 40 miles, as it would be necessary to go first to Bristol, and then some miles below it, in order to get out into flat country, and the expence per mile would be quite as much as the line between Gloucester and Swindon: I make the distance from London to Gloucester, by Swindon, about 112 to 114 miles; viz. 81 miles from London to Swindon, 24½ miles from Swindon to Stroud, and 9 miles thence to Gloucester; and it would be about 38 miles farther by the other Line.——I believe a Railway may be made between South Wales and Gloucester, as I have examined the most difficult part of the intervening country. I am likewise connected with the Iron trade, and am also acquainted with the Copper and Tin-plate trade of South Wales, throughout the whole of which there is a feeling in favour of the Line.——I advocated a Line to Windsor before a Committee of the House of Commons last year, which was a very excellent Line, but it was met by such great opposition that the Directors abandoned it. It was superior to the Great Western of last year, (in consequence of its stopping at a higher level; the Great Western now goes higher;) and is not inferior to the present Line; the levels were very good, but not better than the Great Western. From Reading to the Birmingham Line the Undulations amounted to 6 feet per mile. I think there were two Summits to overcome between Acton and the River, one near Sion House beyond Brentford, the other near Hounslow, which was the highest of the two, but they were not considerable. From Sion House it was a gradual descent, which commenced a little below Hounslow, in order to pass under the Long Walk at Windsor; from thence there was a Line that would have gone to the South of Reading, which Line was quite practicable. It crossed the Long Walk at about ½ a mile up, in a Tunnel which extended about a ¼ of a mile from the walk on each side: we went through the field between the Park and the Turnpike Road, close to the foot of Frogmore Grounds, and we crossed all those fields to the long walk: the Line curved round the Town after it passed the Long Walk by the Barracks; we had not room enough for one Tunnel, therefore we proposed making two, passing between the roots of the trees diagonally. I met Lord Duncannon on the spot, and satisfied him that it might be done without being very objectionable; but the Directors were informed by some Peers that they would not sanction the measure if it went under the Long Walk, as they considered it would be prejudicial to the Royal Residence at Windsor, notwithstanding the entrances were to be planted; it terminated at Paddington, but we did not make use of the Birmingham.—— ——I last year advocated the Terminus at Paddington, in opposition to the one at the Thames. I conceived it would be much more convenient for the distribution of Passengers, also for Cattle and Live Stock. Light Goods carried by the Railway, being for the consumption of London, will require to be distributed throughout same: the Northern Terminus at Euston Square is therefore unquestionably preferable to the Southern at Nine Elms, being more central, and upon a higher level, which renders it more convenient for the loads to descend. I consider a junction of the Great Western with the London and Birmingham advantageous, notwithstanding its having to encounter some extra Tunnels, and a Plane of 1 in 86, which under all circumstances is not very objectionable, as the distance is short, and it is the best that

Remarks upon a Line from Gloucester to Bristol following the Severn.

Ditto by Swindon.

Remarks upon a Line between Gloucester and South Wales.

Obser. on Mr. P.'s Line to Windsor.

It was abandoned on account of the Diffi.

It passed under the Long Walk

by

2 Tunnels.

Comparison of the London Termini.

The advan. of a Junction of the Gt. Western with the L. & B.

can be obtained. I cannot say whether they intend using a Locomotive or a Stationary Engine, but I should prefer the latter. I am not aware of any great difficulty that can arise to the Trade of the two Railways through their having the same Terminus, provided there is ample space for the Depôt; but I prefer, and I think a separate Terminus would be better, which was my reason for not running my Windsor Line upon the Birmingham.

Difficulty of obtain. a Terminus in Oxford Street.

I still prefer the Terminus of the Windsor Line, which was at the end of Oxford Street; but the difficulties were very great, as the property belongs to the Bishop of London, whose objections we could not overcome: we could have obtained a Terminus about ½ a mile North of the Oxford Road, near the New Road, which would have been better than that at Vauxhall Bridge, but not equal to that at Euston Square. I think it would be very advantageous to have two Termini: when the subject was in discussion last year, the Terminus of the London and Birmingham was almost as distant from the principal parts of London as ours at Paddington, and I thought there was very little to choose between them, but now that they come to Euston Square the case is materially altered.——A Bale of Goods sent by the Railway, and intended for the Sea, would be passed by the Canal at the Depôt, and would occupy about 5 or 6 hours, which would not be very disadvantageous. I believe the Bristol Fly Waggons occupy about two nights and a day on the journey; and I expect the Railway will carry the

The Railway would take all the Goods that travel by the Fly Waggons.

bulk of the Goods at present carried by them. I am not aware of any Goods that travel by the Fly Waggons that could not go by the Railway, as they would be carried cheaper, which, coupled with the additional speed, would render it advantageous.——I believe

The Windsor Plan advo. before Par. was not the deposited Plan.

Account of the Deposited Plan.

that the deposited Plan (of the Windsor Line) was not the Plan advocated before Parliament last year in opposition to the Great Western, but a Plan made subsequently, owing to our being obliged to alter the Line, on account of its passing under the walls of Windsor Castle, between it and the River; also as Eton College and the Commissioners of Woods and Forests would not assent to it: there was a reservation with the latter that, provided we did not interfere with the Waterworks, they would assent; and upon an examination being made by an Engineer for the Crown, his objections were such that they declined giving their assent to it. I did not lay down the Line which went so near

Mr. Brunton's Line from Bath to Bristol

Observations upon Messrs. Brunton & Price's Line from Bristol to London.

the Palace, nor ever advocated it in my evidence before Parliament. I have seen the Plan Mr. Brunton projected from Bath to Bristol. I was connected with a Railway Mr. Brunton projected from Bristol to London, and I was a party to the Prospectus of same, which was before the publication of Mr. Bradshaw's Map. Colonel Page gave us the Levels of the Kennet and Avon Canal; therefore, having the Canals to guide me, and knowing the country generally, I looked over it sufficient to satisfy myself the Line was practicable. I proposed to go on through Newbury to Reading, but I never entertained an idea of cutting through St. George's Hill; the Line from Bristol to Newbury was very similar to the Basing Line; we did not take the Levels of the country, but we drew a Line on the Ordnance Map, which Mr. Brunton afterwards

Mr. P. declined giving Evidence on it.

It had no Tunnels.

It had a Com. with the Deep Water at Bristol.

went over; I declined giving evidence upon it last year; it was opposed by the Land Proprietors; it went near the front of Mr. Gore Langton's House, which would have been very objectionable; he avoided having any Tunnel by keeping higher Levels; but I should not wish to omit all the Tunnels on the Great Western, even if there was no objection in point of expence; he also had a Communication with the Deep

Water at Bristol at Redcliff Wharf, (the Goods were to go by a Tunnel and an Inclined Plane) where there is not room for more than 1 or 2 Vessels, (Irish Steam Boats cannot get up to the Wharf, although Cattle and Live Stock from Ireland will be a very important part of the traffic of the Railway. I cannot say whether an advantage is gained over the Great Western by the same, as the latter has a Communication with the Float, the water of which never flows out, it being dammed up to the feeder, and it is deep enough for barges, but not for ships at all times of the tide.

The Gt. Western has a Communication with the Float.

Ex. Mr. ROBERT STEPHENSON, C.E.

I am Chief Engineer upon the London and Birmingham Railway, 80 miles of which, out of 111 miles, have been let for £1,300,000.; my Estimate for the same was £1,325,330. The works are let by Public Contract, and the Directors usually accept the lowest Tender, if the parties are respectable and are able to give security; I am present at the opening of the Tenders, but I have no opportunity of knowing their several amounts until then: in alloting the several Contracts, I have subdivided them pretty equally, and arranged them so that one Contract shall not interfere with another; the Work is measured at the time the Contractor delivers his Tender, which is accompanied by a "Schedule of Prices," upon which his Estimate is founded, and at the expiration of every month the work is measured and priced according to such List, and the amount is paid him, with the exception of 20 per cent., which is withheld until half the contract is finished, the amount retained is then 10 per cent. only, and the Contractor afterwards receives the full amount of his work. The commencement of a Contract is easily and cheaply executed; there have been many cases of a contractor making a profit upon it notwithstanding the drawback; thus he may be receiving a profit throughout the whole of his contract; but although he may possibly clear 20 per cent. upon the first half, he cannot realize more than 10 per cent. upon the whole. I am not aware of any instance of more, as the work will cost him more as he proceeds, and the payments are made so as to equalize his burden. This is one of the Blank Tenders, with a "Schedule of Prices;" which Prices are filled in by the several parties tendering:—

Chief Engineer of the Lond. and Birm.

Mode of Letting and Executing the Works on the same.

The commencement of a Contract is the most Profitable part.

10 per cent. is as much as can be made by a Contract.

Contract, No.

To the Birmingham Committee of Directors of the London and Birmingham Railway Company.

of do hereby propose to make the Railway with all the Excavations, Embankments, Tunnels, Bridges, Culverts, Drains, Fences, and other Works complete, and to keep the same in repair for One Year after Completion, and to find and provide all the requisite Materials, (except the Iron Rails, Chairs, Pins, Keys, Blocks, Sleepers, and Oak Trenails,) according to the Plans and Specifications exhibited to within the Periods, and upon the Terms and Conditions mentioned and contained in the Draft Contract, also exhibited to for the Sum of and have in the First Schedule hereto set forth the Price of the different Descriptions of Work at which the aggregate Amoun of this Tender is computed.

Form of

And further propose to execute the several Works in the said Specifications denominated Extra Works, at the Prices set opposite to each description of Work in the Second Schedule hereto.

And in case this Tender shall be accepted, hereby undertake to execute a Contract and Bond to be prepared by the said Company, according to the Draft before referred to, within a Fortnight from this Date.

And propose of and of as Sureties for the due Performance of such Contract.

Tender,

And do hereby undertake that they shall, within a Fortnight from this Date, execute a Bond, to be prepared by the said Company, conditioned for that Purpose, in a penal Sum equal in Amount to Ten per Cent. on the said Sum of of which Bond the Draft has been also exhibited to me.

And lastly, do hereby undertake and agree that in case the said Contract and Bonds shall not be executed by and said Two Sureties, within the Time above mentioned, the said Company shall not (unless they think fit) be bound by this Tender and Contract, but the same shall be absolutely void, in case the said Company shall so think fit; nor shall they in any Case be liable to any Claim by in respect of Work then already done by upon the said Railway.

Witness Hand, this Day of 183.

First Schedule referred to, containing a List of the Prices at which the above Tender of £ is computed.

£. s. d.

The Price of permanent Fencing, consisting of Ditching, Railing, and Quicks, at per Lineal Yard . . . .

The Average Price of the Whole of the Excavations, with the Slopes of the Excavations and Embankments, completed as described in the Specification, at per Cubic Yard .

The Price of the Tunnel at exclusive of the Facings or Fronts, at per Lineal Yard . . . .

The Price of Brick Work set in Mortar, at per Cubic Yard . . . .

Ditto Ditto in Cement . . . .

The Price of Freestone Ashlar Work, dressed and set, with rustic chamfered Joints, the Face of the Stones being rough, when the Stone is obtained in the Excavation, at per Cubic Foot . . . .

Schedule

Ditto Ditto of Derbyshire Stone, at per Cubic Foot . . . . . .

The Price of Freestone Ashlar Work, dressed and set, with rustic chamfered Joints, the Faces of the Stones Chisel-dressed, when the Stone is obtained in the Excavation, at per Cubic Foot . . . .

Ditto Ditto of Derbyshire Stone, at per Cubic Foot . . . . . .

The Price of String Courses and Coping, dressed and set, when the Stone is obtained in the Excavation, at per Cubic Foot . . . .

Ditto Ditto of Derbyshire Stone, at per Cubic Foot . . . . . .

The Price of Cast Iron Work employed in Tunnels and Bridges, at per Ton . . .

of

The Price of Wrought Iron Work employed in Tunnels and Bridges, at per Ton . .

The Price of paved Crossings referred to in the Specifications, at per Square Yard . .

The Price of the following Culverts, including Foundations, Fronts, Wing Walls, &c.; viz.

Culvert, 2 Feet Diameter, at per Cubic Yard . . . .

Ditto, 3 . . Ditto . . . Ditto . . . .

Ditto, 4 . . Ditto . . Ditto . . . .

Ditto, 5 . . Ditto . . . Ditto . . . .

Ditto, 6 . . Ditto . . . Ditto . . . .

The Price of laying the Brick Drains under the Bridges, at per Lineal Yard . . .

Metalling the Surfaces of diverted Roads, or Roads forming the Approaches to Bridges or paved Crossings, at per Superficial Yard . . . .

The Price of Posts and Rails similar to those described in the Specification for the Tops of the Embankments of Approaches to Bridges, at per Lineal Yard of single Line . .

£. s. d.

The Price of the double Line of permanent Way laid on Stone Blocks, with all necessary Drains, as detailed in the Specification, at per Lineal Yard

Ditto Ditto laid on Wooden Sleepers, at per Lineal Yard

The Price of maintaining and keeping in good Order the double Line of permanent Way for the Period of One Year after the Completion of the whole of the Works, at per Mile

SECOND SCHEDULE referred to, containing a List of the Prices of the Extra Works.

The Erection of Gates, including Posts, Gates, Iron-work, and Fencing, at per Gate

The Price of Fencing, similar to the permanent Fencing described in the Specification, with Ditching, Railing, and Quicks, at per Lineal Yard of single Line

The Price of Posts and Railing, similar to that described for permanent Fencing, at per Lineal Yard of single Line

The Price of cutting a Ditch Six Feet wide at the Top, Two Feet wide at the Bottom, and Two Feet deep, at per Lineal Yard

The Price of cutting a Ditch Three Feet wide at the Top, One Foot wide at the Bottom, and One Foot deep, at per Lineal Yard

The Price of Brick-work in Fence Walls or Retaining Walls, including Foundations, at per Cubic Yard

The Price of Excavation in Marl, Clay, or Sand, when the Lead is a Quarter of a Mile, at per Cubic Yard

Ditto .... ½ of a Mile .... Ditto

Ditto .... ¾ Ditto .... Ditto

Ditto .... 1 Ditto .... Ditto

Ditto .... 2 Ditto .... Ditto

Ditto .... 3 Ditto .... Ditto

Ditto .... 4 Ditto .... Ditto

Ditto .... 5 Ditto .... Ditto

The Price of Brick Work in Tunnels set in Mortar, at per Cubic Yard

Ditto Ditto laid in Roman Cement, at per Cubic Yard

The Price of Brick Work of every Description in Occupation and other Bridges, and Facings of Tunnels, including Labour and all Materials, and all Labour in digging and preparing Foundations, at per Cubic Yard

Prices.

The Price of plain Freestone Ashlar Work, dressed and set, with rustic chamfered Joints, the Face of the Stone being rough, when the Stone is obtained in the Excavations, at per Cubic Foot

Ditto Ditto of Derbyshire Stone, at per Cubic Foot

The Price of plain Freestone Ashlar Work, dressed and set, with rustic chamfered Joints, and the Faces of the Stone Chisel-dressed, when the Stone is obtained in the Excavations, at per Cubic Foot

Ditto Ditto of Derbyshire Stone, at per Cubic Foot

The Price of Freestone Quoins to skew or square Bridges, dressed and set, with rustic chamfered Joints, and the Face Chisel-dressed, when the Stone is obtained in the Excavation, at per Cubic Foot

Ditto Ditto of Derbyshire Stone, at per Cubic Foot

The Prices of String Courses and Coping, dressed and set, when the Stone is obtained in the Excavation, at per Cubic Foot

Ditto Ditto of Derbyshire Stone, at per Cubic Foot

The Paving of Roads crossing the Railway on a Level, at per superficial Yard

The Formation and metalling, according to the Specification, of Roads, at per superficial Yard

The building of Brick Culverts, including Foundations, Fronts, Wing Walls, &c.; viz.

Culverts, 1 Foot in Diameter, at per Cubic Yard

Ditto 1½ .. Ditto .. Ditto

Ditto 2 .. Ditto .. Ditto

Ditto 3 .. Ditto .. Ditto

Ditto 4 .. Ditto .. Ditto

Ditto 5 .. Ditto .. Ditto

Ditto 6 .. Ditto .. Ditto

	£.	s.	d.
Thoroughly seasoned Memel Timber for Bond, laid in the Brick Work, at per Cubic Foot .			
Beech, Larch, or American Oak Piles, from Nine to Ten Inches mean Diameter, hooped and shod, 10 Feet long, and driven, at per Cubic Foot			
Ditto Ditto 15 Feet long, at per Cubic Foot			
Ditto Ditto 20 Ditto			
Oak Sleepers and Planking, spiked to the Piling, including Labour and Spikes, at per Cubic Foot			
Ballasting and laying the permanent Sidings upon Stone Blocks, including cutting the Rails into proper Lengths, fixing Crossings' Points, &c., and making the whole complete, at per Lineal Yard of single Way			
Ditto Ditto upon Wood Sleepers, at per Lineal Yard of single Way . .			
The Price of One Tool Recess, complete			

It is to be understood that only the neat Measurement of the Masonry, Brick Work, or Timber Work will be allowed, notwithstanding any Custom to the contrary.

Tricks of Contractors.

The Company enter into separate Contracts with Iron and Stone Merchants for Rails and Blocks.

Account of the Failure of Jackson & Sheddon in the Willesden Contract.

Reasons of the Company taking it into their own hands.

Jackson and Sheddon's price for the Excav. was 11½d. which is costing the Company 1s. 2¼d. Mr. S.'s Price was 1s. 2d. Average Price on the Line is 1s. 1d.

I have the means of checking the quantities by my Estimate, as it is an usual thing for Contractors to deliver in a higher price than that upon which the Tender is founded, particularly upon such portions of the contract which will be first executed, but I endeavour to make them agree with the Tender; and frequently have them altered in consequence of the discrepancy. We make separate and distinct Contracts with Iron Masters for the Rails and Chairs, and with Stone Merchants for the Blocks, (we are now paying about 6s. for each Block,) which are delivered to the Contractors for executing the Railway, who lay them down.——The parties who took the first Contract from London (Jackson and Sheddon) having failed, the Company took it (the Willesden Contract, which is six miles long,) into their own hands in order to avoid delay; Jackson afterwards applied to me for employment, and as I did not attribute the failure to him, but rather to some impropriety on the part of his partner, (want of capital occasioned it in some degree, and it was likewise grossly mismanaged,) through which he lost all his money, I therefore let some of the brickwork to him, among other persons, by the rod, knowing him to be a very efficient bricklayer; it is a very difficult thing to relet work which has been thrown up by a Contractor, as they know the difficulty you are in, and take advantage of it; the same remark applies to the sale of the stock upon the ground, they are obliged to be sold at a great loss, as the Company in such cases generally seize upon the stock, or the materials upon the ground, (the Waggons, Rails, &c.) but in this instance they waived their right: one or two of the Directors were in favor of negociating with another Contractor, who offered to undertake it for £3,000. above the original Tender, but when he was desired to go over the work with me, he made it £16,000. above his former statement; the Directors therefore took it into their own hands, and it has since gone on very well. Jackson and Sheddon's price for Excavation was 11½*d.* and it has cost the Company 1*s.* 2¼*d.*, including Superintendant and every thing; 1*s.* 2*d.* was the amount of my Estimate. The Contract Prices for the Excavation of ordinary materials are from 1*s.* to 1*s.* 6*d.* per yard: the average price of the Cutting is let at about 1*s.* 1*d.*; the highest price paid upon our Line for Excavation* is 1*s.* 2*d.*, which is clay, sand and marl, with a lead of about 1¼ miles only; if the lead was 3 miles

* There must be some error regarding the Maximum Price of Excavation, as it is stated to be from 11*d.* to 1*s.* 6*d.*: and a few lines further 1*s.* 2*d.* is stated as the highest price upon the line.—*Editor.*

I should make it 1*s.* 5*d.*, or 2*d.* a mile extra for the leading, (we have some Excavation in Northampton as high as 2*s.* 3*d.* owing to its being in Rock,); the average Lead upon the whole Line is perhaps 1¼ miles.——The increase in price since the Company have executed the works, is occasioned by additions to the Stock in order to work the Contract efficiently; the Schedule of Prices delivered in by the Contractor in this case was not a true one, he first put down such prices which he thought would merely cover him, and then added a sum for his profit as he thought, but as the works cost more than he calculated, he did not receive all that was due to him.——The following is a detail of the Estimate—Contractor's Stock,—Rails and Chairs £2,587. Points and Crossings and other Utensils £287., and Waggons £575. making £3,449.;—to which we added, Rails, Chairs and Sleepers to the amount of £1,893., and Waggons £2,013. making £3,906.,—and we shall require 50 more Waggons for the Contract, which I estimate at £1,000.,—thus the Stock applicable to the Excavations only will amount to £8,355. independent of the Stock required for the Tunneling and Brickwork;—the value of the old materials at the completion of the works will be £3,237.; the Rails will be worth about half their present value, Points and Crossings one-third, and Waggons one-fourth, which deducted from the above, leaves a balance of £5,118., therefore as the total number of cubic yards in the Contract is 837,000, the price of Stock is consequently 1½*d.* per yard; from June 1834 to June 1835 the Company have also paid for repairing waggons, for iron wheels, axle-trees, &c., also for timber and other current expences £817.—for keeping the roads in order, purchasing sleepers, &c. £1,054., and for inspection £127.—for various expences, including leading rails and materials to the works, and wharfage, and for compensation for making roads through adjoining lands, £390.—making a total of £2,388. The work, executed within the above time by these materials amounts to 109,000 cubic yards, which brings the Current Expences to 5¼*d.* per cubic yard. The Labour we let to a Sub-contractor, who digs, leads it away, and teams it for 9½*d.* per cubic yard, but I have taken it at 9*d.*, as the price was 8*d.* at the commencement, and my calculation applies more immediately to the heavy part of the work; we supplied the Sub-contractor of the excavation at Chalk Farm with all the materials, and paid him 7½*d.* with a lead of about 6 or 800 yards, until he demanded 8*d.* which we gave him; his next demand was 8½*d.*, which we refused; the work was then let to another, but he would not take it at 8*d.*, and he afterwards left it. I made out a Summary of these expences, deducting 1½*d.* for Gravel, which is unusually expensive near London as it is very scarce, but it is not the case at any other part of the Line, I also took off about £700. from the "current expences," which made it 3¾*d.* per cubic yard instead of 5¼*d*, I then added 1½*d.* for the Interest of £8,000. the capital employed; the works will occupy about 2½ years, but as one-half will be finished before the whole of the capital is advanced, I have not allowed interest upon the whole of that time,—this gives a total sum of 1*s.* 2¼*d.* per cubic yard, which I conceive to be the true cost; this amount does not include the resodding of the Slopes nor any expence which may arise from Slips or Contingencies; certainly we have not had any extraordinary contingencies in the cutting, but something should be added for them; Contractors always consider these points, a separate and specific sum is provided in the Schedule for the expence of "Keeping the Railway in Repair for 12 months after the completion of the works."

Exc. in Rock 2*s.* 3*d.*

Average Lead on the Line is 1¼ Miles.

The reason of the Works costing the Company more than the Contractors.

Detailed Estimate of the same.

Value of the old Materials at the conclu. Works.

Rails - - - ½.
Points and Cross. ⅓.
Waggons - - ¼.

Stock 1½*d.* p. c. Yd.

Currt. Exp. 5¼*d.* yd.

Average Price of Labour, 9*d.* per Cubic Yard.

Average Price for Current Expences 3¾*d.* per Cubic Yd.

Interest of Capital 1½*d.* per Cubic Yd.

Making a Total of 1*s.* 2¼*d.* per C. Yd.

Independent of Contingencies.

The Lead being 1¼ Miles.

Descrip. of the Soil.

The present Lead upon which the Calculation is made does not exceed 1¼ miles, but I should add 2*d.* a mile more if it should increase, it would vary in a small degree according to the material we had to prepare the road with; the Soil is bad, being the upper part of the London Clay mixed with alluvial; the expence is not in the digging but in the loss of time it occasions, as the Contractor cannot possibly work in wet weather. If this system of executing the work was adopted generally throughout the Line, it would increase the expense of Superintendence; I have also had experience of executing works by the method of small contracts upon the Newton and Warrington, the Leicester and Swannington, and upon the Liverpool and Manchester Line, (although I was not officially employed upon the latter,) which were executed precisely similar to the Willesden contract; I consider it by far the most objectionable. I have never known any works executed by day work only, *i.e.* without having recourse to small contracts, neither should I wish to be connected with any Railway executed by this method, as men that are working by the day under such circumstances, will not do half a day's work; there are some works in our contracts called "Extra Works," which we have not the power of letting, and I always find that they approach to nearly double the price of the contract work. Some of the work which we are executing at Willesden Green is in the same description of Soil as that at St. George's Hill; it is the surface of the London Clay, mixed with alluvial matter; at the bottom of the cutting it is genuine London Clay, but not so bad as upon some of the excavations on the Southampton Line. It is an exceedingly good material in fine dry weather, when as much can be worked of it as of any other material; but in wet weather it is as bad as any soil. The Waggons by which it is discharged hold 2 Cubic Yards, or more; but if they are filled too full they are apt to tilt on the road, and cause stoppages, it is therefore desirable not to fill them to above 2 or 2½ cubic yards: some are filled under the arches, the centres of same being made sufficiently high to allow of their passage; the number of Boxes in one Train varies from 8 to 16, accordingly as the men are getting forward, (I never saw more than 17.) The Locomotive which leads them averages 25 Journies a day, and 3 an hour is the very utmost they can do: it is worked from six to six, two hours being allowed out of that for rest: the distance it runs is rather more than a mile, which it does with ease in 4 minutes. I have observed that after the Engine has delivered a load, it invariably returns with another before the first has been teamed; and consequently has to wait perhaps half of an hour on wet days, and a quarter of an hour on fine days, for the empty waggons. I think there are 11 Horses, which lead the waggons to and from the end of the line, after the Engine has left them, a distance perhaps of 600 yards: the Engine goes down upon a permanent road, but the Boxes pass along a temporary road to the end: the Rope being attached diagonally, which does not cause any waste of power; the usual mode is to hook them on in a direct draught.——We used a Temporary Road from the beginning, but not the temporary road we now use; as it has been re-laid two or three time since the commencement of the Cutting, and gravel has been inserted under the sleepers, to render it passable; many of the original Sleepers are gone. I cannot state the number; but all those laid down by Jackson and Sheddon are among them, as they were Scotch Fir, and of small Scantling. They were pressed into the ground, and when taken up were put aside, as they were worn out and useless: those of a

Remarks upon executing Works by small Contracts.

Ditto Day Work.

The rate of Cost of "Extra Works" on the L. and B. was nearly double the Contract Works.

The Soil at Willesden is similar to that at St. George's Hill.

Description of same

The Waggons hold 2 Cubic Yards.

There are 8 to 16 Boxes in a Train.

The Locom. makes 25 Journeys a day or 3 an hour.

Remarks on the Working of the Engine.

Remarks upon Sleepers.

Scotch Sleepers.

larger Scantling will last longer; but in a Cutting like this, with the same quantity of traffic, they will not last above 12 Months; those we are using have not lasted 6 or 8 months; therefore, of whatsoever scantling they may be, they will become rotten in that time, as it is a wet situation.

They will not last longer than Twelve Months.

Having made a purchase of some Larch five or six weeks back, a tree or two of Scotch Fir happened to be among it, which was cut up before the Superintendant saw it, but I believe there are only 10 Scotch Fir Sleepers on our permanent Road; those which were on our temporary road have been obliged to be replaced by Oak or Larch; there may be a few of the Scotch fir remaining. we have increased the Scantling of the Sleepers to a certain extent, but some of the first were not very small. I am not aware that any of our Sleepers have been purchased in the neighbourhood of St. George's Hill. I do not believe that Kyan's liquid will make Scotch Fir as durable as Oak or Larch, although I have not tried it. I believe the effect is confined very much to the surface of the wood. A piece of Teak, for ship timber, was left with me by Mr. Kyan, and I endeavoured by a chemical process to render the mercury visible in the interior of the wood, but failed, which is my reason for supposing that it did not penetrate to the heart, but it might not have been tried a sufficient length of time.

Remarks upon Kyan's Patent Preparation.

A good Scotch Fir Sleeper costs 3*s.* or 3*s.* 6*d.*, a substitute in Stone upon our Line would cost 11*s.*, Larch Sleepers cost 6*s.* or 6*s.* 6*d.*, and Oak Sleepers are 7*s.* 6*d.* at Coventry: it depends very much upon local circumstances, but I do not think you could get Oak Sleepers in any part of the country for less than 6*s.*

A Scotch Sleeper costs 3*s.* or 3*s.* 6*d.* in Stone 11*s.* Larch 6*s.* or 6*s.* 6*d.* And Oak 6*s.* to 7*s.* 6*d*

A mile requires upwards of 3000 Sleepers, and the Saving in a double Line, if Scotch Fir could be used instead of Stone, would be equal to £1200.; the Saving per Mile from the use of Scotch Fir, compared with Larch, would be almost £300., and with Oak £600.

3000 Sleeps. to 1 mi. The saving if Fir was used inst. of Stone would be £1,200. Ditto Larch £300. Ditto Oak £600.

Oak Sleepers will do upon a permanent Road as well as Stone, under certain circumstances. I prefer Oak Sleepers upon Embankments for a certain period; in fact, they are obliged to be used for the first four or five years, after which they may be re-placed by Stone Blocks, if the Embankment is properly consolidated, but upon a very high Embankment I would not take the Sleepers out.——Jackson and Sheddon had executed 89,000 Cubic Yards when it came into our hands, and we have since executed about 109,000 of the highest part of the Embankment.

Sleepers better than Stone upon Embank. and should be used for the first 5 yrs.

Jackson & Sheddon exc. 89,000 C. Yds. The Co. have excd. 109,000 Cubic Yds.

We are now laying down our permanent Railroad to the extent of 600 yards of the termination of the Embankment. The portion upon which we are laying the permanent road has been made nearly 6 months. The Cutting is laid on Stone Blocks, which extend beyond the same, until the Embankment gets 8 or 9 feet high, when Sleepers are used; the part we have put Stone Blocks upon has been executed 9 or 10 months, and is that part which was first done. The Embankment to which I have referred as not having been made more than 6 months, has wooden Sleepers upon it. We let all the work now executing at Willesden, and the Contractor pays the men by the piece, *i. e.* so much for tilting and teaming each Waggon; the men at the other end are paid 4½*d.* a yard for filling the Waggons, which is a higher price than he would pay in open Excavation. We could not proceed more expeditiously, unless at an increased expence.

Description of the Execution of the Works at Willesden.

They work from six to six; we cannot work 16 hours a day without 2 Shifts, and some time is always lost in changing the Shifts, the men also have higher wages (or contract price) for working 8 hours than for 12; they would rather work 12 hours and get 4*s.* than 8 and get 3*s.*; they

Double Shifts are attended with extra Expence.

are now doing 800 or 900 yards a day, they may do 1000, which is the utmost that can be economically worked in one day; we have many applications for employment from Agricultural Labourers, but we have not found them useful men generally speaking; as many different classes of men are required in excavating, the Agricultural Labourers of the country may supply one class, which is not of great extent, perhaps one-fourth of the whole.——I went to St. George's Hill on Saturday, (the 4th of July, 1835,) accompanied by Mr. Bidder, my assistant, (this transpired since the visit of Mr. George Stephenson and Mr. Palmer). I found the Cutting was principally in Plastic Clay, which is unfavorable and easily acted upon by water, and it is a dearer soil to work than sand or gravel. I cannot form an opinion, taking the works as a criterion, when they will be finished, as they were merely commenced. It is very uncertain whether the clay will be cut through, or whether it will increase; but some opinion may be formed from the several wells, they cannot sink more than 10 or 12 feet without finding water, and it is impossible the common gravel can hold water. I do not think the clay on the line is stratified, as the face of the Cutting is covered by a mass of alluvial, but off the line, towards the high end of the hill, there are some Brick Yards open, where it is good clay for bricks, and decidedly stratified: it is yellow sand, and no doubt belongs to the lower part of the Bagshot sand. I therefore think there will be a great quantity of Clay, particularly from the circumstance of their Borings being in water, which is indicative of clay or some impervious strata at the bottom of the cutting; (we sunk shafts on the London and Birmingham Railway at a great expense.) I calculate the Cutting will cost at least 12½*d.* per cubic yard if it is in Sand with a portion of Clay; if the Clay prevails it will be more, exclusive of risk of slips.——Supposing they team an average of 800 cubic yards per day, and equal to 250 days per annum, it will give 200,000 cubic yards per annum, (1,000 per day gives 250,000. per annum, which is the max.) thus it will occupy about 7 Years in the execution. The Cutting has been reduced from 116 to 96 Feet, which diminishes the quantity of cutting, but it increases the embankments, the excavation instead of being thrown to spoil, as intended, will have to be led a considerable distance to form the embankment: I do not conceive this will shorten the time for executing it, as the embankment is the proper measure of the time that it will take.——They were preparing to lay down a teaming road upon Scotch Fir Sleepers of the same description and scantling as those which failed upon our line; their soil was better than ours, but Jackson and Sheddon endeavoured to make it as good by laying down gravel, (which we have to a depth of from 6 to 12 inches), but they sunk, which will also be the case with them: I observed our Sleepers were sinking about 4 or 5 months after Jackson and Sheddon took the Contract, which was before their failure of 6 or 9 months back.——I cannot say whether the Works on the London and Birmingham were commenced within 1 year after the passing of the Bill, I am sure they were not within the first 9 months, as we were occupied in staking out the line and drawing up the plans and specifications apportioning the line, purchasing the land, and other preliminaries; and I do not think any time was lost by so doing; the Works were commenced in June 1834: we began in much the same manner as they have at St. George's Hill, *i. e.* with a small number of men, and gradually increasing them; I cannot say how many we had for the first 2 months, but there were several

Remarks on the inefficiency of Agricultural Labourers.

Account of his Visit to St. George's Hill, 4th July, 1835.

Nature of the Soil, &c.

He cal. St. Geo. H. if in Sand and Clay will cost 12½*d.* per Cubic Yard.

800 to 1000 C. Yds. the utmost that can be teamed in a day.

St. Geo. Hill take 7 Years to execute.

It is reduced from 116 to 96 ft. Cutting.

Mem. Team Road.

The Works on the L. and B. did not commence until 9 Months after the obtaining of the Act.

Account of how the Time was occupied.

hundreds; by the last return we have upwards of 4000 men at work; a certain portion of an Embankment can be worked with Barrows and Planks, (as we did part of ours) without a Railway. The Works as far as Watford are in a state of execution, by which we shall open Communications between certain points before the whole Line is finished. We have not commenced any Excavation that would occupy less space of time than 18 months or 2 years, as it would be of no service to complete the lighter portion until the heavier are finished. Our last Contract was made about 4 months back. St. George's Hill is stated to have been began on the 6th of October, 1834, and we had an Embankment commenced on the 1st of November: I estimate the quantity removed at St. George's Hill to be 25,000, and upon our Embankment 82,000 cubic yards have been teamed. If we had such a work as the former we should not consider it important to commence any other works until it was in a very forward state, on account of its being the largest, by which we should save the interest of the money that would be expended in the smaller works; and there cannot be any advantage in finishing that part between St. George's Hill and London, as the country appears very thinly populated. As St. George's Hill is the longest work upon the Southampton Line, the fact of their being an Embankment of half a mile long at Shapley Heath will not affect my calculation. The amount of Work that can be done on our Embankment is limited. I have seen various expedients used to facilitate the progress of them, but never knew the average to exceed 800 Yards per day; it is possible to team more over an Embankment, but I am alluding to the average progress of the work. 1,000 yards would be the utmost you could average per day, even with exceedingly favorable material, and it may be possible, by working 24 hours, (which is not often done) to team 1,000 yards of the Clay at St. George's Hill; it would also be attended with great additional expence; great speed in Embankments and economy in Cuttings are not synonymous, the reverse is equally applicable. I have seen a Plan tried for facilitating the progress of an Embankment at Watford upon the London and Birmingham Railway; it was a scheme of the Contractor Mr. Noel, (or of his Foreman,) but it was abandoned; it was not exactly like Mr. Grahamsley's Plan, of which I had a drawing made, but I have never seen it practised. I also persuaded the other Contractor at Watford to go to Hartlepool to see Mr. Grahamsley's machine, or something similar to it, by which (as was reported) they were teaming 1,500 yards a day; but he returned quite convinced that the old plan was the best.——I have seen Locomotives partially used upon an Embankment to remove earth, and I may add successfully, as proper precaution was used, the Engines were not allowed to move too near the end of the Embankment, where it is impracticable to keep it in proper order to receive them, the Road should be nearly equal to a Permanent Road: a new Embankment is very bad, as it requires constant alterations. I used a Locomotive at Liverpool in the formation of the Railway: we used a Locomotive upon our Permanent Road upon the London and Birmingham, as soon as circumstances permitted, the Embankment extended over a space of about 1,100 or 1,200 Yards, and had been consolidated about 8 months including a winter, which renders it less difficult to keep in order. The Contractors upon different parts of the Line are preparing to use them, but I conceive that they cannot be advantageously used for a less distance than 1 mile or 1¼

They began the Works with Wag. Planks, as at St. George's Hill.

Description of the System of Form. the L. & B. in comparison with the System of Forming the Works upon Southampton Railway.

The Form. of an Embank. is limited

800 to 1000 c. Yds. the utmost that can be Teamed in 1 Day

Method of Teaming used at Watford.

Mr. Grahamsley's Plan.

Locomo. may be used in the Form. of an Embank.

A Good Road indispensable for same.

The Locomo. at Willesden is upon a Length of 1100 or 1200 Yds.

They should not be used upon less than 1 or 1¼ Mile.

Perm. Rails should not be used in the Form. of a Line.

miles.——The Permanent Rails should not be used in the formation of a Line unless upon a good road and good blocks, as they become bent, and even when straightened the injury is not removed: the great cause of the failure of the Rails upon the Liverpool and Manchester was their having been used in the execution of the work, in the course of 1 or 2 years the whole Line will have to be replaced with heavier Rails; they were 35 lbs. per yard, Rails and the Engines were proportionably light, about 5 or 5½ Tons: the Wear and Tear upon these rails, although very slight, is very easily detected by heavy weights passing over them, perhaps it is ⅓rd of a lb. per yard per annum, or $\frac{1}{50}$ or $\frac{1}{100}$ of an inch, and in some places more. The Rails at Willesden are 50 lbs. to the yard, and the Engine we are now using to execute the work is 10½ Tons. The Rails usually adopted at the present time are from 50 lbs. and upwards.

They were 35 lb. Rails, and the Engine 5 or 5½ Tons.

The Willesd. Rails are 50 lbs. and the Engine 10½ Tons.

Rea. of some of the Sleep overhang. the Road on the L. and M.

——In the execution of the Works upon the Liverpool and Manchester, it was necessary to get 3 or 4 lines in the width of the Embankment, (3 Lines would be narrower than any now executing), and some of the Sleepers projected beyond the Embankment, which was quite a casualty during the execution of the works, but I do not think the Rails were laid upon the projecting part: the same thing has occurred at Willesden, on that portion of the line which is not traversed by the Engine; the Rail even projects over the Embankment, which consequently could not be so well supported as where the Sleepers lay entirely on the Embankment, the position also produces a greater tendency to bend; the weight of the Engine bent both Rails in the above instance on the Liverpool and Manchester Railway, as it was considerably heavier than the waggons at Willesden. I cannot say whether the Rails were or were not afterwards used for the Permanent Way, but there were many broken and thrown aside before the Line was executed.

Ditto at Willesden.

Consequen. of same.

Calcula. of the Ex. of using a Locomo. Eng. in the Formation of an Embank.

——I have made a Calculation of the Expense of using Locomotive Engines to remove earth, with a view to ascertain whether it would reduce the expense of the Cutting of St. George's Hill, and I found it amounted to about the same expense as Horses. A Locomotive used upon a short distance, as at Willesden, is working at great disadvantage, as the distance is short, and the loss of time at the end of the journey is extremely objectionable, on account of the expense; but where it is of any length, a considerable saving will be effected, if the road is fit to receive them; but it would not affect the Estimate to a considerable extent; it is not the economy effected by it, but the expediting which influences us in adopting it. I have allowed 4½*d.* per cubic yard for Locomotive power at St. George's Hill, (1½*d.* per cubic yard per mile,) it is nearly 6*d.* by a Cart and Horse, and I have allowed ¾*d.* for the distance to the end of the Embankment, (3 or 400 yards) that a horse would have to lead it after the Locomotive had left. Our Engine at Willesden is at present new; when repairs become necessary, the total cost will average about £1,500. per annum, including Capital and Wear and Tear, the calculation of which is as follows: first, we have two persons to pump the water, at 6*s.* per day; an Engine Man at 7*s.*; and a Fire Man at 3*s.*; 2 chaldron of coke is consumed per day, which is about 50*s.*, and I estimate the repairs at about 15*s.* per day; the cost of the Engine is about £1,000., the interest of which is equal to 4*s.* a day, and the depreciation on the Engine is equal to 10*s.* a day, which altogether amounts to about £5. or £6. per day; the expense of the Locomo-

Cost of Lead. by Locomo. at St. Geo. Hill 4½*d.* per C. Yd. it is 6*d.* per C. Yd. by a Cart and Horse.

Details of the Expence of the Engine at Willesden.

which amounts to £5. or £6. ℔ Diem.

tive power at Willesden for leading materials from Excavations to Embankments is therefore about 2*d*. per Mile for a Cubic Yard, (1½ Tons.) I believe the Locomotive Power is let at $\frac{4}{10}$ths of 1*d*. per Ton per Mile upon the Stockton and Darlington Railway, but the fuel upon this line cost the engine man but 3*s*. 6*d*. or 4*s*. per ton. at Willesden it is 26*s*. which adds materially to the expense; and upon the Liverpool and Manchester it is $\frac{6}{10}$ths of a 1*d*. The difference is caused by their being finished, and their arrangements perfectly completed, which is very different to the leading of clay or sand from a Cuttting and discharging it over the Embankment; besides they have but one stoppage in 30 miles, whereas our Locomotive Engine is stopped at every mile, in fact the Engine is not going half or quarter of the day: if a waggon should be upset, or a wheel broken, or any accident happen to the Engine, the whole is thrown aside for a day or two, (a greater risk of accidents and extra expences would arise by laying down the rails upon an unsettled embankment), therefore, there is no analogy between them. I have taken the expense of keeping the Line in order from our books, of the Cost of the Works at Willesden, where I have no doubt every exertion is used to do it as cheaply as possible: as the Embankment at St. George's Hill is only 300 yards long, Locomotives cannot be used economically upon it at present, although they could hereafter. The extreme height of the Embankment at Willesden Green is about 13 feet. (I consider the difficulty of using a Locomotive would increase with the height of the Embankment.) If the bank had been 25 feet high, which is the average height at St. George's Hill, I could not have set the Locomotive at work so soon, unless it had been exposed to much wet weather, which contributes much to the solidity of an Embankment, and has been laid out about 6 months: the Slope is 3 to 1, and we have 6 Roadways upon it at the end; the same number may be worked upon an Embankment of 1½ to 1, provided it is the same width at the top as ours, but it is not usually done: if the Embankment is made wider at the top than it is ultimately intended to be, it has to be trimmed down afterwards, it would be impossible to have 6 Roads if the Embankment was not more than 31 feet wide; 4 would be the utmost it would take, which would essentially diminish the amount of the day's work: when we had 4 at Willesden, which was for some time, we could not average 700 yards a day. I may apply the same remark to all parts of the Line, but they are at present working the whole 24 hours at Watford. (Working on dark nights costs nearly double what it does in the day) and discharging upwards of 1,1 00or 1,200 yards in one day, yet by the last returns not more than 16,000 yards was done in a month, which is little more than 700 yards per day: the Contractor finds that he cannot do 1,200 yards for more than 2 or 3 days together, even with the best possible arrangements: the work also increases with the quantity teamed, as fresh rails are required to be laid.——

Or 2*d*. per C. Yd. (1½ Tons) per Mile.

The Loco. Power on the Stock. and Darl. is Let for $\frac{4}{10}$ths of a 1*d*. per Ton. And on the L. & M. for $\frac{6}{10}$ths of a 1*d*.

Remarks upon same.

Remarks on the Formation of Embankments.

I consider Mr. Giles's Prices for the Basing Line, as given by him last year, are totally inadequate, which I stated in my evidence at the time. I have since found, by my experience upon the London and Birmingham Railway, that the Cost of executing Works in this part of the country is more expensive than I anticipated, I therefore consider his prices still more preposterous. Mr. Giles having given evidence regarding his system of executing that Railway, a short time previous to the first letting upon our Line, the difference was so great between our Estimates, that the Chairman and some

He cons. Mr. Giles's Prices inadeq.

Works near London cost more than he anticipated.

Remarks upon Mr. Giles's Prices.

The South. Ry. con. 15 to 16 Mill. Ex. for which Mr. Giles has alwd. £365,495.

The L. and B. con. 12½ Milln. for which Mr. S. allo. £779,000. at 1*s*. 2*d*.

Fencing.

Southampton 1*s*. 6*d*. per Yard.

On the L. and B. it is 4*s*. 5*s*. and 6*s*. Tunn. £15. per Yd. on the South. and £30. on the L. & B.

Further Comparison of the Prices.

The L and B. goes near the Schools at Harrow and Rugby.

The intention of the Great Western to Join the L. and B.

His reas. for appro. of the Junction.

Acct. of the Route of the G. W. along the L. and B. Line.

The Lev. of the L. and B. at Cam. Town is 11 or 12 ft. above the Lev. of the Can.

other Directors made some remarks upon it; I recommended them to make enquiry, which they did, and the result was that they were perfectly satisfied with their original plan.——The Southampton Railway is 76 Miles long by the Section, and from the evidence given has between 15 and 16,000,000 cubic yards of excavation upon it, for which Mr. Giles has allowed £365,495.——The Birmingham Line is 111 Miles long, and contains 12,500,000 cubic yards of excavation, for which I have allowed an average of 1*s*. 2*d*. per cubic yard, or a gross sum of £779,000., and I have paid every attention to the economical execution of it, yet my Estimate is twice as much upon 12 miles as Mr. Giles's for 16 miles.——The Fencing upon the Southampton Line is stated at about £16,000., and for a similar distance on the Birmingham I have allowed £76,000., leaving an actual difference of £50,000.: good fencing cannot be put up for 1*s*. 6*d*., but it depends upon the nature of the adjoining closes, and local circumstances: in grazing country, (which the Birmingham principally runs through) it certainly could not, but this cannot make a difference to any great amount: the price on the London and Birmingham is 4*s*. per double yard; in some places I have taken it at 5*s*. and 6*s*. The price for Tunneling on the Southampton is £15. per Yard, and on the London and Birmingham it is £30.: the Ballasting, Laying the permanent way, and Rails, are estimated by Mr. Giles at £2,223. per mile; I estimate the same at about £4000. per mile: and there is no allowance made in Mr. Giles's Estimate for the permanent working of the Railway, which would amount to £500. per mile. I have no doubt the Southampton is more favourably situated for Gravel than we are; the 5 or 6 miles next London costs us about 3*s*. per cubic yard for ballasting. We have purchased a great quantity of Gravel for the permanent way at Willesden, it is brought both from the Paddington Canal and the Thames; that which is brought from the Canal is for an Embankment across the Brent, about 5 miles along the Paddington Canal, and as there is Gravel at the bottom of that Valley, we have it dug out, as far as the quality will admit, to ballast the road: we have plenty of Gravel and Chalk, both at and beyond Watford; we then get into the Iron Sand country, where it is quite as plentiful as with them. The London and Birmingham Railway passes about 1½ miles from the Harrow School, and also about a mile from the Rugby School, and we have not met with any opposition from either. I believe the promoters of the Great Western intend to join the London and Birmingham, but there has not been any terms entered into regarding it; indeed, I believe there is a Prohibition in their Act (the Great Western) against it. I approve of the junction, as it will save the enormous expense of 2 entrances into London: the Directors of the former having asked my opinion upon it, I stated that I did not consider it would interfere in any respect with their interests, as it was possible to make arrangements which would prevent any confusion. I have been present at all the Meetings where it has been discussed, and the majority of the Shareholders always remained neuter; they will join us at nearly three-quarters of a mile beyond the Tunnel at Willesden: they will have to pass through the Tunnel under the Harrow Road, and a Bridge under the Kilburn Road; there is also another Tunnel, 1100 yards long, under the Finchley Road; after which there will be no more Tunneling until they get to Camden Town. The Elevation above the natural ground where we terminate at Camden Town, is exactly the height of the present Canal Bridges, or a

Rise of 11 or 12 Feet; from thence there is a Fall of 1 in 86 to Euston Square, a distance of perhaps three-quarters of a mile, which the Carriages would descend of themselves. They would acquire considerable velocity before they came to the bottom, but we have sufficient power to check it, by Breaks, and backing the Engine. It is not determined whether this Plane will be worked by assistant Locomotive Engines or a Stationary Engine. I have no doubt an extra sum is to be paid for going to Euston Square, as it is stipulated in the Act; but I do not recollect the price.——The Birmingham Railway crosses about half a dozen Roads upon a Level, (there is one about a mile from Harlesden Green;) Gates and Lodges are provided for security, there are none crossed between London and the point where the Great Western joins us, and I understand Mr. Brunel does not cross any road on a level.——I have had the Sections of the competing Lines laid before me, but I have not taken any Levels, neither have I been over the Lines, I made my observations from the Sections in both cases, presuming they were accurate. I consider the Great Western is the best; in fact, I do not know a Line that has Gradients equal to the Great Western, and I think it would be difficult to get such another Line from London, considering it is 115 miles long, 4 miles longer than the London and Birmingham. The alternative suggested, instead of the Box Plane of 1 in 107 for 2½ miles, (which is partly in a Tunnel) was a Plane of 1 in 333 for 9 miles, upon which the regular Engines of the Line could run. I would adopt a Tunnel to avoid a steep inclination, and a certain curve, *i. e.* if they were considerable; as I prefer the concentration of the inclinations, by which a fixed Engine may be applied, as it is the most economical; the remainder of the distance is thereby rendered extremely favourable for the working of Locomotives, which is a great saving of power, also of wear and tear; it has also the advantage in point of speed. I am not aware of any Railway Tunnel in the middle of the Line that is laid with as great an inclination as the Box; there are some Planes of a greater inclination at Liverpool, but they are situated at the commencement of the line. I do not see any objection to there being a curve at the upper end of the Tunnel.——We have nearly 4 Miles of Tunneling upon the London and Birmingham: there are 4 between London and Tring, 2 beyond Willesden, one about a mile, the other a quarter of a mile long. I have incurred one at Watford 1 mile in length, to accommodate a Landowner of importance. A Tunnel is better for a Landowner than any description of Cutting, as it preserves the surface to a certain extent; if it was a Cutting of 100 feet deep it would be still more objectionable; it is also injurious in point of severance, and generally speaking I prefer a Tunnel in point of expence. We have a Tunnel in Cashionbury Park which falls 6 feet a mile, or 1 in 880, which is a very good Level; Locomotives pass through it with ease. There are several different methods of working a Stationary Engine, all of which I have seen act very efficiently.——The Engines that work upon the Liverpool and Manchester Line would work very well upon the inclined planes upon the Bath and Basing Lines, provided the loads were proportionably reduced, but unless they were reduced it would be impracticable, *i. e.* with the power that would be sufficient for the rest of the line; I refer comparatively in all my statements, being aware that Locomotive Engines will work, perhaps efficiently, at 16 feet in a mile, and some even at a greater

There is a fall 1 in 86 to Euston Sqr.

Remarks upon the same.

Paved Crossings.

He considers the gradients of the G. W. the best.

Remarks on same.

Expedi. and Desc. of the Box Pla.

There are 4 Mi. of Tun. on the L. and B.

Account of same.

Advt. of a Tunnel over a Cutting.

The Eng. upon the L. and M. would not work adv. on the Basing Line.

A Locomo. will work at 16 feet per Mi.

Remarks on the same.

inclination, but they will work to more advantage and carry heavier loads at 11 feet in a mile; all the several items of expence (which occur to me at the present moment) are

The full Power always used upon the Incl. Planes of a Railway.

increased with the inclinations, but not in the same ratio: the Full Power is generally used upon the inclined planes of a Railway; at Rain Hill and Sutton Hill, upon the Liverpool and Manchester Line, the Passenger Train is generally, and the Goods universally,

As on L. and M. &c.

helped up by Assistant Power; they are also obliged to limit the Passenger's Train to 3 Carriages instead of 8; the Warrington and Newton is another line where the full power is

Comparison of the Gradients of the competing Lines.

used.——I have made a calculation of the Proportionate Loads that could be drawn on each line by Locomotive Engines: the measure of the weight and power of an Engine upon a great Line of Rail-road is the greatest inclination over which it has to pass, provided the length of the inclination is within the Reach of any Concentration of Power which the Engine possesses upon the adjoining more favourable Planes; a Plane of four miles would regulate the power of an Engine, (it allows of it in 1½ miles upon the Liverpool and Manchester,) in other words, an ordinary Engine with a full load would come to a "stand still" immediately after it had passed that distance, unless the load was reduced. Excluding the Box Plane, the second Plane in height on the Great Western is in the Depôt at Bath, which is 1 in 308, but being very short it will not affect the power of a Locomotive; the third Plane in height is 1 in 473, upon which I have formed my calculations; I have taken the highest upon the Basing, (as the Line is to be worked entirely by Locomotives,) 1 in 202, which is 6 miles 54 chains long, it therefore exceeds the length which the Engine would be capable of concentrating any power to overcome, as it must have come to the speed which it would continue throughout the Inclination

Excl. the Box Pl. the Gradi. of the G. W. are 50 per cent. more econo. than the Basi.

long before it reached the end of the 4 miles: I allowed 10 lbs. per Ton for Friction, although I am aware it is a little below that, but I took the same in both cases; the effective Load on this Plane would be 42 Tons, and the gross Load 52, allowing 10 Tons for the Engine; the effective Load on a Plane of 11 feet 2 inches per mile on the Great Western would be 64½ Tons, which is in the proportion of 3 to 2, or nearly 50 per cent.

Excl. the Box Pl. and the steepest Pl. on the Basing, and it is 30 or 33 in favo. of the G. W.

in favor of the Great Western. If an Assistant Engine was used upon the Plane of 1 in 202, which would be very disadvantageous, the comparison would then be in proportion of 4 to 3, which is about 30 or 33 per cent. in favor of the Great Western, the 2nd steepest Inclination upon the Basing Line traversed by Locomotives being 1 in 250,

Memo.

I have not taken into consideration the undulations upon either line, as they would not

On the Mode of Calculation.

affect it; I take the highest Inclined Planes as the basis of my calculations, which I consider the practical way of doing it; (the undulations would of course affect a comparison of the summit Levels;) my calculation is upon a supposition that the Box Plane is worked by a Stationary Engine placed at the top of the Tunnel, and there will be no occasion

Working on the Box Plane.

to unhitch the Locomotive, as I should prefer taking it up with the Load, and allowing it to work all the time; whether a Stationary Engine or an Assistant Locomotive was used would not affect my Estimate, if the latter was used it would go up with the

The Eng. on the L. and M. have 11 to 14 inch Cylin.

Locomotive, and then wait for another load.——The diameters of the Cylinders of the Engines upon the Manchester and Liverpool Railway are 11 or 12, some are 14 inches;

Length of stroke 18 to 24 inches.

And 5 ft. Wheels,

the length of the stroke varies from 18 to 24 inches, and the diameter of the Wheels are generally 5 feet, some have all their wheels of the same size, and many have their wheels coupled together for the purpose of increasing the adhesion; the weight of an Engine with the water will vary from 9 to 11 Tons, including coke and every other necessary; the Tender when loaded with Coke and Water will weigh about 5 Tons: the Adhesion of an Engine upon a Railroad, excluding foggy weather, varies from an 11th to a 15th of its weight, I would reckon on a 15th, but it depends upon the state of the weather, adhesion is as good in wet weather as at any other time, but any sort of damp renders the Rails greasy: the Engine at Kensall Green has a 12 inch cylinder, a 20 inch stroke, and I think 4 feet 6 inch wheels, and I calculate its Tractive Power is equal to 1,000 or 1,100 lbs, *i. e.* it will draw that weight over a pulley: a Locomotive with two 10 inch cylinders, with Water, &c. would weigh about 8 Tons; I made one with 14 inch cylinders which weighed about 13 Tons, the price of which was £1,000: we intend using larger Engines upon the Line than the Kensall Green Engine, but I have not yet determined the size; a 16 Tons Engine is the Maximum that the Road is calculated to carry, and I do not anticipate using Engines of such weight. I believe the Coaches that carry Passengers generally weigh 2 Tons or 2½, and about 8 are usually attached to an Engine: the number of Passengers in each Carriage varies from 4 to 20 or 30. ———We have two or three Inclinations on the Birmingham Railway full as much as 1 in 330, perhaps ⅓ of our Line is 1 in 330; I do not mean that an Engine starting from London will have to pass over 33 miles at 16 feet in a mile, as that amount includes both sides of the summit, one half will probably be in one direction and the other half in the contrary; there is not a very great difference between a Plane 1 in 300 and another 1 in 330. I allowed about £70,000. for the item of Locomotive Power, Waggons, &c. for the working of the Railway when completed; £57,000. is allowed upon the Great Western, owing to its being more favourable as regards Inclinations, (the difference will be in the expence of the Engines); I take the Stationary Engine at the head of the Box Tunnel into consideration; the expence of the Carriages, &c. &c. forth will be the same upon both lines; we have contracted for 8 miles of 50 lb. Rails, and I think the price is £9. 10*s*. per Ton; my Estimate was £10. 10*s*., but fluctuations take place in the trade; the Chairs are £8. or £8. 10*s*. per Ton, and we have one in every yard, for which specifications were drawn up and advertized, we also wrote to several Iron Masters, and we accepted the lowest Tender that was eligible.———There was an idea of carrying a Railway from the London and Birmingham at Tring to Oxford, but it has not been seriously entertained; I have no doubt of the possibility of it, as the country is favorable.———I do not know the contemplated Line between Swindon and Gloucester, nor have I drawn my attention to the Gradients, but I should judge from the Canal passing over that height, that there would be no great difficulty in keeping the same levels; it may be taken as a general principle, that a Railway can be carried much upon the same level as a Canal. I think that the offset to Oxford from the Great Western would form a better communication between London and Gloucester, in point of level, than an offset to Oxford from the Birmingham at Tring.———I have frequently gone over the road between Oxford and Gloucester, and can state that the country appears as bad as can be selected, and quite as bad as that between Sheffield and Manchester.———Cowran Hill upon the Newcastle and Carlisle Railway was stated (in the evidence given last year) to have cost 5*d*. per cubic yard, which I thought impossible, and from inquiries which I have since made, I am induced to consider it

Weig. of Eng. 9 to 11 tons.
Tenders 5 tons.
Adhesion of an Engine, is 11th to 15th of its weight.

The Ken. Green Eng. has 12 in. Cy. and a 20 in. stroke, and 4 ft. 6 Wh. the Tractive Power equ. to 1000 lbs.

Eng. with 10in. Cy. will weigh 8 T. Ditto 14 in. Cy. will weigh 13 T. and cost £1000.
Coach 2½ T. and carry for 4 to 30 Pass.

⅓ of the L. and B. is 1 in 330, half one way and half the other.

Mr. S. allowed £70,000. for the future working of the L. and B. £57,000. will be suff. on the G. W.

The Price of the Rails is £9. 10*s*. per Ton. Chairs £8. or £8. 10*s*. per Ton.

Remarks on a Line from Tring to Oxford.
Ditto Swindon and Gloucester.

Ditto Oxford and Gloucester.

Cowran Hill is stated to have cost 5*d*.

even more preposterous, although the Soil is evidently a loose sand, and ¼th only of the Cutting is led to embankment, the remainder goes to spoil, and the hill is very abrupt, which is favorable for Spoil Banks. I should be very much surprised if the aggregate cost of the excavation, taking a length of 5 miles upon this railway, amounted to no more than 6*d.* per cubic yard, including the lead and every expence, I last year put it at 10*d.*; but admitting that it cost only 5*d.*, it does not follow that St. George's Hill can be done for the same price, as the lead is greater, also the length of the cutting and the quantity of earth to be removed. I know Whilom's Scars very well, at which place there is a considerable quantity of work executed which is in rock, but I have not been there since it was executed; I do not recollect the opinion I expressed upon it when the bill was in Parliament, but I might have said it would cost 2*s.* per yard, and if Mr. Grahamsley has executed it for 1*s.* it cannot be the line laid down on the section, as the material was to be conveyed away to form an adjoining embankment, and I can state that the whole character of the line has been altered near Newcastle; and by a deviation of the line at Whilom's Scars it may be brought so near the river that the precipice has merely to be hewn down, and Rock of this description may be executed at 1*s.* per cubic yard, whereas the Parliamentary Line was in Rock cutting on both sides; an Engineer is perfectly right in availing himself of the 100 yards deviation, by which he may lessen the expences considerably, but when the Bill was in Parliament, the opposition of Beaumont restricted it to the identical line laid down on the map: 100 yards made a great difference on this line, where it was in a deep cutting it would have changed it into a high embankment.——I have not saved much by Deviations on the London and Birmingham Railway, but I have otherwise improved the line, although I have not substituted open cutting for tunnels unless the change required it, with the exception of Oxhey Lane Tunnel, which is made into an open Cutting on account of the material turning out better than I expected; neither have I, nor do I intend to shorten any of the tunnels by adopting deeper cuttings at the two extremities.——I do not consider myself at liberty to state whether I am meditating any plan to carry the Birmingham Railroad down to the River; I certainly suggested it about six or eight months back to one or two of the Directors, but it has never been discussed by the Board, as I had not considered of it sufficiently to justify my decided recommendation: I think it very desirable, but it depends entirely upon the expence whether I should recommend or disapprove of it, if it can be made for £200,000. it would be desirable, but I state that sum with the greatest possible limitation, as it is a question that involves much consideration; I have no doubt it will eventually be carried into effect, but I am not aware of any negociations, (or of letters having been written to proprietors of property in the direction of Hatton Garden or Saffron Hill;) it would be particularly advantageous to get the goods and passengers more into the heart of London, as it would render the terminus complete; I thought of going to Waterloo Bridge, which certainly is preferable to the Regent's Canal, but the latter Depôt is not much complained of; if we went to the former, we should most likely continue to make use of the Canal, as it embraces London in so many directions, and has possession of all the avenues to business; we expect our

Mr. S. put it in at 10*d.*

Remarks on the line at Whilom's Scars.

Inst. the great diff. the 100 Yards Devi. will sometimes make.

Remarks on the Deviations on the L. and B.

Remarks upon the Extension of the L. and B. Railway to the River.

If the same could be done for £200,000. it would be desira.

Mr. S. thought of carrying it to Waterloo Bridge.

Advantages of the Regent's Canal Depôt.

principal Profit to arise from Passengers and Light Goods, and Euston Square is the most favorable site that could be obtained for the disposal of the same throughout London. I am not aware of any prohibition against the goods coming to Euston Square, but if there is any, they will be conveyed to their several directions from Camden Town. The originators of the London and Birmingham Railway proposed to have four Lines of Rails, until I explained the absurdity of such an arrangement; I consider that two Lines are capable of conveying six times as much traffic as they will have.

The prin. Prof. antl. on the L. & B. is the Carr. of Passen. and Light Goods.

Abs. of hav. 4 Rails on the L. & B. as propo., 2 will carry six times the traffic.

Ex. Mr. CHARLES VIGNOLES, C. E.

I have had considerable experience in my profession, and have assisted in the setting out and construction of several Railways, among others the Liverpool and Manchester. I laid out and constructed the St. Helens and the Wigan, and I have lately completed the Dublin and Kingstown, I am also constructing the North Union. I was consulted upon the Birmingham and Grand Junction, and I have given many opinions upon different lines of Railroads in America, and on the Continent; I laid out a line through Hanover, and I was consulted upon a line between Leipsic and Dresden, which was commenced last year.——I have considered the filling up of the space between Reading and the terminus at Bath, regarding these as fixed points; I examined the Valley of the Kennet and Avon Canal, and the Valley of the White Horse, also the ground between Newbury forward nearly across to Bradford, and I think the Line should pass to the North of the Marlborough Downs, on account of the superiority of the levels, the greater facilities for the execution of the works, and particularly as it is connected with Gloucester and South Wales; I entertained this opinion in my evidence of last year; I contrasted the Line North of Marlborough Downs with the Southern Line from Basing to Bath through the Vale of Pusey, and although I have not examined the latter with the same critical accuracy as the former, I am satisfied it is the best, the principal objections to the Basing being very apparent; as the great difficulty of crossing from the Valley of the Kennet and Avon to the Vale of Pusey, and then into the Vale of the Avon, and from the Vale of the Avon between Bradford and Bath to Bath; the obstacles to a Railroad from the Vale of Pusey to the Vale of the Avon is the great rise which is concentrated in a small distance. The country between Bradford and Bath is very rugged and difficult, if you follow the line of the Avon to Bath it will make the Railway exceedingly circuitous and difficult, and judging from an inspection of the Ordnance Map, the curves would be very objectionable, almost impracticable for high velocities, as 25 miles an hour; yet the Line must either follow this valley or cut through the Claverton Hill; I have examined the Line which passes by a Tunnel, and I consider it very formidable, as the Shafts would be exceedingly deep, thereby involving very

His experience upon Railroads in England, Ireland, America, and the Continent.

Regarding a communication between Reading and Bath.

He prefers the Line North of the Marlboro' Downs.

His Reasons for the same.

Difficulty of a Line between Bradford and Bath.

great additional expence.——I certainly would avoid a Tunnel if possible, but not at the expence of curves and levels; I consider it absolutely necessary for a junction with South Wales, Gloucester, and that portion of England, that the Railway should take to the North of the Marlborough range to Chippenham, and I consider the Line selected is under all circumstances the best, upon an assumption that I get to a certain point at Bath, and then pass on to Bristol; I have not examined the country sufficiently to enable me to say whether there is a better, not having taken any levels: I object to going round by Bradford on account of the ascent from Chippenham, and as the Line must pass through the vallies at very sharp curves, and at considerable expence, *i. e.* supposing good levels can be found, and the termination would be at a much higher level at Bath than convenient; they cannot get at our Terminus without taking very inferior gradients, I therefore consider the Line of the Great Western by Chippenham and Box Hill much the cheapest; it would also be highly objectionable to take the Trunk Line from Chippenham to Bradford, and turning our Branch into a portion of the Main Line, (taking the line of a bow instead of the bow-string); and although there is a considerable elevation to Bradford, which is very good for a Branch line, it is quite incompatible with a Main line. A good practical locomotive Line between Bath and Bradford could not be obtained without tunnels, or without curves of less than half a mile radius, and if you carried it through by a tunnel, or passed through the valley in whatever curves you could get, and disregarded the termination, the inclination would not exceed 1 in 500, which is the level shewn upon the Plans of the Basing Line, which terminates 120 feet higher than our Depôt at Bath; cutting off the brow of the hills for 50 or 100 feet would improve it, but much depends upon the point of termination at Bath. A curve of ¾ of a mile is very objectionable, but if I could not get a better I should be content with it; we consider that a radius of less than a mile retards the progress of an engine.——The Line between Chippenham and Bradford is very good as regards the communication between Bradford and Bath, and the junction of the former with the main trunk, but it is the contrary between Bradford and the West of England, as you must pass over the summit, which is 1 in 250. (this is very different to the Concentration of the levels at the Box Tunnel; it is a great rise of several hundred feet in the case of the Basing, spread over perhaps 5 or 6 miles, and making the rise equivalent to a diminution of the power of more than one half the whole line; in the other case, it is a concentration of a great rise in a short distance, and by the help of an additional Locomotive Engine, or other assistant power, upon this short plane, the Engines are enabled to work upon the remainder of the Line to the full extent of their power). A great Line of Railway should be laid down upon a principle of being the greatest advantage to the whole district through which it passed, affording the easiest communication with those places with which it is not in direct communication; and upon comparing the Great Western Line with the Line from London to Basing, and from Basing to Bath, I consider the Box Tunnel a trifle compared with the advantages which the trading community would derive by the former. I consider that the Great Western Depôt at Bath is not capable of any improvement, both in reference to the local facilities and to the continuation of the Line to Bristol, (a good termination could not be obtained

Margin notes:

The North Line ab. necessy. for a junc. with South Wales and Gloucester.

Rem. on the Line round by Bradford, (the Basing.)

Difficulty of the same.

Basing Depôt Bath.

Curves of less radius than 1 Mile retard the Engine.

Observats. regardg. the Basing Line at Bradford contd.

Comparison of the Plane of 1 in 250 with the Box Plane.

Memo. regarding laying out a Railway.

The G. W. cannot be improved.

Conven. of the G. W. Depôt at Bath.

at Bath for both Companies upon the same level) neither can the Line generally be very materially improved, although 5 Tunnels in 12 miles may appear a great number, but it must be taken in reference to the remaining portion. If a line could be formed without Tunnels, but similar in every other respect, "*cæteris paribus*," I should prefer it, but as regards the local accommodation of Bath I doubt whether you could find a more suitable elevation.——I went over Mr. Brunel's Line between Bath and Bristol last year, and I did not see any great objection to it: I understand the Line is made by consent of the Land-owners: the several alterations which have since been made have been pointed out to me, and I consider they are decided improvements: the Gradients are very good.——I have had my attention called to the subject of the Box Plane for the Great Western, and I consider it can be worked with the greatest ease, as I know there are planes of equal or even greater inclination upon existing railways; and I do not consider a Curve of ¼ of a mile radius objectionable on account of being near the end of an inclined plane and tunnel, (there is a curve of similar radius at the termination of the line at Manchester;) if the inclined plane is immediately before it, the curve becomes the landing place. I have one ¼ of a mile radius at the foot of an incline of 1 in 36 upon the St. Helens. There is a Plane upon the Liverpool and Manchester 1 in 90, which is in the middle of the line, and worked by Assistant Engines; the passengers travel up it at least 20 times a day; we have 2 Planes of 1 in 30 or 40 upon the St. Helens, the longest of which is ¼ of a mile, they are situated in the middle of the line, that next the collieries (about 1 in 36 or 40) is worked by a Stationary Engine and an endless rope, and is very easily worked; the other, in the opposite direction by the river, is a self-acting Plane, and to the eye of a common observer, it scarcely appears to rise: it is principally a Colliery railway, but there is a Passengers' Coach once or twice a day. I furnished the original Estimates for this Railway, which were rather under the mark, although I put down good prices, but several branches were made which were not originally contemplated, and a great number of clauses were added, concerning Turnpike Roads, which increased the expences, (I am still their Consulting Engineer.) A Plane of 107 is not a very bad plane, although it requires great additional power; it is more economical where the inclination varies much to concentrate it in one spot, and have a Stationary or Assistant Engine to overcome it, than proceeding with the same power throughout, both over the steeper and flatter portions of the Line. I have also considered the Box inclination with reference to its being in a Tunnel, and I am of opinion that it will not affect the facility of working the Plane; I certainly do not think the tunnel any advantage.——With respect to Accidents, I think it is possible to stop the Carriages by Breaks, I have seen them stopped upon Inclined Planes frequently: they often stop the Train upon the St. Helen's, at 1 in 36, with 6 loaded coal waggons. The distance in which they can be stopped depends upon the velocity with which they are travelling; they run until the "friction of the Break is greater than the gravitating power of the Engine." I am certain that an Engine upon a Plane of 107 could be stopped in 150 or 200 yards. The consequences would be serious if an Engine was to get loose, and run down at full speed, without any check whatever, but the Break is always

Mr. Brunel's Bath and Bristol Line.

Box Plane may be easily worked.

Effect of Curves near Inclined Planes.

The Plane of 1 in 90 on the L. and M. is wor. by Ass. Eng.

Passengers travel by the Planes on the St. Helens Ry. 1 in 36 or 40.

Des. of same.

Explanation of the Working of the Box Plane and Box Tunnel.

Observations upon Accidents upon Inclined Planes and Breaks, &c.

applied; in going down the inclines upon the St. Helen's, we have not had an instance of the men neglecting it; there is generally a Break to every third Waggon, besides the Tender, and a Train can be brought to a stand still by the Break of the Tender only: they do not generally apply the break to every third carriage, as there is no occasion for it, but they carry it as a precaution. In the event of an accident, or of the engine getting off the rails, the steam flies off immediately by the safety valve, and the velocity is also increased, which are sufficient signals for the breaksman. A Waggon running down with a load of passengers, and half a load of Goods, would go at a very high velocity; but an accident would not occur, unless the Engine got off the Rails; or, if it was worked by a Stationary Engine, and the Rope was to break, the Break would be sufficient to check it; but I am not aware of an instance of a Train escaping from the break, or an accident from the breakage of the rope.——We have a Plane 1 in 96, upon the St. Helen's, in open cutting, and there are very seldom more than 5 carriages in each train, for which there are 2 guards, independant of the men who apply the break of the tender, and I have travelled by them for weeks together, and have had occasion to stop at the Junction of the line with the Liverpool, and they have let me down, although they were going at full velocity; the Guard probably did not apply the Break until within 100 or 200 yards of the spot, according to the velocity of the engine.——If an Engine was to get off the Rails, I would certainly rather be in a Tunnel than upon an Embankment, and it would be less liable to get off when pulled by a Rope than when drawn by an Engine; perhaps it would be driven against the wall and crushed, and the Tender also; but this would be quite sufficient to stop the progress of the Carriages, which would consequently run up against each other, but the buffing apparatus would almost obviate the shock: and I doubt whether they would be damaged. If such an accident should occur on an Embankment, the probability is that the Engine and Train would go over. An accident happened upon an Embankment on the Liverpool and Manchester, owing to the axletree breaking, and the Engine, Tender, and first Carriage, went obliquely down the bank, but the other Carriages remained upon the Line, and only one or two got off the rails.——I think a Tunnel upon an incline would have a greater draught through it, which would be an advantage in point of ventilation; the same number that are considered sufficient to work a Tunnel, are in practice sufficient to ventilate it: if there was any deficiency in ventilation, a fire lighted in one of the shafts, or any thing that created a current of air through it, would remedy it.——The Gradients of a Line form one of the most important points, as the "Consumption of Power" is the greatest expence upon a Railway: the cheapest Line is that which has the Levels brought down to the flattest pitch, and not that which costs the least.——I have looked at the Gradients on the Bath and Basing Line, and I find a Plane of 1 in 202 for 6¾ Miles, which an ordinary Engine could not ascend without assistance, unless it had only half its usual load; and as it could not take any more on the Line than it could take up this Plane, a great loss of power would be sustained. The Cost of Fuel is very small, in comparison with the other expences. I have made a Calculation of the Power required on each Line, and I make it 50 per cent. in favor of the Great Western; the transport would, consequently, be much cheaper upon the latter; as the long inclinations of 1 in 202, 250, and 264, will retard the

Observations upon Accidents upon Inclined Planes and Breaks, &c.

Power the most imp. con. on a Rail.

An ord. Eng. could not asc. the Plane of 6 miles 1 in 202.

The Exp. of Fuel is very small comp. with other Exp.

The Gradients 50 per cent. in favor of the G. W.

progress of the Engines, and the slightest accident or derangement of them would increase it; and they certainly will not be able to travel up these Planes with the same velocity as upon a Level, unless they reduce the loads considerably.

Mem. Grand Junction Railway.

———The first Engineer of the Grand Junction, whose name I am not acquainted with, laid out the Gradients upon that line. (I am not aware whether it contained a Plane of 10 miles at 1 in 210.) I was the next Engineer employed, and then came Mr. Rastrick: the Engineer who laid out the Line for Parliament, was not the party employed in constructing the work; the levels, therefore, might have been altered.

A Communication with Wales must pass by the North. Line by Swindon & Stroud.

———I was applied to regarding a Communication between the Metropolis and Wales, previous to my examination of the Great Western, and I stated that it must pass North of the Malborough Downs, as there is no way of getting out of the Vale of Pusey towards Wales. I recommended a Communication by Stroud and Gloucester, up to which place I took the levels, starting from the vicinity of Swindon, the line could pass either by Malmsbury, Minchin, Hampton, and Maresfield, or by Cirencester and Stroud. I did not examine the latter much at the time, as it was not pushed, but I consider it is the best Line.

Country between Oxford and Cheltenham.

———I have not examined or levelled the Country between Tring, Gloucester, and Cheltenham, but I have a general knowledge of it from frequent travelling. I know the points between Oxford and Cheltenham very well; a very high ridge, called the Cotswold Hills, is between them. A Line between the above places, through Burford and Witney, would pass entirely through agricultural land; the manufacturing part of the Line would be by Cirencester and Stroud, at Maresfield and Minchin Hampton: the Witney blanket manufactories are situated to the South and West of the Valley of the Thames.

He was the Enginr. of the pro. Line to Windsor.

Description of same.

He considers it was superior to the Great Western.

———I was Engineer of the proposed Line to Windsor, which tunnelled under the Long Walk; it skirted the Town of Windsor near the Cavalry Barracks, and then went on to Reading under St. Leonard's Hill on the Windsor side of the river; it was within 1 mile of the centre of Eton: the College objected to the Line at first, but I believe they afterwards removed their objections, and were better satisfied with it than they were with the Great Western, as the River and the Town of Windsor was between Eton and the Railway, but the Line of the former passes ½ a mile further off from the body of the buildings than we did. I consider the passage through Windsor, and the general features of the line up to Reading, were superior to the Great Western; it would also have been more advantageous to the Town, and I do not consider it would have injured its beauty. A Petition was presented and the Plans were lodged, but we did not take any further proceedings on account of the opposition of the Great Western Railway.

Terminus of the G. W.

Gradients of the G. W.

———I last year advocated the Northern Terminus at Paddington (for the Windsor Line) as it was upon a higher level than the Great Western, but the latter is now the highest, as it joins the London and Birmingham, which is also much more central: the Gradients for the first 80 miles have consequently been improved by the same. The Line from London to Reading, by the Valley of the Thames and the Valley of the White Horse up to the Swindon summit, presents very favourable gradients; I have not seen any Line of the same distance that could equal it in this respect. (A Line through Lincolnshire might possibly equal it.)

Mr. V.'s Line from London & Brightn.

———A party of gentlemen, merchants of London, some years back conceived an idea of a communication between London and Paris, and I was

Description of same.

engaged and went to Paris to negociate with the French Government the terms upon which a Railway might be made, and certain promises were obtained; upon receiving which the party ordered a Survey to be made to Brighton, but as the French Government did not fulfil their promises, or encourage them as was expected, it was abandoned, and the Survey was endeavoured to be made available as a distinct line to Brighton; the plans were therefore deposited and notices given, but it did not find sufficient favor with the public to fill the subscription list, and was consequently abandoned: the sum subscribed was not more than sufficient to pay for the survey, as it was not brought before the public; such notices as the standing orders of the House required were complied with, but it was never advertised. It passed over very difficult Country, including 2 high summits, but the rugged nature of the ground about them was the most objectionable. I was glad to get a Passenger Line at such little comparative expense;

It passed over very difficult Country.

Description of the Tunnels upon it.

I think the least Curve upon the line was 1 mile radius: there were 3 Tunnels on the line amounting to nearly 3 miles, one was 2¼ miles long at an inclinaiton of 1 in 330, which is my greatest inclination in a tunnel; there was another at Brighton (at the back of the town) ¼ of a mile long, which was level. I found it was impracticable to get better levels than 1 in 330, on account of the extreme difficulty of the railway; the power upon an inclination of 1 in 330, including gravity and friction, is exactly ½ what is required upon an inclination of 1 in 100. I afterwards altered the Gradients, to enable the Engines to continue at the same velocity and with the same load throughout the whole Line; thus whatever load the Engine carried upon 1 in 330 it would take up 1 in 100 by the assistance of an assistant Locomotive Engine.——The following are theT ables of the Gradients, and I have also annexed the Gradients of another Railway to Brighton, (Sir John Rennie's) which they were compared with at the time:—

Explanation of the Gradients.

GRADIENTS on the LONDON and BRIGHTON RAILWAY, as originally proposed by Mr. Vignoles.

Gradients of Mr. V.'s Line from London to Brighton.

1	Mile	56½	Chains	-	Horizontal or Level Line.
0	—	43	—	-	Rise 30 Feet or 1 Foot in 95 Feet.
0	—	20	—	-	Do. 3 Feet or 1 Foot in 440 Feet.
0	—	60½	—	-	Do. 40 Feet or 1 Foot in 99 Feet.
0	—	21	—	-	Do. 3 Feet or 1 Foot in 462 Feet.
1	—	19	—	-	Do. 66 Feet or 1 Foot in 99 Feet.
9	—	59½	—	-	Do. 132 Feet or 1 Foot in 390 Feet.
3	—	1	—	-	Fall 40 Feet or 1 Foot in 398 Feet.
5	—	58½	—	-	Do. 91 Feet or 1 Foot in 332 Feet.
6	—	47½	—	-	Rise 105 Feet or 1 Foot in 331 Feet.
4	—	19½	—	-	Fall 68 Feet or 1 Foot in 329 Feet.
0	—	59½	—	-	Do. 40 Feet or 1 Foot in 103 Feet.
0	—	41	—	-	Do. 8 Feet or 1 Foot in 330 Feet.
0	—	78	—	-	Do. 50 Feet or 1 Foot in 103 Feet.
1	—	2	—	-	Do. 16 Feet or 1 Foot in 361 Feet.
0	—	77½	—	-	Do. 52 Feet or 1 Foot in 96 Feet.
1	—	65½	—	-	Do. 28 Feet or 1 Foot in 96 Feet.
9	—	14	—	-	Level.

GRADIENTS on the revised LINE from LONDON to BRIGHTON, as proposed by Mr. Vignoles.

1¾ Miles	-	Level.		
0½ Mile	-	1 in 100	-	Rise.
0½ Mile	-	1 in 440	-	Do.
0¾ Mile	-	1 in 100	-	Do.
0½ Mile	-	1 in 460	-	Do.
1¼ Miles	-	1 in 100	-	Do.
9¾ Miles	-	1 in 400	-	Do. First Summit.
3 Miles	-	1 in 400	-	Fall.
5¾ Miles	-	1 in 330	-	Do.
6½ Miles	-	1 in 330	-	Rise.
4½ Miles	-	1 in 330	-	Fall.
0¾ Mile	-	1 in 100	-	Do.
0½ Mile	-	1 in 330	-	Fall.
1 Mile	-	1 in 100	-	Do.
1 Mile	-	1 in 360	-	Do.
1 Mile	-	1 in 100	-	Do.
1¾ Miles	-	1 in 340	-	Do.
9½ Miles	-	Level		Kingston Lighthouse, opposite Entrance to Shoreham Harbour.
3¾ Miles	-	1 in 450	-	Rise.
1 Mile	-	Level		To London Road, W. St. Peter's Church, Brighton.

Gradients of Mr. Vignoles' Revised Line from London to Brighton.

GRADIENTS on the LONDON and BRIGHTON RAILWAY, from Kennington Common to the London Road at Brighton, as laid down by Sir John Rennie.

M.	CH.					M.	CH.
4	22½	-	Rise 82 Feet or 1 Foot in 275 Feet	-	-	4	22½
1	43	-	Do. 5 Feet or 1 Foot in 1623 Feet.				
8	18	-	Do. 168 Feet or 1 Foot in 258 Feet	-	-	8	18
1	41½	-	Horizontal or Level Line.				
4	60	-	Fall 84 Feet or 1 Foot in 298 Feet	-	-	4	60
2	56½	-	Do. 8 Feet or 1 Foot in 1786 Feet.				
2	54	-	Rise 18 Feet or 1 Foot in 785 Feet.				
3	4½	-	Do. 50 Feet or 1 Foot in 322 Feet	-	-	3	4½
0	50	-	Do. 2 Feet or 1 Foot in 1650 Feet.				
4	56	-	Fall 80 Feet or 1 Foot in 310 Feet	-	-	4	56
2	20	-	Do. 48 Feet or 1 Foot in 247 Feet	-	-	2	20
2	7	-	Rise 4 Feet or 1 Foot in 2755 Feet.				
2	34	-	Do. 46 Feet or 1 Foot in 278 Feet	-	-	2	34
1	19	-	Horizontal or Level Line.				
4	33	-	Fall 90 Feet or 1 Foot in 259 Feet	-	-	4	33
46	39	-	Say 46½ Miles Total Rise 90 Feet			34	8

Gradients of Sir John Rennie's Line to Brighton.

Gradients on the Branch to Shoreham.

M.	CH.				M.	CH.
2	0½	–	Fall 43½ Feet or 1 Foot in 243 Feet	– –	2	0½
0	46	–	Do. 6½ Feet or 1 Foot in 467 Feet.			
2	39	–	Do. 30 Feet or 1 Foot in 437 Feet.	For Miles	36	8½
0	75½	–	Horizontal or Level Line.			
6	1	-	Say 6 Miles Total Fall 80 Feet.			

The following is also a copy of the Estimate:—

LONDON AND BRIGHTON RAILWAY.

Abstract of Estimate.

Details of Mr. V's Line from London to Brighton.

Earthwork, 11,134,042 Cubic Yards, of which upwards of One Third would go to Spoil, at 9*d*. - - - - - - -			£417,527
Masonry in the Bridges and Culverts, exclusive of the Viaducts out of London and at Brighton, and also exclusive of the Tunnels, at 2,250*l*. per Mile (Cost of Liverpool and Manchester Railway), for 48¾ Miles - -			109,688
Viaducts out of London and for a short Distance at the Brighton End, 1¾ Miles, at 17*l*. per Yard (being the Cost as proved in Committee on Great Western Railway by the concurrent Testimony of Four or Five Engineers) -			52,360
Tunnelling 6,000 Yards, at 22*l*. 5*s*. per Lineal Yard (being the large Contract Price of the Tunnels on the London and Birmingham Railway lately entered into) - - - - - -			113,500
Fences, 48¾ Miles, at 4*s*. 6*d*. per Lineal Yard (usual Price) - -			17,305
Railway laid complete, for 54 Miles, at 48*s*. per Lineal Yard - -			228,006
Land, 487½ Acres, at 100*l*, per Acre - -	£48,750		
Purchase and Damages to House Property -	25,000		73,750
			1,032,226
Locomotive Power.			
20 Engines and Tenders, at 1000*l*. each - -	£20,000		
60 Coaches, in the Average at 200*l*. each - -	12,000		
100 Waggons, at 50*l*. each - - -	5,000		37,000
Stations.			
Six Water Stations, at 500*l*. each - - -	£3,000		
Ground and Building at each End of Line - -	10,000		
Commercial Station at Shoreham - - -	4,000		
20 Passenger Stations, &c. along Line, at 150*l*. -	3,000		20,000
			1,089,226
Contingencies - -			108,922
		Total	£1,198,148

Explanation of a Gradient System of successive Short Inclined Planes and Levels.

These papers have never been placed officially before any Committee constituted by a body of Subscribers. The Gradients upon the Line subjoined (Sir John Rennie's) are very similar to those upon the Basing and Bath, which have been so much objected to, as they have Planes of 1 in 240, 1 in 243, and 1 in 259, which would have prevented the Engines taking more than ½ a load, whereas they could carry a full load the whole way upon my line by the help of Assistant Locomotives, which I had in 3 places, 2 in one direction and 1 in the other.——I have first an Incline of ½ a mile 1 in 100; ¼ of a mile 1 in 440; ¼ of a mile 1 in 100; ¼ of a mile 1 in 460, and 1¼ of a mile 1 in 100, all of which I consider as 1 plane; the assistant engine would travel the 3 planes as well as the ascents, which certainly may be called a waste of power, but every line must be judged of by its own particular merits, and the probable nature of the traffic, which as it would be passengers, I placed two intermediate levels of 1 in 400 between the planes, by which an engine would be able to take up nearly a full load, by having intervals of a more favorable slope it would recover breath and steam, and be enabled to proceed up the next (if not too long); in the case of a full load of passengers, or a load of goods, the Assistant Engine would go up the whole way; but by having these Inclines interposed or broken I should not require the assistance of an Engine for light loads; I should probably travel up these planes at a rate of 15 miles an hour, and the Engines would lose steam towards the end of the ¼ of a mile plane; but if the momentum which the engine possessed at starting should be lost at the end of the first ¼ mile it would cause a stoppage, but the Momentum of the Power and the Steam of the Engine would carry us up our Plane, although it would not up a long Plane, even at 1 in 200; and the Trains would be more frequently stopped, particularly if the engine was out of order, and the plane very long, supposing similar engines and loads as those upon the Liverpool and Manchester, (say 6 carriages); therefore this system could not be applied on the Basing Line, on account of the great length of the Planes.——In the case I have mentioned upon the St. Helens, of an Engine being stopped in the middle of a Plane of 1 in 96, upon resuming its progress it went very slowly. An Engine travelling up 1 in 200 with a load of 6 Carriages, after going some distance, will very often go slow, or come to a stand; but I do not speak from experiments. If Mr. Geo. Stephenson stated he could surmount an Inclination of 1 in 210 by an Engine of ordinary power I cannot agree with him; he had such a gradient originally upon the Grand Junction, which he afterwards altered to 1 in 100 for a shorter distance. I doubt whether any of the Engines on the Liverpool and Manchester Railway would take a train of passengers up an ascent of 1 in 200 for 6 miles, as the velocity of a Locomotive Engine depends upon the rapidity with which it generates Steam, and the inclination being long, it would lose breath, and would be unable to exhaust sufficient steam into the chimney, consequently it would not generate it sufficiently rapid, it would lose speed before it was half way up. There are some Planes 1 in 200 upon the Darlington Railway.——I calculated upon performing the journey to Brighton in 2 hours, the distance being 54 miles, although there were 5¼ miles 1 in 100. I had a Cutting of 80 feet at Wickham Lane, I likewise had an Embankment rather more than ¼ a mile long of 50 feet average height and 80 feet in the highest part, and I am making one very like it upon the North Union; there were several other cuttings of 70 feet at the summit point; the Tunnel of 2¼ miles is about 16 miles

The Reason why it is not applied upon the Basing.

I cannot agree with Mr. G. S. if he intend to work 1 in 210 without ex. Ass.

He does not con. one of the L. and M. Eng. would not travel up 6 Miles at 1 in 200.

Des. Mr. V.'s Brighton Line continued.

from London, and has a 54 feet cutting at each end, but the average is not more than 40; there was a very short Cutting 11 miles from London 106 feet deep at the summit, it was the knob of a hill: also an Embankment of 60 feet for 1 mile at the Croydon Canal, and 68 feet for a very short distance in the deepest point of the hill, the extreme length of the cutting for same is nearly 2 miles, but it is principally shallow.———(The longitudinal Section only, of a Railway will not enable an Engineer to measure the quantity of earth-work upon it, he must be acquainted with the transverse section also; certainly he could assume the latter.)———I think the Line would have taken 5 years to execute, as the sum total of the Excavation amounts to 11,134,042 cubic yards, say 11,000,000 in round numbers, the Embankment would require 7 millions, leaving 4 millions to go to Spoil; my average price upon the whole quantity was 9*d.*, (the quantity put to Spoil being much more than usual lessens the expence,) and I think the average lead was 1 mile; if the lead had been 3 miles I should have increased the price; the Estimate of the Line was £1,000,000., independent of the carrying establishment and contingency fund of 10 per cent. which made a total of £1,198,148.; and amounts to very nearly £24,000. per mile. (I never made any estimate for it at £14,000. per mile; the Wigan Line amounted to £12,000., but the cutting and embankment upon it was very slight.)———I was consulted upon the Liverpool and Manchester Line before it was laid down; the period of my first visit was about 10 or 11 years back, when I assisted Mr. Rennie, for whom I conducted the whole business until the obtaining of the Bill. I have written a report upon the Northern line between Liverpool and Manchester, (which was projected a year or two since,) which I do not consider a better line, nor are the levels better; it has inclines of 1 in 200, and 1 in 180, spreading over many miles. The Angle of Repose of an iron bar upon a Railway is 1 in 250, (at which angle an engine would descend by Gravity,) which retards the progress of an engine considerably; (the angle at which the material of an embankment will stand is likewise called the angle of repose).———In consequence of the interposition of the 2 inclines upon the Liverpool and Manchester in the middle of the line, it is not usual to load the passengers' train with more than the Engine, and Assistant, will take up these planes, therefore as a Plane of 1 in 96 requires 3½ times more power employed upon it than upon a level; the Engines (with passengers) consequently carry only ½ a train.——— On the Leigh and Kenyon, (a line 2½ miles long, joining Bolton to the Liverpool line, of which Mr. Rastrick was the Engineer,) they proportion the passengers to the goods rather than the goods to the passengers, and they travel at a rate of 20 miles an hour, their engines being loaded to the maximum; the inclination is 1 in 440, or 12 feet in a mile, which is not very good, but it is the best that could be obtained, the loss of power upon such an inclination is nearly ⅓rd.——— I have perambulated the Southampton Line as far as Basing, having understood that they were not at work beyond that distance, and I found the works were proceeding at 8 or 10 different places, independant of St. George's Hill, which is the principal spot where any quantity of work is done, the men were not employed in cutting at the time, but were laying down the Rails, and ballasting the Road; and if they commenced it on the 6th of October, they certainly have made but very little progress.———

Both the Trans. and Long. Sec. are nec. in mak. an Est. of a Ry.

Mr. V. put his Brighton Line, with 1 mile lead, at 9*d.*

It amounted to £24,000. per Mile.

The Wigan was £12,000. per Mile.

He con. the L. and M. up to ob. the Bill.

Rem. on the Northern line to Liver.

250 is the Ang. of Repose.

Meth. of Work. the Planes upon L. & M.

1 in 96 req. 3½ more power than a Level.

Do. upon the Bolton and Leigh.

Loss of Power upon a Pla. 1 in 440 is ⅓d.

Description of the Works upon the Southampton Railway.

(Upon the North Union, not more than 4 Weeks were occupied, from the time of our breaking ground for the great embankment, up to the period of our teaming 800 yards a day, by which time we were in full work.)——Next to which is the work at Shapley Heath, and at Battersea. Regarding the formation of the Railway generally, I think it would be better to throw all the force upon St. George's Hill, on account of the great quantity of materials to be removed, and the difficult nature of the work; as the whole of the work upon the other parts of the line must stand still until it is finished, which is a direct loss of the interest of the capital employed, and I am not aware of any town within 1 or 2 miles of the neighbourhood. An average number of 800 Yards may be teamed per day, at St. George's Hill, with 4 teaming places, provided it is well managed, and great exertions are made, the men would be required to work 12 or 15 hours; and they could not exceed 1000 Yards, (on account of the great probability of the Clay increasing, and the great height of the Embankment;) taking the average throughout the year, and allowing 5 days to the week, including hours of daylight, (Night-work adds, at least, 25 per Cent. to the cost, particularly in the getting) and taking slips and accidents into consideration. The length of the Lead will not affect the teaming.——I did not consider, at the time I visited it, that proper arrangements were made for rapid work, although some allowance must be made, as they are merely commencing; neither do I think the method by which the points were carried into the Cutting was well arranged, or the order in which the Teams were filled; but Shapley Heath, where there is probably about half as much work to execute as St. George's Hill, was the only place where I saw any of this description of work. At Elvetham there is about 12,000 or 14,000 cubic yards to do, 6,000 of which is done.——You cannot conveniently increase the face of an Embankment 1½ to 1, for teaming, unless it is very high, and it is done in 2 Lifts.——I have heard of a Machine being used upon the Carlisle and Newcastle Railway which discharged a greater quantity than usual, viz. 2 or 3000 per diem; I therefore sent a Contractor to look at it, who did not think it worth adopting; had he made use of it, he would have applied it upon the Wigan Railway, where the Embankment was 16 feet high and 1½ miles long. If it was practicable to use it upon the works at St. George's Hill, it would very much reduce the time for completing the same.——I doubt the economy of using Locomotive Engines, in preference to Horses, in a Line of 2 or 3 miles; neither would there be a great saving of time, because it is limited by the teaming places, the irregularity of the road, and the long time it takes to subside, the stoppages which occur, the wear and tear of the Engines, the fuel consumed, and the liability of accidents in getting in and out of the points; therefore, it is not advantageous to use them until the road is consolidated, and the Leadis of some length.——The Levels of St. George's Hill have been raised, by making the Cutting less, and the Embankment more; but I do not think it will make any difference in the length of time necessary to complete it; if any thing it would lengthen it. The Hill contains 3,700,000 cubic yards, of which 1,500,000 go to the Embankment, and are teamed from one end, which would occupy 7 to 8 Years, at the rate of 800 cubic yards per day: this calculation applies to the work at one end of the Embankment only; I presume they would work at both ends, which would not affect my calculation.

The N. Union was in full work within 4 Weeks after they com.

St. Geo. Hill.

800 to 1000 c. yds. one day's work.

Night Work adds 25 per Cent. to Exp.

Works at Shapley Heath and Elvetham.

Remarks upon Mr. Grahamsley's Ma.

Rem. upon using of Locomot. Engines in form. of works.

St. Geo. Hill con. 3,700,000. c. yds. and would occ. 7 or 8 Years to exec.

Scotch Sleepers. ——I have never known Scotch Fir Sleepers used upon a Railway, I should decidedly object to them; I also object to the mode of fixing Rails with Chairs, at intervals of 15 feet, when laid down for Locomotive Engines; I have adopted it in Quarries, and upon a Railway which I laid down some years back for temporary purposes, but I found the rolling of the waggons over were constantly disturbing the Rails.

Rem. upon Chairs 15 ft. apart.

Observ. upon Mr. Giles's Estimates. ——I am not acquainted with Mr. Giles's Estimate, although I know his prices generally. I have heard that his Estimate of the Excavation upon the Southampton Railway, (upon the whole) amounting to 16,000,000 cubic yards, was £365,495., which is little more than 6*d*. per cubic yard, which I cannot conceive is sufficient. I think the Lead was 2 or 3 miles, (the Counsel stated it to be 3 or 4.) I consider the Southampton Line will cost much more than £12,000. or £14,000. a mile, on account of the great quantity of cutting, and the distance it has to be removed. According to the opinion of most Engineers, and the experience of all previous and present Companies, Contract is the cheapest mode of executing public works; I prefer it on account of the ultimate economy.

The Southamp. was taken at 6*d*. with 2 or 3 Miles Lead.

Des. of the Method of Working the North Union Rail. ——The Newton obtained their Act in 1830, and was a separate Line until last year, when it was extended to Preston, and joined the Wigan Line. There was a dispute between the Companies, which lasted 2 or 3 years; at length they were incorporated, under the name of the North Union, in the Spring of last year; the works were not commenced until about 8 or 9 months after the Act was obtained, (the separate Act from Wigan to Preston was obtained in 1831, and the first meeting of the Proprietors was in July last year,) as much preparation was required before we could break the ground, as staking out the Line, &c.

Cost of the Works, &c. ——This is a Schedule of the Prices of Contract No. 3, upon which the Contractor made his Estimate.

No. 3.

To the Directors of the North Union Railway Company.

Schedule of Prices.

	£.	s.	d.
I, the undersigned, do hereby propose to construct that portion of the "North Union Railway" between a Footpath in Field marked No. 11 on the Ground Plan, in the Township of Euxton, and the Head of the Incline in the Township of Coppall, in a Field marked No. 6 on the Ground Plan, being the Third Lot as at present laid out, in Length Three Miles Forty-three Chains, with all the necessary Fencing, Earth Work, Brick Work, Masonry, Drains, Formation and other Works for the entire Completion thereof, and to uphold and maintain the same for Two Years after the public Opening of the same, finding all Materials of every description, (except Railway Bars,) and the ordinary double and single Chairs, and Pedestals,) according to the Maps, Plans, Sections, and Drawings, within the Period, and in the Manner, and upon the Terms and Conditions mentioned and set forth in the Specification, and contained in the Draft Contract exhibited to me, at and for the Sum of £51,589. 8*s*. 9*d*. Sterling, say	51,589	8	9
And in case the Directors should elect not to allow the Use of their Rails and Chairs in the temporary Roads, according to the Specification, and should only deliver them as they might be required to be permanently laid down, I do similarly propose as aforesaid at and for the Sum of Pounds Sterling, say . .			

Should the Directors require the Work to be completed within the Period of Two Years and a Half from this Date, instead of within the Period of Three Years from the 31st January, 1835, according to the Specification, I do similarly propose as aforesaid at and for the Sum of Pounds Sterling, in addition to either of the above Sums, say . . . .

Should the Directors decide to adopt Blocks in the Cuttings and on the Low Embankments, to be Twenty-four Inches square and Twelve Inches thick, to contain Four Cubic Feet only, instead of Five Cubic Feet according to the Specification, but not to be otherwise inferior, my Proposal would be diminished in that respect Pounds Sterling, say . . . .

And I have in the Schedule hereunto annexed set forth the Prices of the various Descriptions of Work at which this Proposal has been estimated, and by which I do hereby propose that any Deductions from the Works (as specified, but which may not hereafter be executed,) shall be made, and at which I will execute any further or additional or extra Works.

And in the event of this Proposal being accepted, I do hereby undertake to execute a Contract, according to the Draft referred, within Sixteen Days from this Date.

And I do propose of in the County of and of in the County of to be my Sureties for the due Performance of such Contract.

Witness my Hand this 18th Day of April, 1835.

William M'Kenzie,
Liverpool.

Schedule of Prices, Contract No. 3, North Union Railway.

Details of the Cuttings.

Quantities.	Average Distance to be waggoned.	Price per Cubic Yard.
Cubic Yards.		
199,527	1¼ Mile	11*d.*
38,040	900 Yards	8½*d.*
339,611	1¼ Mile	11*d.*
25,000	To be run into temporary Spoil, and afterwards carried to Ballast.	4*d.*
8,342	220 Yards.	6½*d.*
610,520		

Schedule of Prices referred to in the annexed Proposal.

THIRD LOT.

No.	Description of Work.	Dimensions.	Prices. £.	s.	d.
1	Larch Posts and Railing, Ditches and Quick Mounds, including both Sides	per Lineal Yard	0	3	6
2	The Average Price of the whole of the Excavations, with the Slopes of Cuttings and Embankments trimmed, but exclusive of Soiling	per Cubic Yard	0	0	10½
3	Soil-stripping and Resoiling (each)	per Superficial Yard	0	0	1
4	Retaining and Breast Walls, of the Stone of the Country	per Cubic Yard	0	10	0
5	Boundary and Fence Walls, of the Stone of the Country	per Cubic Yard	0	12	0

Schedule of Prices, Contract No. 3, North Union Railway.

No.	Description of Work.	Dimensions.	Prices. £. s. d.
6	Bridges and large Culverts of Brick, including Pointing, Centres and Foundations	per Cubic Yard	0 16 0
7	Small Culverts of Brick, including Fronts, Ends, and Wings, Pointing, Centres, and Foundations; viz.		
	Culverts of 1½ Feet internal Diameter	per Lineal Yard	0 8 6
	Culverts of Two Feet ditto	per Lineal Yard	0 12 0
	Culverts of Three Feet ditto	per Lineal Yard	0 16 0
	Culverts of Four Feet ditto	per Lineal Yard	1 16 0
	Culverts of Five Feet ditto	per Lineal Yard	3 0 0
	Culverts of Six Feet ditto	per Lineal Yard	5 10 0
8	String Courses, Quoins, Pedestals and Copings, dressed and set	per Cubic Foot	0 1 3
21	Paved Crossings of Roads upon and near Railway	per Superficial Yard	0 7 0
22	Formation and Metalling of Roads and Approaches	per Superficial Yard	0 1 3
23	Field Gates, including Posts, Iron Work, and Painting, fixed complete	each	3 0 0
24	Road Gates, including Posts, Iron Work, and Painting, fixed complete	per Pair	10 0 0
25	Gatekeepers Houses or Station Lodges	each	150 0 0
26	Double Line of permanent Railroad, laid complete on Stone Blocks of Five Cubic Feet, with all the detailed Drains, Ballasting, and Boxing, and so pro rata for single Lines of Sidings, clear of the Crossing Plates	per Lineal Yard	1 3 1
27	Ditto on Blocks of Four Cubic Feet ditto ditto	per Lineal Yard	1 1 3
28	Ditto on Sleepers of Larch Timber ditto ditto	per Lineal Yard	1 5 0
29	Crossings and Sidings of single Line of permanent Railroad, laid complete, on Stone Blocks of Five Cubic Feet, including Cast and Wrought Iron Points and Crossing Plates, Points, Tongues and eccentric Gearing, and Machinery, measured from Point to Point of Crossing, and from Point to Crossing of the Sidings to Outside of the Main Line	per Lineal Yard	0 14 6
30	Ditto on Blocks of Four Cubic Feet ditto ditto	per Lineal Yard	0 13 6
31	Ditto on Larch Sleepers ditto ditto	per Lineal Yard	0 15 6
32	Railroad Material or Ballasting and Boxing laid down and spread	per Cubic Yard	0 1 7½
33	Transverse and central longitudinal Rubble or French Drains under and among the Ballasting	per Lineal Yard	0 0 3
34	Longitudinal covered Drains at the Bottom of the Drains of the Excavations	per Lineal Yard	0 1 6
35	Open Drains and Retaining or Breast Walls at the Bottom of the Slopes of the Excavations, when substituted for the covered Drains	per Lineal Yard	0 2 6
36	Stone Blocks of Five Cubic Feet	each	0 2 6
37	Ditto of Four ditto	each	0 2 0
38	Larch Sleepers	each	0 3 0
39	Cast Iron Work of Points, Crossings, and Turnplates	per Cwt.	0 9 0
40	Wrought Iron Work for ditto	per lb.	0 0 4
41	Turnplates fixed complete with Stone Curbs and Wells	each	40 0 0
42	Protecting Mounds on Embankments, including Pipe Drains	per Lineal Yard	0 1 0
43	Maintaining Railway Crossings and Sidings, when laid on Stone Blocks of Five Cubic Feet, for the First Year	per Mile	150 0 0
44	Ditto on Larch Sleepers, for the First Year	per Mile	200 0 0
45	Maintaining Railway, &c. on Blocks as above, for the Second Year	per Mile	80 0 0
46	Ditto on Larch Sleepers, for the Second Year	per Mile	120 0 0

The Items of the Schedule are priced to be executed as per Specification, and only the Neat Measurements to be claimed, allowed, or deducted, notwithstanding any Custom to the contrary.

WILLIAM M'KENZIE, Liverpool.

It is 3 miles, 43 chains long, and amounts for one Line only to £51,580., which is £14,700. per Mile, and any increase or decrease of the work made during the execution is to be added to or deducted, according to these prices. We had several other proposals, some of which were higher and some were lower than this; but they did not bear an exact relative proportion to each other; we accept those with the most satisfactory securities, and the lowest prices. In the Schedule referred to, the average Price of Excavation, with the slopes and banks trimmed, is 10½d., which is exclusive of soiling, for which 1d. per superficial yard is added, making the Price of the Cutting 11½d.; the Lead is from 1 to 1½ miles. There is a Hill near Preston, where the average Lead is 1½ miles, for which we pay 11d. (in one of the Contracts, which is smaller than the above, and of shorter distance, the Price of Excavation is 8d., but the Lead does not exceed half a mile;) the Contractor pays 7½d. for it, and runs all risks of Slips, and finds the men under him all materials, as Waggons, Rails, and Sleepers, but the men find Horses. We generally reckon 1d. a yard for Waggons (such as ours) upon a Cutting; the oiling of them is additional, and is very expensive; the laying of the Rails, and also the continual re-placing of them, makes the whole of the Contractor's extra outlay, probably amount to 2d. or 2½d., which added to the 7½d., leaves him a profit of ¾d. or 1d.; Sand would perhaps cost less, on account of the difference between the filling and cutting, (this work being in Clay;) the general rule is, after filling the Waggons, to add 3d. per yard a mile, to cover the expences of loading and contingencies; but the first mile would cost more, and ¼ or ½ a mile yet more in proportion, than 1 mile, as the expence of filling is the same; the first mile would be about 9d. in Clay, and 8½d. in Sand, including cutting, filling, and leading; I should add 3d. more for the second mile, and probably 2d. for the third mile, according to circumstances, as the nature of the soil, and the state of the weather.——The price of Ballasting varies for "Railroad forward, and Ballasting laid down and spread," the price is 1s. 7½d. per cubic yard, and a yard forward of the Railroad takes at least 5 cubic yards, it therefore costs about 8s. 9d. I have not been able to get any Ballasting done under 1s. 6d. a cubic yard, some has cost 2s. although we have it out of the Cutting, or immediately adjacent, we have no Ballasting so low as 5d. a yard forward, including the labour of spreading it, but much depends upon the quantity used. The chief expense of Ballast, when found in the excavations, is the labour of laying it along the side to select, and the bringing and laying of it down at different times. I generally spread it 15 inches thick, or 12 inches, under the sleepers. They were laying it down upon an embankment at St. George's Hill only 7 or 8 inches deep; but I am not aware whether it was for the permanent railway, I inferred the contrary on this very account, if they intended to adopt the mode in which they were fastening the rails, generally, they need not relay it, unless they wished, as they might take it up and pack ballast under; the quantity of ballast required depends upon the nature of the subsoil, as the object is to prevent the latter rising up. Clay requires more ballast than other soils, I also find it necessary to put a greater number of drains in the ballasting of it when upon an embankment.——Upon comparing the Soil of the North Union Railway with the Sand of Bagshot Heath, I am inclined to add 1d. or 1½d. more for the latter, and it is then a small average price, considering the lead is 3 miles, but I think it will do it; I would not put my name to a less Estimate even if the soil was favorable, considering the

£14,700. per Mile.

Price of Cutting, 11½d. Incl. 1½ Miles Lead.

Details of the same.

Mater. 2½d.
Labor 7½d.
Makg. 10d. pr. c. yd.
indep. of Slips, &c.

Method of Prising the Lead.

Ballasting.

1 yd. run. of Ballst. the Railway costs 8s. 9d.

15 inches thick.

Remarks upon Clay Soils.

Comp. of the Soils of the N. Union and St. Geo. Hill.

Description of the Soil at St. George's Hill.

quantities and length of the Line. The upper part of the Soil at St. George's Hill, about 2 or 3 feet, is Sand and Gravel, the next 4 feet are loamy Sand in long veins, and the lower part, measuring 15 feet, is getting into the London Clay, the peculiar nature of which is that upon being exposed to much wet it melts away like soft soap; it is similar to the cuttings at Highgate Hill and at Barnet: the depth of the Cutting was 20 feet, which is merely the commencement, therefore if the Clay should increase, which I think will be the case, the cutting and leading cannot be done for 1*s*. The length of the Cutting is 2 miles, and the Embankment at the London end or the Mole Embankment must be brought from the further end, and it is near 4 miles across the valley of the river Mole.——

Remar. on Fencing.

Fencing is generally regulated by the material that can be most easily obtained upon the line, but quicks are generally employed, although it depends upon the nature of the country; for instance, on the High Peek Railway the fences are all stone, as there is a plentiful supply of it.

Post Rail and Quick Fence is the cheapst.

Costs about 3*s*. 6*d*. both sides of Railw.

I could not make a cheaper permanent fence than a Post Rail and Quick fence; I do not think Furze, Beech, or dwarf hedges would be as cheap, whatever the nature of the country may be. I am paying 3*s*. 6*d*. for Fencing on both sides, consisting of a Ditch, the material of which is thrown up and forms the quick mound; good soil is then selected either from the top of the ditch or adjacent to it, and a quick border is trenched in 12 inches deep and 18 inches wide; the quicks are planted 18 to the yard, and a Larch Post and Rail Fence, having 3 rows of rails and a standard at intervals of from 6 to 9 and 12 feet; in the latter case we have a post in the middle called a prick post; the above fence answers the purpose better than any I am aware of; the post and rails are merely temporary, having to be maintained about 4 or 5 years, which they are lasting; we find sawing is better than splitting the rails. We are bound to have proper Fences to keep the Cattle off the Railway; sheep, for instance, frequently trespass. The Larch is brought from the northern parts of Lincolnshire and the southern parts of Cumberland.——

Description of same.

Exp. of Workg. the Line should be inc. in the Estimate.

It has lately been the practice in forming an estimate of the cost of a Railway to include a sum for Locomotive power in working the line, but I have not yet practised it.——

Expense of Paved Crossings.

I have had several Lodges and Gates built where there are level crossings, the cost of which varies from £150. to £200., the gatekeeper is generally paid 14*s*. a week, independant of his house; it is not the practice to erect Gates across roads without Lodges.——

Tunnels cost about £30. or £40. per yd. run.

I have not laid out any Tunnels, but I designed one on the Liverpool and Manchester Railway; their average cost upon a large scale, is £30. or £4 per yard. I believe the Tunnel through Chalk upon the Thames and Medway Canal, which is about the same in sectional area as a railway tunnel, was nearly £30. a yard.——

Dublin and Kingstown.

I have recommended the use of lighter Engines on the Dublin and Kingstown Railway, (and not heavier, as enquired by the Counsel for the Opposition,) we have ascents 1 in 440 at each end, or about 17 feet in a mile, and the remaining portion of the line is level.

Ex. Mr. EDWARD DRIVER, *Land Surveyor and Valuer.*

I have had considerable experience in my profession generally, and have been engaged for the Crown for all purposes, buying, selling, &c.——I have been employed by the Directors to value the Land, from the river at Maidenhead to the junction with the London and Birmingham Railway. I went over the Line this as well as last year, occupying 4 days each time; the several quantities were furnished by Mr. Brunel, and amounted to about 190 Acres, the sum total of the valuation of same amounts to £39,446. I have allowed £22,341. for Land, and £17,105. for Compensation. I have taken it at different prices, according to the situation, from £5. which is a very high price, (there is but a very small portion higher) down to £1. 10*s*. per Acre. I commenced from the Maidenhead Bridge across the valley of the Thames, the valley of the Colne, and the valley of the Brent. I noticed several mills, but we do not pass near them, with one exception, which is at Drayton near Uxbridge, and about a ¼ of a mile from the line; compensation is allowed for all buildings interfered with, also all sub-tenants and lessee's claims for severance, and consequential damage of every kind. There is but one house actually upon the line, which is at Ealing Common, and I am not aware that it will be interfered with, as they may shift the line a trifle; but there are one or two labourers' cottages in other parts.——I consider it the highest Estimate I ever made, and I have been engaged to value for the owners of property against the London and Birmingham Railway; as the ground is very level there is not much to be apprehended from slips, but I have allowed for them. I have not valued each individual's property separately by the acre, but I divided each district into quarters of a mile; if the land in one quarter varied part was at one and part at another price. I also assumed an average compensation throughout the whole line; it is possible that the cuttings may endanger the springs in the adjoining neighbourhoods, but I have not considered the effect likely to be produced upon them, therefore have not allowed for the same; I consider that they would be provided for by the Engineer, but I do not think the compensation will amount to the sum which I have allowed.

Land from Maidenhead to the Junc. with the L. and B. 190 Acres, valued at £39,446.

Allows £17,105. out of it for Compen.

The Land is valued from £1. 10*s*. to £5. per Acre.

Description of the Property on the Line.

Mode of Valuing it.

Remarks upon Springs.

Ex. MR. DANIEL LONSLEY, *Land Surveyor, at Blurbury in Berkshire.*

I have had 20 years experience in my profession; I also farm between 1,200 and 1,300 Acres.——I valued the land from the Bridge at Maidenhead to Reading, and from the River Thames in Cholsey Parish down to the River Cole near Shrivenham; the land required was pointed out to me by Mr. Brunel and Mr. Hennett, it amounted to 142 acres 1 rood 29 poles from Maidenhead to Reading; and 241 acres 1 rood 38 poles for the latter distance. The sum at which I valued the land from Maidenhead to Reading was £14,221. 5*s*. 2*d*., and the latter distance £25,785. 17*s*. 10*d*.; this valuation is very high, and includes Compensation of every kind: the Land amounts to £20,468. 11*s*. 6*d*. and for Compensation £19,538. 11*s*. 6*d*. (Enclosed Land is worth about 28 years purchase, and unenclosed about 27 years, my valuation averages 56 years purchase in order to allow fully for every compensation.)——There are large Corn and Cattle Markets Abingdon and Wallingford; also a similar but smaller market at Wantage. ——The Railway will be of great benefit to the Agricultural Poor in our neighbourhood, as we have no wood of any consequence for several miles, and it costs the poor more for the carriage of wood for Fuel than the original price of it; we

He Farms about 1,200 Acres.

Land from Maiden. to Reading, 142 A. 1 R. 29 P. valued at £14,221. 5*s*. 2*d*.

Land from Cholsey Pa. to Shrivenham, 241 A. 1 R. 38 P. valued at £25,785. 17*s*. 10*d*.

Particulars of same.

Almost as much all. for Comp. as for Ld.

Markets in the Line, &c.

The G. W. will be great benefit to the Neighbour. Poor.

Memo. Manure.

are therefore obliged to give them the Haulm or Stubble, or they would not cut down the Corn. Ours is generally a poor agricultural country, and requires more manure than we can procure, and the Haulm is generally considered to make the best, which the Farmers would then be able to secure; some people use rags, which are expensive, as they are sent from London, we also use a great quantity of Ashes, which we get from Newbury and Thatcham, and the expense of carriage is more than their original cost. There are also great quantities of Coals sent down, which is very expensive on account of the carriage; they would get it by the Railway for half the present price.——— Our Barley generally goes to the Bristol Market from Abingdon, and a cheaper conveyance would be a great benefit to the farmers; I have known the time when Barley has been unsaleable at Abingdon, even at any price, the Canal being frozen and the Barges having all their sacks full. I have known a difference of 4*s.* or 5*s.* a quarter in it, owing to the same cause, in the course of a market or two.

Barley Unsaleable for want of a Conv.

Ex. Mr. F. HAWKES, *Land Surveyor, of Reading.*

Land from the Meadows at Reading to Southstoke, 115¼ A. 9 P.

Val. at £12,583. 5*s.*

Compens. £5,713.

Particulars of same.

Land Pr. from £35. to £250. per Acre, indep. of Compen. Arab. Lnd. 28 Yrs. Com. Fld. 30 Yrs. Purchase.

I valued the Land at Reading, taking the Meadows immediately opposite the Town, and extending as far as the River, where it separates Cholley from Southstoke. The quantities were given me by the Engineer, and amounted to 115¼ acres 9 poles, which I value at £12,583. 5*s.*, this sum includes Compensation of every kind, Severance, &c. and I have put it at very high prices compared with what I should put it at was it for sale. ———Of the above amount £5,713. is for Compensation. The lowest that I priced the Land (from King's Mead to Reading) was £35. per acre, and highest £250. (independent of compensation which amounts to almost as much more.)———I have taken 28 years purchase upon common arable lands, and I have taken some common fields we pass through at 30 years' purchase. (I have let the meadows at Reading for £5. per acre.)

Ex. Mr. H. E. GOODRIDGE, *of Bath, Architect and Surveyor.*

Values the Houses and Land at Bath, at £40,864.

Particulars of same.

Some of the Ld. val. at £1000. per Acr.

I valued the Land and Buildings upon part of the proposed Line, from Oak Street at Bath (No. 221) to the entrance point of the Parish of Bathwick, at £46,345. 16*s.*; against which a credit of £5,481. 16*s.* is to be set, being the saleable value of the materials; the net amount is therefore £40,864., which I consider a full and ample valuation, including all compensation to tenants and others. Some of it is priced quite extravagantl y, compared with what it would fetch in the market. The Land I valued, unconnected with buildings, was about 9 acres, and is included between 234, in the Parish of Lidcombe and Widcombe, and 142, at the extremity of the Parish of Bathwick; I valued some as high as £1000. per Acre, including compensation. The

Land on the other side of the River, when required for the new Road, was valued at £200. per acre only. I have allowed compensation to both Landlord and Tenant, not only for Land that actually comes in contact with the Railway, but whenever I conceived any damage likely to occur.——Sydney Gardens a place of public entertainment (about 15 acres) is within the district; the Line runs below the Canal and cuts right through the Gardens, where it will be covered over, for which I have allowed a highly liberal sum; some of the Proprietors, to whom I hinted the amount, were quite astonished. ——The Line also passes in a Tunnel under Bathwick Terrace, consisting of good respectable Houses, which are faced with Ashlar, the partitions also are of Ashlar; I have allowed for some of the Houses entirely. The Line also passes by Raby Place, which consists of a row of good Houses, some having been sold at £1200. each; whenever the line touches the area of a house, I consider it best for the Proprietors to purchase it, which is the system I have pursued.——I have seen the Drawings of the Bridge over the Avon, and I consider that Mr. Brunel has allowed an ample sum for same.——I know the proposed Depôt at Ham Gardens; the level of the former is considerably above the highest flood in the river, as shewn by a mark on Widcombe Poorhouse, made in 1809; the Level of Ham Gardens is something below it.—— I understand the Bath and Basing Line will interfere with Prior Park, upon which a considerable sum has lately been expended, it would pass near an ornamental sheet of water of considerable beauty, and also destroy the terrace, which would be highly injurious; this Park is the most ornamental property in Bath, and belongs to Dr. Bates, who consequently objects very strongly to the line: there are also a great number of Gentlemen's Houses and Pleasure Grounds near it: it also interferes with Colonel Wrench's and Mr. Tugwell's pleasure grounds, and I conceive that the necessary compensation will amount to considerably more than ours at Raby Place I have been much connected with the property in the neighbourhood of Bath, particularly on the Southern side of the river; the river of Bath is liable to be flooded; the meadows are often flooded to a considerable extent. The lower Bristol Road, which is the same level as the meadows, is occasionally overflowed. When the Bridge at Bath was built, it was expected to have had a great effect upon the meadows; a scale was therefore attached, to ascertain the result, and it was found to have very little effect upon them; it consists of 1 arch, 100 feet span, and was erected under my direction.

The G. W. passes through Sydney Gardens.

Bathwick Terrace.

Raby Place.

Rail. Brdg. at Bath.

Level of the Bath Depôt.

The Basing Line interfer. with Prior Park.

And var. other Pro.

Particulars of same.

Floods at Bath.

Bath Bridge.

Ex. Mr. YOUNG STURGE, *Land Surveyor, Bristol.*

I valued the Land (principally with Mr. Townsend) and Buildings between Bristol and Bath, also between the East end of Bath and the County of Berks, (through Chippenham and North Wiltshire.)——The amount of my valuation between Bristol and the Parish of Lidcombe and Widcombe, adjoining Bath, exclusive of any Houses in Bath, is £33,592. 10*s*., being £15,690. 14*s*. for Buildings, and £17,901. 16*s*. for

Valuations between Bristol and Bath, Land £17,901. 16*s*. Hous. £15,690. 14*s*.

Betw. East of Bath and Shriv. Land and Buil. £47,770. inc. the two Branches.

Part. of same.

Land.——My valuation of the Land and Buildings, from the East end of Bath to the Parish of Shrivenham, amounts to £47,770., viz. from Bath to Chippenham, Land £18,941. 5*s.*, and Buildings £1485., making £20,426. 5*s.*; from Chippenham to Shrivenham, Land £16,726. 10*s.*, and Buildings £500., making £17,226. 10*s.*; the Branch line to Bradford, Land £9,397. 15*s.*, and the Buildings £200., making £9597. 15*s.*, and the Branch to Trowbridge, Land (no buildings) £2704.: the total amount of my valuation is £83,547. 1*s.* In which amount I have allowed for compensation of every kind; I have priced some of it full twice as much as I should if it was going to sale.——The Land around Bath is very valuable.——The country generally about Christian Malford is Dairy Land, and very spongy.

Ex. Mr. JOHN HAMMOND.

I was employed last year by Mr. Brunel to superintend the Borings on the Line from London to Reading; upon which I gave Evidence in the House of Commons. This is a Statement of them.——The figures refer to points upon the section:

Account of the Borings from London to Reading.

No. on Specimens.	Parish.		Depth of each.	Depth from Surface.
			Feet.	Feet.
1	Ealing, No. 74.	Fine Gravel	8	
		Coarse Gravel	2	
		Yellow Clay, hard and dry	2	10
		Blue Do. Do.	16	12
		Total Depth bored	28	
2	Ealing, No. 68.	Vegetable Soil and Yellow Loam	2½	
		Large hard Gravel	15½	2½
		Yellow Clay, hard and dry	2	18
		Blue Do. Do.	20	20
		Depth bored	40	
3	Ealing, No. 62.	Vegetable Soil	2	
		Rough Gravel and Sand	18	2
		Yellow Clay, hard and dry	2	20
		Blue Do. Do.	18	22
		Depth bored	40	
4	Ealing, No. 61.	Vegetable Soil	2	
		Greenish Gravel, moist	0½	2
		Mottled Loam, dry	3½	2½
		Blue Clay, hard and dry	24	6
		Depth bored	30	

No. on Specimens.	Parish.		Depth of each.	Depth from Surface.
			Feet.	Feet.
5.	Ealing, No. 59.	Vegetable Soil	3	
		Coarse loamy Ground . . .	9	3
		Yellow Clay	2	12
		Blue Clay, firm and dry . . .	1	
		Depth bored .	15	
6	Hanwell, No. 9.	Rough loamy Gravel . . .	9	
7	Hanwell, No. 9.	Yellow Clay, with thin Vein of Gravel near the Top	20	
		Blue Clay, hard and dry . . .	20	
		Depth bored .	40	
8	Hanwell, No. 2.	Vegetable Soil	2	
		Coarse Gravel, moist . . .	2	2
		Yellow Clay, hard and dry . .	22	4
		Blue Ditto	5	26
		Depth bored .	31	
9	Norwood, No. 76.	Made Ground	4½	
		Rough Gravel	5½	4½
		Coloured Loam	6	10
		Blue Clay, hard	14	16
		Depth bored .	30	
10	Iver, No. 13.	Loamy Soil	9	
		Rough Gravel	11	9
		Quicksand	3	20
		Yellow Clay	1	23
		Blue Do. hard	12	
		Depth bored .	36	
11	Iver, No. 2.	Vegetable Soil	4	
		Dry rough Gravel	13	4
		Sand	3½	17
		Yellow Clay, hard	1½	20½
		Blue Do. Do.	8	22
		Depth bored .	30	
12	Taplow, No. 4.	Vegetable Soil	2	
		Peat	9	2
		Rough Gravel	9½	11
		Chalk	1½	20½
		Depth bored .	22	

Account of the Borings from London to Reading.

No. on Specimens.	Parish.		Depth of each.	Depth from Surface.
			Feet.	Feet.
13	Taplow, No. 2.	Vegetable Soil	5	
		Rough Gravel	4	5
		Fine Do.	11	9
		Chalk	4	20
		Depth bored	24	
14	River Thames, Maidenhead,	Water, when Eight Feet below the Top of Towing Path.*	5¾	To Bottom of River.
		Gravel	12¼	
		Chalk	2	18
		Depth bored	20	
15	River Thames, Maidenhead.	Water, when Eight Feet below the Top of Towing Path.*	5½	
		Gravel	12¾	
		Chalk	2	18¼
		Depth bored	20¼	
16	Bray, No. 27.	Loamy Soil	1½	
		Rough Gravel	12	1½
		Coloured Clay and dry Sand	9½	13½
		Sand, with Water	3	23
		Coloured Clay and dry Sand	11	26
		Chalk	2	37
		Depth bored	39	
17	Bray, No. 23.	Loamy Soil	2	
		Coarse Gravel	6½	2
		Loamy Sand	2	8½
		Coloured Clay, with a little Sand	12½	
		Chalk, with Flints	14	23
		Depth bored	37	
18	Bray, No. 13.	Sandy Soil	2	
		Coloured Loam and Clay with Sand	13½	2
		Chalk	1½	
		Depth bored	17	
19	Waltham St. Lawrence, No. 10.	Vegetable Soil	3	•
		Chalk mixed with Gravel	9	3
		Chalk	14	
		Depth bored	26	
a	Ealing Haven.	Yellow Clay	11	
		Coarse Sand	½	
		Gravel	1½	
		Depth bored	13	

* I bored in the River, and in the Towing Path, and found Water at the same depth in both places. I found Water on the shore, as soon as I came down to the level of the River. (The spot was about 30 feet West of the River.)

Account of the Borings from London to Reading.

No. on Specimens.	Parish.		Depth of each.	Depth from Surface.
			Feet.	Feet.
b	Ealing, on Road from Apperton.	Yellow Clay, hard	13	
		Sand	1	
		Sand, Clay and Gravel mixed	1	
		Depth bored	15	
20	Ruscombe Church.	Gravel	7	
		Variegated Clay	16	
		Depth bored	23	
21	Vale of the Loddon.	Silty Ground	6	
		Gravel	12	
		Chalk	2½	
		Depth bored	20½	
22	Sonning, No. 40.	Gravel	8	
		Sand	2	
		Gravel	1	
		Variegated Clay	24	
		Silt	1½	
		Variegated Clay	15½	
		Depth bored	52	
23	Sonning, No. 38.	Gravel	16	
		Variegated Clay	12	
		Silt	1	
		Coloured Clay	5	
		Red Sand	4½	
		Clay, very rough	15	
		Clay, with Silt	6½	
		Sand and Silt	2	
		Blue Clay	2½	
		Depth bored	64½	
24	Sonning, No. 27.	Gravel	6	
		Clay with light Sand	5	
		Sand, very dry	3	
		Silt, with Clay and Sand	2½	
		Sand, sharp and dry	4	
		Do. streaked with Blue Clay, and a Vein of Shells	2½	
		Silt	1	
		Variegated Clay and Green Silt	2	
		Chalk	2½	
		Depth bored	28½	
25	St. Lawrence's Parish, No. 12.	Mould	2	
		Soft Ground	6	
		Gravel	8	
		Chalk	2	
		Depth bored	18	

Account of the Borings from London to Reading.

No. on Specimens	Parish.		Depth of each.	Depth from Surface.
			Feet.	Feet.
26	Ham Gardens.	Mould	7	
		Clay	$10\frac{1}{2}$	
		Blue Silt	1	
		Thin Lias Stone, with Beds of Clay . .	3	
		Depth bored .	$21\frac{1}{2}$	
27	Broad's Ferry.	Clay	15	
		Blue Clay and Gravel	5	
		Lias Stone	3	
		Depth bored .	23	
28	Weston, No. 45.	Loam	11	
		Clay and Gravel	5	
		Gravel	6	
		Depth bored .	22	
29	St. Philip's Parish.	Silt	14	
		Peat	3	
		Silt	14	
		Depth bored .	31	

Ex. Mr. JAMES OTTO HEISE.

Account of the Borings between Reading and Bath.

I superintended part of the Borings on the line between Reading and Bath. This is a Statement of them:

No. 1. In No. 7. Tilehurst Parish.

Alluvial Soil	1 Ft.	6 In.
Chalk and Flints	38	6
Total .	40	0

No. 2. In No. 4. Purley Parish.

Loose Stones	1 Ft.	0 In.
Chalk	2	6
Total .	3	6

No. 3. In No. 4. Purley and Whitchurch Parish.

Hard Gravel, rather small	10 Ft.	0 In.
Coarse Kind of Sand	3	0
Total .	13	0

Account of the Borings between Reading and Bath.

No. 4. In No. 49. Goring Parish.

	Ft.	In.
Clay and Silt	6	0
Soft Gravel	9	0
Hard Gravel	4	0
Chalk	4	6
Total .	23	6

No. 5. In No. 2. Basildon Parish.

	Ft.	In.
Clay	2	6
Silt	2	6
Soft Gravel	8	0
Hard Gravel	4	0
Chalk	2	6
Total	19	6

No. 6. In No. 114. Cholsey Parish.

	Ft.	In.
Clay	3	0
Silt	13	0
Chalk	4	0
Total .	20	0

No. 7. In No. 86. Wootton Bassett Parish.

	Ft.	In.
Blue Clay	9	0
Gravel	1	0
Blue Clay	16	0
Total .	26	0

No. 8. In Grittenham Wood (Brinkworth Parish).

	Ft.	In.
Yellow Clay	3	0
Dark Blue Clay	61	0
Total .	64	0

No. 9. In No. 14. Box Parish.

	Ft.	In.
Yellow Clay	14	0
Dark Blue Clay	36	0
Total .	50	0

No. 10. At the Tanners Poundpill, Corsham Parish.

	Ft.	In.
Yellow Clay	7	0
Hard Stone in thin Layers	4	0
Blue Clay	0	6
Stone	4	6
Blue Clay	0	2
Stone	3	4
Blue Clay	5	0
Stone	8	0
Total .	32	6

Account of the Borings between Reading and Bath.

No. 11. In No. 50. Corsham Parish.

Yellow Clay	4 Ft.	6 In.
Stone	3	6
Brown Sand, wet	3	0
Stone	1	0
Clay and Sand	0	6
Stone	4	0
Blue Clay	10	0
Stone	2	6
Brown Sand, dry	0	3
Stone	1	0
Clay	0	3
Stone	2	6
Clay	0	10
Stone in thin Beds	11	2
Blue Clay	0	10
Stone in thin Bed	8	2
Blue Clay	1	6
Stone	0	6
Total	56	0

No. 12. In No. 14. Corsham Parish.

Blue Clay	18 Ft.	6 In.
Dark coloured Stone	13	6
Blue Clay	0	3
Dark coloured Stone	17	9
Very hard Stone	10	0
Soft Stone, Bath Stone	18	6
Total	78	6

Remarks upon the Floods at Gate Hampton.

We made the Borings during the flood season, and when at Gate Hampton the Floods filled the hole we were boring; the water did not proceed from the boring but from the river, which rose above the fields to about 6 inches above the banks, and a large portion of the country was flooded; the Railway will be on Embankment at this part and consequently above the floods. I do not think the Embankment would impede the water from passing off; but upon the floods subsiding, it would run off by the bed of the river (there would be land arches at the bridges to allow of it passing off.)

Ex. Mr. RICHARD CREED.

Sect. L. and B. Railway.

The Shares are at £15. premium.

The works are let by Contract.

I am Joint Secretary to the Board of Directors of the London and Birmingham. I have received letters this morning from Liverpool, stating that the Shares were selling at £50. although £35. only has been paid upon them.——We take every precaution to get the Work done cheap, and we have every reason to believe it is being executed as economical as possible.——The works are let by Public Contracts, which are advertized in the London and certain Country Papers; Mr. Stephenson is called in at the time the Tenders are opened, (who is paid a Salary, with an express understanding that he is not to derive any pecuniary advantages from the Contracts).——We accepted the lowest Tenders for the 3 first Contracts, and with the exception of Contract No. 1 B, the average price delivered in for Excavation is 1*s*. per cubic yard; No. 1 B was commenced by Contract, but it is now under the immediate superintendence of Mr. Stephenson, and we have a Clerk of the Works, who is appointed to check every bill connected with the works. We have built an Engine House, and it is erected as economically as possible consistent with solidity, I am not aware that it has any ornaments, being merely a brick and mortar shed of the most simple construction. The Directors have never expressed an intention of making a Branch from our line at Tring to Oxford, nor have any directions been given to Mr. Giles to make such a representation to the Cheltenham people, (as it is not their intention at present to engage in forming Branch lines, although they would be advantageous to a Great Line of Railroad); he paid me a visit, probably within the last three months, upon which occasion I stated that a line from Tring to Oxford might be advantageously formed, as it was perfectly practicable; the expediency of the Line would of course depend upon the increased Traffic it brought our Line; but I think a Branch from the Great Western, down the Valley of the Thames, would be a better line, as upon ours you must first ascend out of the basin of the Thames, (over the Tring summit,) and then descend into it again.

Ave. pr. of Exc. is 1*s*. per cub. yard.
The Com. are exc. Contract No. 1 B.

Observations upon a Branch from Tring to Oxford.

A Branch from the G. W. would be better.

Ex. Mr. HENRY ROWLES.

I have practised as an Architect, but I have retired from business many years; I have also been engaged as a Contractor. I have had great experience in the construction of Public Works, having been employed in the execution of them under government; I have been engaged in every way, and have had 40 years experience.——The Surveys and Estimates of the London and Birmingham Railway were made by Mr. Robert Stephenson, the Superintendant Engineer, who is paid a Salary, and an allowance for travelling expences, (he has no perquisites of any kind to my knowledge,) under the instruction of the Directors.——About 40 Shares were forfeited of people dead and gone, which were sold by auction and realized £47. although £35. only is paid upon them.——I advised the Company to employ one sole Contractor as far as laid in their power, as it was the most economical plan, and at my suggestion the Board of Directors (of which I am one) determined to let the works by Contract.——The Engineer prepares the Specifications and furnishes an Estimate before the Tenders are delivered.——There is a Committee for each half of the line, and of the 57 miles of the line under the management of the London Committee, 37 have been let, amounting in the whole to £584,438., the aggregate of our Engineer's Estimate was £573,467., the works have therefore exceeded his estimate about £11,000.; his Parliamentary Estimate for the above distance amounted to £522,519.; the increase was occasioned by additional works for the better security of the Railway. We always require Security of the Contractors, (there have been only two exceptions,) we also retain a portion of their payments as they become due, thus we

His Experience.

The Surveys of the L. & B. got up by Mr. R. Stephenson.

Is Director of the L. & B.
He advised the Board to let the works by Contract.

37 Miles let of the London end for £584,438.
Mr. S.'s Est. for it was £537,469.

Security required of Contractors.

keep back 20 per cent. upon our Engineer's first certificate until half the work is executed, which is equal to 10 per cent. upon the whole, until the completion of the contract.*——Mr. Stephenson's Estimate for Contract No. 3 C. was £45,224, which

Acct. of the failure of Con. No. 3 C.

a Contractor took at £39,720. but was unable to proceed with it, we were consequently obliged to take it off his hands, and relet it at an excess of £9,000. above the original tender. We received 8 Tenders for the Contract No. 1 B., the first out of London, (we received 5 Tenders for No. 2 B.) and it was let to Jackson and Sheddon, the highest was £139,285. and the lowest £119,987., our Engineer's Estimate amounted to £120,668.; the amount of the contract appears enormous, particularly as the circumstance of being near London does not make any difference, but it is difficult work. We

Acct. of the failure of Con. No. 1 B.

also took this contract into our own hands, and are finishing it ourselves, which I think we shall find more expensive; we employ a Superintendant (Mr. Birkinshaw) to look after the men and see that the work is executed properly, to whom we pay a salary,

Acct. of the compl. of the above Cont. by the Company.

independent of the Chief, the Assistant, and the Sub-Assistant Engineers, we also find all the materials; the Contract comprehends the line between Camden Town Station and the River Brent, a distance of 6 miles, upon the whole of which we are at work; the

Primrose Hill Tun. in London Clay co. £50. per yard.

Primrose Hill Tunnel is 1700 yards long, which cannot be taken at less than £50. a yard, as it is through London Clay; there are 1,100 yards of it to excavate, which is a great quantity; there is an immense quantity of cutting at the terminus over the Brent, quite as heavy as at any part of the line; the latter cutting is through the ordinary soil of the country, and very expensive to remove, the lead is 2 miles

20 A. at the Depôt has to be raised 11 ft.

long; the clay from the tunnel and cuttings is used at the Depot, about 20 acres of which has to be raised 11 feet. I am quite satisfied the work is being executed as economically as possible consistent with efficiency, as the Directors are desirous that they should be properly constructed, or there would be enormous expences for repairs, the liability of accident upon the line would also be increased.——We had several reasons for taking this Contract into our own hands, amongst others the

Acct. of the failure of the Highgate Tunnel.

fact of Mr. Rennie having tried to make a Tunnel through London Clay at Highgate some years back, and failed; I acknowledge that I did not think the preparations at Primrose Hill were sufficient, considering the extent of the work; the tunnel was

Primrose Tunn. was int. to be 2 Bks. tk. but was exe. 3 thick and in Rom. Cem.

intended to be 18 inches thick, but upon executing a few yards the internal pressure was inclined to flush the brick-work, by swelling and compressing, a quality which the London Clay possesses upon the first introduction of air: we therefore directed the tunnel to be made 2 feet 6 inches thick, and to be executed in stronger bricks, and in

Reason for the same, and Remarks on the London Clay.

Roman Cement, which made an addition of £10,000. The difficulty of working in London Clay is well known; we cannot travel a road without perceiving the slips which have arisen from working in it, and we took this precaution both for the sake of the Public as well as ourselves, particularly as the failure of the Contractors gave us a good

The Difficulty of re-lett. the work.

opportunity: the re-letting it to another Contractor would have also occasioned a great loss of time, as the work must have stood still for a time, whereas we proceeded with it at once. The employing of Superintendants, in preference to Contractors is more

* This does not agree with Mr. Robert Stephenson's statement of the Drawback, as will be seen by reference to his Evidence.—*Editor.*

expensive, as they have not the intelligence, ingenuity, or personal interest of a Contractor.——I was present at a meeting at Birmingham of the Proprietors of the London and Birmingham Railway, in February last, (there was also a meeting the previous February) at which the propriety of a Junction with the Great Western was discussed, and a paper was sent round, stating that some of the Proprietors were hostile to it, and requesting us not to give our assent; there was also a special Deputation from Manchester, which advised us to remain neuter, which we did, but I do not consider the Deputation expressed the opinion of Lancashire in a collective body. (I cannot say whether any of the parties were holders of shares in the Southampton Railway.) There were also meetings at London and Manchester upon the subject, which many of the largest Proprietors did not attend, therefore they did not express the feelings of the Proprietors generally. I am not aware whether these meetings were called by circulars, but I have every reason to think they were called privately, as there is no authoritative body that could send them but ourselves. I divided upon the question that the Company should be neuter, but was in a minority, as were also the London Committee. If our Railway and Depôt were contracted, or any difficulty was apprehended in regulating the 2 railways, in that case the junction would be injudicious, but as our Depôt is very extensive, (38 or 40 acres) this could not possibly arise; they would be off our 4 miles very soon. I am also quite satisfied it would be the most economical plan for the Great Western; their saving would be enormous, notwithstanding their having to pay for 6 miles, (we have a regulation that all parties coming upon the line shall all pay for at least 6 miles) although they would only run 4 upon it.——We have obtained a Bill to Extend our line to Euston Square, where we shall form a Depôt for Passengers; the heavy goods will stop at Camden Town Depôt, which is upon the the Regent's Canal; the conveyance of them by the canal to the river will not be as economical as having a Terminus at the Thames, but I do not think any mode of reaching the Thames can be as economically formed as our depôt. The works at the termination of a line are by far the most expensive; thus, "the Extension," which is only 1 mile in length, will not cost less than £130,000., and if we went to Vauxhall Bridge, through the neighbourhood of Cadogan Square, Chelsea, we should find it equally expensive, as it is all building ground, and even more crowded with buildings. I am not aware that a Depôt upon the South side of the Thames would be more economical, but it would depend upon whether it was upon building ground.——We adopted the Terminus at Euston Square on account of the situation being convenient for the distribution of passengers over the City, the distance to London Bridge being about 2¾ miles, to Covent Garden 1¼ miles, to Charing Cross within 1½ miles, to the top of St. James's Street about 1¾ miles; to the Houses of Parliament 2 miles, and to the Bank 2⅜ miles. I do not know any spot equally convenient without incurring very great expense; it is more convenient than a terminus at Vauxhall Bridge, as it is more central, and it is 2 miles nearer to Newgate and Leadenhall Markets; its position in reference to the Regent's Canal is also advantageous, as it communicates by the river with the great Docks of London; we should put the goods on barges at the Depôt, pass them through several locks, and float them down to the ships, whereby there would be no further carriage. Passengers form the great Returns upon the Liverpool and Manchester Railway; we also depend upon them

Observations upon the Junction of the L. and B. with the G. W.

The L. and B. Depôt covers 38 or 40 Ac.

A Bill is passed for the Exten. to Euston Square.

Comparison of the Depôts.

Advantages of the Euston Sq. Depôt.

Passengers form the great Returns upon the L. and M.

He con. the Canal is better suited for Heavy Goods.

for our returns, as the conveyance of goods to be exported is not very profitable. I should wish the Canal to take all heavy articles, as they would tear our Railway to pieces, and occasion great expense: some Goods would certainly travel by us; muskets would in the event of a war, if they were much required, but the Canals being open in Summer, when great expedition was unnecessary, they would travel by the latter.

Ex. Mr. W. W. SUTHERLAND.

I am a Clerk in the Stamp Office. This is an account of the number of Coaches between London and Bristol, and other places in the west of England, taken from the Stamp Office returns, which have been on the increase during the last few years.

Account of the number of Coaches between London and Bristol.

An Account of the Number of Coaches and the Number of Journeys licensed by the Commissioners of Stamps and Taxes between the under-mentioned Places:—

No. of Coaches.	From what Place.	To what Place.	No. of Journeys per Week.	No. of Coaches.	From what Place.	To what Place.	No. of Journeys per Week.
6	London	Bath	40	1	London	Great Marlow	12
20	—	Bristol	136	1	—	Maidenhead	12
4	—	Cheltenham	26	Nil.	—	Monmouth	Nil.
3	—	Devonport	14	4	—	Newbury	24
17	—	Exeter	80	10	—	Oxford	64
1	—	Farringdon	6	11	—	Reading	80
6	—	Gloucester	38	4	—	Stroudwater	26
1	—	Harlington	16	3	—	Taunton	12
2	—	Henley on Thames	24	7	—	Uxbridge	92
4	—	Hereford	24	2	—	Wantage	12
1	—	High Wycombe	12	1	—	Wallingford	12
1	—	Marlborough	6	11	—	Windsor	144

Ex. Mr. WILLIAM SHEARMAN.

I am Clerk in the Stamp Office, in the Coach Department, and I have prepared the following account:—

Account of the No. of Coaches.

An Account of the Number of Coaches and the Number of Journeys licensed by the Commissioners of Stamps and Taxes between the undermentioned Places:—

No. of Coaches	From what Place.	To what Place.	No. of Journeys per Week.
2	Windsor	Reading	24
2	Bath	Reading	12
4	Bath	Oxford	24
1	Bath	Farringdon	6
2	Bath	Chippenham	18
1	Reading	Oxford	12
13	Bath	Bristol	188
1	Bristol	Trowbridge	12
4	Bristol	Portsmouth	26
1	Bristol	Brighton	6
4	Bath	Clifton	48

Ex. Mr. JOHN SCHOLES.

I am the Agent of Mr. Chas. Ward, who is a Common Stage Van Owner; the rate at which the Vans travel is about $2\frac{1}{2}$ miles an hour. I am also Agent to Mrs. Basing, who is an Owner of Waggons that travel Westward.——I calculate that each Waggon Horse draws 15 cwt. of Goods, each journey, exclusive of the dead weight (the waggon, corn, &c.) which they take with them.

Vans travel about $2\frac{1}{2}$ miles an hour.

A Waggon Horse draws 15 cwt. of Goods each journey

Ex. Mr. THOMAS JONES HOWELL.

I am one of the Inspectors of Factories appointed by Government. This is a Table shewing the number of Factories in the Counties of Gloucester, Wilts, and Somerset, and the number of Persons employed therein:—

Official Statement of Number of Factories, and of Persons employed therein.

	Factories.	No. of Persons.
County of Gloucester	120	8,057
County of Wilts	67	3,721
viz. Bradford, Calne, Chippenham, Melksham, Malmesbury, Kingswood, Christian Malford, Trowbridge, Devizes, Westbury, Wilton, Mere, Heytesbury, Warminster, and Salisbury.		
County of Somerset	65	4,285

Account of the Factory Population.

Ex. Mr. JOHN PUGH.

I have examined the list deposited in the Parliament Office of Owners and Occupiers on the Line, and have made an Analysis of the Assents, Dissents, and Neuters; I have distinguished the Owners only.

Analysis of Assents, &c. as deposited in the Lords.

COUNTY.	Assents.	Neuter.	Special Answers.	Dissents.
Gloucestershire, including Bristol .	35	6	—	—
Somersetshire, including Bath . .	145	13	7	20
Wiltshire, including Branches . .	131	31	10	29
Berkshire, including Parts of Oxon and Wilts intermixed	143	86	33	69
Oxfordshire	30	7	2	4
Buckinghamshire	15	7	4	31
Middlesex	32	32	10	11
	531	182	66	164

Analysis of the Assents & Dissents.

N.B.—Those Special Answers which are tantamount to a Dissent are put in the latter Column.

Analysis of the Assents & Dissents.

SUMMARY.

Assents	531
Neuters	182
Special Answers	66
Owners not dissenting	779
Dissents	164
	615

" Declines answering," is put among the special; there are only 3 special answers that are tantamount to a Dissent.

Ex. MR. R. J. VENABLES.

Analysis of the Shares, in reference to the Places from whence they proceeded.

I have made an analysis of the Deeds that have been executed and proved of the £100. Shares of the Great Western, with regard to the locality from which they proceeded.——I have also prepared a similar analysis of the £100. Shares of the Basing and Bath line, taken from the list deposited in the Private Bill Office, which are as follows:—

Analysis of Shares, Great Western Railway.

Bristol	6,522
Bath	1,186
Gloucestershire, Devon, Somerset, and Cornwall	881
South Wales	1,421
Worcestershire and Herefordshire	111
Ireland	378
Wilts and Dorset	388
Berks and Oxon	797
London	6,261
Various other Places	691
Making a Total of	20,087

Analysis of Shares, Bath and Basing Line.

Somersetshire—consisting of		
Bath	65	
Frome	5	
High Littleton	2	
Making		72
Wiltshire—consisting of		
Bradford	5	
Trowbridge	80	
Staverton	30	
Other Places	65	
Making		180
Berkshire—viz.		
Newbury and Speenhamland	70	
Making		322
Then Manchester	960	
London	660	
York	10	
Making		1,640
Making a Total of		96,200

The Cap. of G. W. is £2,087,000. Do. the Basing £196,200.

The amount of Capital of the Great Western is £2,500,000., of which £2,087,000. is subscribed: the amount of Subscription of the Bath and Basing line is £196,200. ——I have also prepared a Statement of the anticipated Traffic upon the Great Western line, having had access to the returns delivered in by the Witnesses from the Stamp Office:—

STATEMENT of PASSENGERS calculated for RAILWAY TRAFFIC between LONDON and BRISTOL.

Coaches (from Stamp Office Returns)—Average, Mails, Five Passengers; Four-horse Coaches, Nine; Two-horse Coaches, Six Passengers.

Places.	Weekly Journeys.	No. of Passengers.	Places at which Passengers will join or leave the Railway.	Total No. of Passengers to or from each Place.	No. of Miles by Railway.	Aggregate No. of Miles travelled Weekly by Railway.
London & Bristol	136	1,168	London or Bristol	1,168	116	135,488
— Bath	40	304				
— Devonport	14	70	— or Bath	1,146	104	119,184
— Exeter	80	664				
— Taunton	12	108				
— Cheltenham	26	234				
— Gloucester	38	286				
— Hereford	24	188	— or Swindon	940	74	69,560
— Stroudwater	26	178				
— Marlborough	6	54				
— Farringdon	6	54	— or Wantage	162	58	9,396
— Wantage	12	108				
— Oxford	64	576	— or Steventon	576	54	31,104
— Wallingford	12	108	— or Moulsford	108	46	4,968
— Newbury	24	216	— or Reading	915	34	31,110
— Reading	80	699				
— Maidenhead	12	72				
— Great Marlow	12	108	— or Maidenhead	504	22	11,088
— High Wycombe	12	108				
— Henley-on-Thames	24	216				
— Windsor	144	1,296	— or Slough	1,296	16	20,736
— Uxbridge	92	828	— or West Drayton	972	11	10,692
— Harlington	16	144				
Bath & Reading	12	72	Bath or Reading	72	70	5,040
— Oxford	24	144	— or Steventon	144	50	7,200
— Farringdon	6	36	— or Shrivenham	36	31	1,116
— Chippenham	18	108	— or Chippenham	108	13	1,404
Reading & Oxford	12	72	Reading or Steventon	72	20	1,440
— Windsor	24	144	— or Slough	144	18	2,592
Bath & Bristol	188	1,128				
— Clifton	48	288				
Clifton & Trowbridge	12	108	Bristol or Bath	1,756	12	21,072
— Portsmouth	26	178				
— Brighton	6	54				483,190

Account of Passengers, between London and Bristol.

It has been admitted that the Speed, Cheapness, and Security of a Railway will more than double the ordinary Traffic of the Road. On the Stockton and Darlington Railway the Proportion of Increase has proved Twenty to One; on the Liverpool and Manchester Railway more than three to One. Assuming it in this instance, as was allowed in the London and Birmingham Railway, at Two to One, the foregoing Statement exhibits an annual aggregate Number of Miles 50,251,760, which at 2*d.* per Head per Mile, will yield 418,764*l.* 13*s.* 4*d.*

The Trav. has incr. S. and D. 20 to 1, L. and M. 3 to 1, He assumes 2 to 1.

I have assumed the same number of Passengers by coach as was proved on the London and Birmingham, and Southampton Railways, viz. 5 for the Mail, 9 for a four-horse Coach, and 6 for a two-horse Coach; I have taken the distances from the Ordinance Map, and compared it with the Section.——I have considered that persons going to

5 by the Mail, 9 by 4-horse Coach, 6 by 2-horse Coach.

Explanation of the Table.

Newbury would go by the Railway to Reading, (34 miles,) and proceed by coach to Newbury, (17 miles); and I have considered persons going to Cheltenham would go by Railway to Swindon.——The first column is extracted from the papers delivered in from the Stamp Office, and in the third column are the names of the places at which persons would join or leave the Railway, and then the number of miles multiplied by the number of passengers will give you the aggregate number of miles travelled weekly by the Railway: I have taken the same rate per mile as on the London and Birmingham, viz. 2*d.* per head; I have only taken the traffic by Stage Coaches, excluding Posting, also caravans, waggons, and carts with manure, lime, coals, and corn; I have calculated upon an increase of Passengers in the ratio of 2 to 1, but a greater proportion has been taken in other cases.——I have also made another Statement, shewing the probable amount of Merchandize and Cattle between Bristol and London:

Allows 2*d.* per mile per Passenger.

MERCHANDIZE and CATTLE between BRISTOL and LONDON.

Account of Merchandize and Cattle between Bristol and London.

	£.	s.	d.
Vans and Waggons observed on the Road between *Bath* and *Bristol*, drawn by 962 Horses, Average proved at 15 Cwt. each Horse, will draw 14,430 Cwt. per Week=37,518 Tons per Annum for 12 Miles at 4*d.* per Ton per Mile .	7,503	12	0
Ditto between *Reading* and *Bath*, drawn by 377 Horses, at 15 Cwt. each=5,655 per Week=14,703 Tons per Annum for 70 Miles, at 4*d.* per Ton per Mile	17,153	10	0
Ditto between *London*, *Gloucester*, and *Cheltenham*, drawn by 192 Horses, at 15 Cwt. each=2,880 Cwt. per Week=7,488 Tons per Annum, 74 Miles, at 4*d.* per Ton per Mile	9,235	4	0
Ditto between *London* and *Falmouth*, 2,496 Tons, by Russell and Co.'s Waggons, and to *Taunton*, 2,184 Tons, by Alexander and Co.'s Waggons; together 4,680 Tons per Annum for 104 Miles, at 4*d.* per Ton per Mile . .	8,112	0	0
Ditto between *Reading* and *London*, drawn by 968 Horses, at 15 Cwt. each=726 Tons per Week=37,752 Tons per Annum for 34 Miles, at 4*d.* per Ton per Mile	21,392	16	0
Ditto between *Maidenhead* and *London*, (in addition to the above) 147 Horses =2,205 Tons per Week=5,733 Tons per Annum for 22 Miles, at 4*d.* per Ton per Mile	2,102	2	0
The Annual River Traffic between *Bath* and *Bristol* has been proved to consist of 79,740 Tons of Goods; the Annual River Traffic between *Reading* and *London* has been proved to consist of about 109,230 Tons of Goods; it may therefore be assumed that more than the lesser quantity passes the whole Line. Of the former quantity, 41,759 Tons consist of Goods adapted to transport by Railway. The Rate for such Goods, at 2*d.* per Ton per Mile, is less than ½ the present Charge. Assuming, from the general Evidence, that the Cheapness, Speed, and Certainty will secure that Traffic to the Railway, as a very low Estimate, the Revenue at 2*d.* per Ton per Mile on 41,759 Tons for 116 Miles is	40,367	0	8
One hundred and twenty-two Beasts from *Bristol* to *Bath*=6,344 per Annum for 12 Miles, at 3*d.* per Head per Mile	951	12	0
Two hundred and eighty-four Sheep=14,768 per Annum for 12 Miles, at ½*d.* per Head per Mile	369	4	0
Eight hundred and seventy-nine Pigs=45,708 per Annum for 12 Miles, at ½*d.* per Head per Mile	1,142	14	0
Carried forward	£108,329	14	8

Account of Merchandize and Cattle between London and Bristol.

	£	s.	d.
Brought forward	£108,329	14	8
One hundred and ninety-two Beasts between *Reading* and *London*=9,984 per Annum for 34 Miles, at 3*d.* per Head per Mile . . .	4,243	4	0
Five hundred and seven Sheep=26,364 per Annum for 34 Miles, at ½*d.* per Head per Mile	1,867	9	0
Six hundred and twenty-two Pigs=32,344 per Annum for 34 Miles, at ½*d.* per Head per Mile	2,291	0	8
Probable Increase of Traffic from South Wales and South of Ireland to London, now conveyed by Sea, Butter and Bacon, (being perishable Articles,) from Cork, Waterford, and Limerick; 39,605 Tons for 116 Miles, at 3*d.* per Ton per Mile	57,427	5	0
Copper, Tin Plates, and Iron, upon a very low Estimate, 20,000 Tons per Annum, at 2*d.* per Ton per Mile for 116 Miles	19,333	6	8
20,000 Beasts from Devon and Somerset from Bath to Reading, 70 Miles, at 3*d.* per Head per Mile	17,500	0	0
104,000 Sheep from Wilts and Devon, 40 Miles, at ½*d.* per Head per Mile .	8,666	13	4
40,000 Pigs from Bath to Reading, 70 Miles, at ½*d.* per Head per Mile .	5,833	6	8
	£225,492	0	0

PARCELS and PACKAGES by COACHES.

Parcels and Packages by Coaches.

				Weekly Amount £.	s.	d.	£.	s.	d.
London and Bristol .	136	Weekly Journies	@ 40*s.*	272	0	0			
— Bath . / — Devonport . / — Exeter . / — Taunton .	146	. .	@ 30*s.*	219	0	0			
— Cheltenham / — Gloucester . / — Stroud .	90	. . .	@ 25*s.*	112	10	0			
— Newbury . / — Oxford . / — Wantage . / — Wallingford / — Farringdon / — Reading	198	. .	@ 15*s.*	148	10	0			
Bath and Bristol . . / Bath and Clifton . .	236	. .	@ 2*s.*	23	12	0			
				775	12	0			
			Annual Amount £				40,331	4	0

Merchandize and Cattle . . .	£225,492	0	4
Packages and Parcels	40,331	4	0
	£265,823	4	4

GENERAL SUMMARY.

	£.	s.	d.
Statement of Passengers, A	418,764	13	4
Statement of Merchandize, Cattle, and Parcels, B.	265,823	4	4
	684,587	17	8
Statement of the Annual Charge upon the above Traffic, calculated on the Annual Expences of the Liverpool and Manchester Railway for Three Years. Average Expences, £87,305. per Annum, (after deducting Interest for Money borrowed for 30 Miles,) same Proportion for 116 Miles .	337,579	6	8
Net Revenue .	£347,008	11	0

The Returns give £13. 17s. per Cent. Profit.

Or £13. 17s. 0d. per Cent. upon a Capital of £2,500,000.

15 Cwt. taken by each Waggon Horse.

He takes the Carr. at 4d. per Ton a Mi.

Charg. by the Canal is 2d. per Ton a Mile.

The Traffic by Public Vans and Waggons only have been taken on the different lines of roads, exclusive of all carts and waggons with manure, corn, lime, and coals. I have assumed 15 cwt. as the average weight of goods carried by each waggon horse, as proved by Mr. Scholes, and I have taken the rate of carriage at 4d. per ton, but Mr. Wilkins proved it to be 10d. or 11d., and the same description of traffic was estimated at 6d. per ton per mile in the Southampton. We got the number of Waggons and Cattle from observations on the line, I allowed a fortnight day and night watching. The Traffic proceeding from the Westward is founded upon the returns of the Traffic through the Locks, these are calculated at 2d. a ton per mile, (which is less than the present charge). I calculated the annual charge upon the Railway at a similar rate to the Liverpool and Manchester Railway, adding the increased extent of this line; If the Basing line was made and executed, and the Great Western was not, all the traffic would go by the former Railway, perhaps the bulk of the traffic to Devonport and Exeter goes through Basingstoke, and not through Reading; the first town we include is 15 miles from London.——I also delivered the answer to the following Proposition of the Bath and Basing Company:

"To the Chairman and Committee of Management of the Great Western Railway Company.

"Gentlemen,

Proposition of the Bath and Basing Company to the Great Western.

"The Committee of Management of the Basing and Bath Railway Company have instructed me to submit to the consideration of the Committee of Management of the Great Western Railway Company the following Proposition. Ths Committee of Management, the Solicitors, the Engineers, and all the other Officers of the Basing and Bath Railway Company are willing to give up the further prosecution of that Undertaking to the Great Western Railway Company, provided the latter will adopt the Basing and Bath Line in lieu of the proposed Great Western Line. The Line between London and Bristol, by Basing, is shorter than the Great Western Line by nearly Three Miles; it is, in a great measure, free from the objection of Tunnels; it is capable of being completed at little more than one half the cost; it combines the advantage of a Communication with the River Thames at Vauxhall, and of a connexion with the Docks at Bristol; it unites the British with the Bristol Channel, and both with the Thames; and by the Tables of Traffic it appears to be likely to yield a larger amount of probable Profit to the Shareholders than the Great Western.

I have the Honour to remain, Gentlemen, your obedient Servant,

W. ROBERT BIDDULPH, Chairman."

"Office of the Basing and Bath Railway Company,
No. 9, John Street, Adelphi, 6th June, 1835."

THE ANSWER.

Answer of the Great Western to the Bath and Basing Proposition.

" Sir,

" The Proposition from the Committee of Management, Solicitors, Engineers, and other Officers of the Basing and Bath Railway, conveyed to me in your Letter of the 6th instant, and at the same moment extensively circulated in Lithograph among the Peers and to the Public, has been submitted to the consideration of the Directors of the Great Western Railway. Although they are led, as well from the nature of the Proposition and the Time chosen to make it, as from the unusual course of giving publicity to it, almost before it was received by me, to presume that the real object of it has been already attained upon the Second Reading of the Bill, I am requested to inform you definitively, that the Directors decline to avail themselves of the Offer, for the following Reasons:— From the first Formation of this Company their leading purpose has been to select such a Line of Communication between Bristol and the Metropolis, with reference to Levels and other important Considerations, as might appear under all circumstances the best, and which might at the same time interfere as little as possible with the convenience and wishes of the Landed Proprietors. After many anxious Inquiries, and repeated Surveys, the Directors were satisfied that the Line through Reading, and passing to the North of Marlborough Downs, embraced these objects in an extraordinary degree, affording, at the same time, a facility of communication with the Manufacturing Districts both of Gloucestershire and Wiltshire, the Port of Gloucester, South Wales, Devonshire, Somersetshire, and Cornwall. The Great Western Line was accordingly adopted, as embracing the greatest Public Interests. In the meantime the Directors of the Southampton Railway, in order to benefit exclusively their own Undertaking, used their utmost Influence and employed every Means to defeat the Great Western Railway, or to induce the promoters of it to join their Line at Basing. These Efforts were met by a distinct Proposal from this Company to refer the matter to impartial and competent Engineers to determine which Line would be best calculated to serve the Interests of the Public and the Subscribers of both Undertakings, which, however, was declined by the Southampton Railway Company. Nothing has since transpired to shake, in the slighest degree, the firm conviction entertained by the Directors of this Company, as to the Superiority in almost every respect of the Line they had selected; and they have succeeded in raising the necessary Subscriptions for its completion. These were obtained to a great extent at Bristol, in South Wales, Gloucester, Stroud, Cheltenham, and other places, from persons who advanced their money on the faith of the adoption of the Great Western Line; not, as has been said, from great Capitalists only, but from the most numerous Proprietary ever established in the present stage of the Undertaking. It is almost superfluous to add, that, under these circumstances, the Directors of the Great Western Railway cannot so far abandon the duty which they owe to those who have placed so important a Trust in their Hands as to adopt a Line which has been reported by the most eminent Engineers, and which they conscienciously believe, to be decidedly objectionable. It must not be forgotten also, that the Basing Line is objected to by many of the most influential Landed Proprietors on the Line, is scarcely supported by any local Subscriptions, does not extend beyond Bath, and is moreover dependant for its communication with London on the Southampton Railway, which is subject to Estimates almost universally considered fallacious, and to unprecedented difficulties of Construction, as exemplified in the immense Cutting of 116 feet deep at St. George's Hill. As the Merits of the Great Western Railway are now to be proved by Evidence, on Oath, before a Committee of the House of Lords, I have merely to observe, with reference to the Assertions in the concluding paragraph of your Letter, First, that the comparative Distance of the Two Lines cannot be ascertained, unless some common Starting Point in London be assumed; but that if the difference were even as great as alleged, (which is by no means admitted,) the Superiority of the Levels on the Great Western Line would more than counterbalance any such supposed advantage. Secondly, that the Impossibility of making Shafts to give Air or Light to the Tunnel on the Basing Line near Bath, exceeding One Mile in length, has been admitted by Mr. Giles the Engineer, which renders it far more objectionable than the proposed Tunnel near Box, in the Great Western Line. Thirdly, the Cost of the Basing Line must exceed that of the Great Western Railway if the Prices are estimated on a uniform Scale, in consequence of the immense quantity of Excavations and Embankments on the former. The Engineer of the Southampton Railway calculates the several Prices, and particularly that for the Earth-work, at less than one half of the Estimates of Mr. Brunel for the Great Western Railway; the latter having adopted the Rates at which the London and Birmingham Company are actually executing the works, which circumstance alone will account for the apparent difference of Capital. Fourthly, the Basing Line cannot be said to unite either the Bristol Channel or Docks with London, as it terminates near Bath, at a distance of Twelve Miles from Bristol. The Great Western Railway crosses the Floating Harbour at Bristol, and can be extended to any part of the Docks with the same facility as any other Line approaching that City. Fifthly, the Tables of Revenue by the Bath and Basing Railway, instead of a greater Profit, must show, if fairly made, an actual Loss to the Proprietors; the whole Traffic from Gloucestershire, South Wales, Oxford, Reading, and Maidenhead is excluded from access to the Line; and the only Profit to be derived from the Extension of the Southampton Railway will be acquired by that Company, who are notoriously the real promoters of the Scheme, and almost the only Opponents of the Great Western Railway.

"I have the Honour to be, Sir, your obedient Servant,

BENJAMIN SHAW, Chairman."

"Great Western Railway Office,
12th June, 1835."

Ex. Mr. ROBERT PODMORE CLARK, *Shipbroker, of Bristol.*

Import. W. Indies, N. America, Brit. America, Portugal, The Baltic, E. Indies, Africa.

Bristol is the natural port of Bradford: it imports from the West Indies, colonial produce, sugars, and rum; North America and British America, wood, timber, tobacco, and rice; Portugal, wines and other produce; from Russia and the Baltic, tallow, hemp, and deals; and in the article of tea with the East Indies; with Africa in iron, tin plates, refined sugar, glass ware, hard ware, stone ware, &c.——I attribute the decline of

Cause of the decline of Trade in Bristol.

Bristol to the rise of other ports, Gloucester for instance, not to any fault in its geographical position.——As cargo Shippers we complain of the want of a commodity, that will bear a higher rate of freight than those goods can bear which are to be had in Bristol, light goods being scarce, (we call refined sugar light goods.)——A communi-

The Railway would remedy it.

cation with London would have a tendency to bring them, and the Railway would improve the trade at Bristol. Supposing cargoes were more readily disposed of,

Freights from Bris. to America & other places.

Captains would be more likely to come for export again.——The Freight of iron from Bristol to America at this time, is to New York £1. per ton, to Quebec and equal distances 10*s.* per ton, Mediterranean £1. per ton; the Freight of iron at Liverpool is

Freights at Liverp. and the reason of their being less than at Bristol.

occasionally as low as 4*s.* or 5*s.* per ton, and 7*s.* or 8*s.* to America; always 50 per cent. less than from Bristol, because they bring merchandise which will bear a higher rate of freight, such as bales of measurement goods, woollens, which they pay by measurement.

The Carrying Trade

——The carrying trade is an unimportant part of the Bristol trade.——I cannot

Bristol Depôt.

say I should prefer a railroad terminating at the Temple Meads in preference to Redcliffe Wharf, as the latter is nearer to the shipping, but it would not make much difference,

The Dock.

as it must be done by lighters or barges in either case. The whole river is a dock, the

Mode of Shipment, &c.

ships laying perhaps a mile and a half from the terminus: the goods must be taken by barges to the ships; the heavier goods being at present conveyed by same, and the lighter goods are brought to and from the manufacturer's houses in carts or sledges, to avoid noise.——I have seen ships lay alongside of Redcliffe Wharf, but not any drawing 14 feet: shipments might be made immediately off the wharf into them.—— I do not think the navigation up and down the Bristol Channel, and up the river to Bristol, is more difficult than Liverpool, (but I am not a nautical man;) there is the expense of pilotage between Pill, and afterwards down the Severn out to sea. I should say that the expense of navigating ships at Bristol is not worth more than at Liverpool, and by referring to documents, it will be seen more ships are lost going to Liverpool than coming to Bristol: it is a rare thing for a ship to be lost in coming to Bristol.—— At Liverpool, the light goods being assorted cargoes account for the lesser expense of freightage. Goods going from Frome, Bradford, and Trowbridge, are frequently sent to Liverpool for exportation, because we have no opportunities of communication to the ports to which those goods would have to be sent, which embraces almost the whole world.——An impediment to Gloucester with the United States is that the Severn is dangerous; large ships cannot go without danger: if we had the capabilities of distributing what we get in Bristol, the trade of Gloucester would find its limit. ——Goods sent from Bristol to London by sea are such as painter's colors, occasionally lime, and rough goods, for the use and consumption of London: we often ship 60 tons of lamp black for Day and Martin.——The majority of produce from Ireland is consumed in this country: the salted meats, such as beef and pork, which are inconsiderable, are intended for exportation. I should think the principal trade from Bristol to London would be cattle and Irish provisions: Barilla is often brought from London to Bristol, and sometimes sent back again, according to the state of the market: sugars are often sent, and sometimes brought back; sometimes they are sent for the purpose of refining.

Redcliffe Wharf.

Navigation to the Port of Bristol safer than that of Liverp.

More Ships lost going to Liverpool.

Reason of the lesser Freight at Liverpl.

Advant. of Liverp.

Trade of Gloucester

Navigation of the Severn dangerous.

Trade to London.

Trade and Produce of Ireland.

Cattle and Irish Pro. is the princip. Trade to London.

State of the latter Market.

Ex. Lieutenant NICHOLAS CHAPMAN, R.N.

I consider the Port of Bristol more easy of approach than either Southampton or Liverpool, and the access is good, and to ships homeward bound from the Atlantic the Port of Bristol is more ready of access than that of London. Two ships, striking soundings in equal situations, one bound to the Northern Channel, and the other to the Southern, taking equal departure, the ship going to Bristol would lie in the dock at Bristol ready to discharge, when the other would be no further than the Isle of Wight, and a ship going to London with a westerly wind, would find it foul upon getting round the North Foreland: a vessel coming from the Isle of Wight to London, under favorable circumstances, makes the voyage in about 3 days; I have known them to be 3 months; as an average I should say 6 days.——In the event of a good communication between Bristol and London, ships with London cargoes might discharge at Bristol, and would be in London 6 days sooner than going by sea in the average of cases; and the advantage would be greater in time of war than in peace, inasmuch, as in the English Channel we have the enemy's coast in view, and that will give security to those at

Port of Brist. better than either Southa. Liverp. or London.

Advantages of the Port of Bristol.

Voyage from the I. of Wight to Lond. averages 6 Days.

Cargoes would be delivered in Lond. 6 Days sooner by discharging them at Bristol.

In times of War the advantage is considerably greater.

enmity with us to shelter their ships of war, and the steam vessels could go clear of ships bound up or down the Channel. Their ships of war could come out by the efforts of the steamers, and secure any convoy there might happen to be. In the Bristol Channel it is different; you enter that channel protected on all sides by your own country, the English coast to the Southward, Wales and Ireland to the Northward and North-west, and you would sail with a prevalent wind blowing right in. In time of war a ship London-bound from the West leaves her escort at Portsmouth; but a Bristol-bound ship leaves her convoy off Scilly.——London-bound ships frequently fall into the hands of Bristol Pilots, but Bristol-bound ships seldom fall into the hands of Channel Pilots, owing to a Northerly current, that sets in when approaching the land with a prevalent wind; these winds prevail eight months in the year, from the Southward to the South-west, to West and North generally, to the Westward, which gives great facilities for approaching the Harbour of Bristol.——The floating Harbour would contain ten times the number of ships that resort to it, and is capable of having wharfs on each side.——A ship drawing 14 feet of water is between 200 and 300 tons burthen, by far the greater part of the vessels using Bristol are about 300 tons burthen.——The Export Trade from Bristol to Ireland is sugar of two kinds, raw and refined, (they are in the habit of making two weekly voyages) oil, tallow, tin plates to a considerable quantity, manufactured copper, plain copper, and brass of all descriptions, manufactured and plain, raw hides in salt, woollen cloths, cheese, and wooden hoops in very large numbers, wheels, spokes, and iron of all descriptions, manufactured colors, seeds for the purpose of seed, and young trees for planting.——Fresh butter cannot be brought to a London market, to compete with Holland, for want of a ready conveyance, and it is the same with eggs.——The Freight by steam, of fresh butter from Cork to Bristol, was 10*d.*, but it is now 6*d.* per firkin; cured butter can always find its way to London.——We make it a rule to receive no goods on board our Packets out of barges (except iron and other handy things;) we prefer having them brought by carts, being more expeditious.——The Railway at Liverpool does not communicate with the water; all goods must be taken out of the waggons, and removed across the road.——The trade between Newport and Bristol is very considerable; Newport supplies large quantities of coals to Bristol.

The facilities of approaching the Port of Bristol.

Capabilities of the Harbour.

Tonnage of Bristol Traders.

Exports to Ireland.

Freights from Irel.

Steam Pack. prefer Goods delivered to them by Cart.

The L. & M. Depôt does not touch the Water.

Newport brings Coals to Bristol.

Ex. Mr. WILLIAM DONNE BUSHELL, *Merchant, of Bristol.*

The Port of Bristol is much impaired for want of a cargo of vessels outwards. ——Its trade is principally with the West Indies.——I trade chiefly with the United States, and import articles used in soft soap, oil and colour manufactories, and turpentine distilleries.——The Harbour is a long floating dock, or a river, stopped with locks. Redcliffe Wharf and the floating Harbour are in fact two rivers.

Trade of Bristol.

Trades with the United States.

Harbour and Redcliffe Wharf.

Ex. Mr. CHARLES LUDLOW WALKER.

I am engaged in a large Brass and Copper concern in Bristol. I was a member of the Provisional Committee of 1832, and I am likewise a Director of the present Company, and hold shares to the amount of 16. We sent letters to almost every firm (taken from the Directory) in South Wales and Ireland, and a variety of other places connected with Bristol and London, on the subject of this railway, and 90 out of 100 were in favor of it: a very great number of the middling class of shopkeepers have subscribed to the undertaking.——I am also an owner of Steam Packets, and export large brass articles to Africa.——In sending goods to London, I can depend on the punctuality of land carriage, but not on canal navigation, as they are liable to interruption from frosts, droughts, and floods: I have known the canals stopped between 3 and 4 weeks together: the floods are principally between Bristol and Bath: the average time from London to Bristol per canal is from 7 to 10 days; it is sometimes from 3 to 4 weeks. I have had goods lying in London 12 months, which I had lost the sale of, owing to their not having been delivered in time.——Goods are sent from London to Bristol, to be exported to Africa, instead of going direct, because the African cargoes consist of assorted goods, and they are obliged to have a particular manufacture of ours to make up their cargoes, as brass rods and neptunes, brass articles, which go to the gold coast of Africa.——There is a great trade between Bristol and Ireland, principally in provisions and cattle.——I do not recollect Mr. Brunton having made a survey previous to Mr. Brunel.——I think a Branch from the Great Western Line to Gloucester would be an advantage, the navigation of the Severn being very difficult and hazardous; I have known a trow (a barge) lost: they generally tide it up in 2 days. I am not certain whether goods shipped for the North (Birmingham for instance) are unshipped at Gloucester or go up to Starport in the same barge, but they change them at Starport.

Is in the Brass Trd.
Is a Dir. of pro. Line
Letters were sent on the subj of this Line
90 out of 100 were in favor of it.
He is also an Owner of Steam Packets.
Irregularity of the Water Carriage, and consequent Losses in Trade.
Reason of Goods being sent from London to Bristol, to be exported for Africa.
Trade with Ireland.
Branch to Glouces.
Navi. of the Severn
Goods for the North

Ex. Mr. JOHN HARLEY.

I am Manager of the Iron and Tin Plate Works of Mr. Lee, at Ponty Pool. —— ——The amount of boxes of tin plate of the manufacture of South Wales is about 78,000 annually. (A box of tin plate weighs from ¾ of a cwt. to 3 cwt.); a small portion only of it goes to London, which is conveyed by sea from Newport, occupying 2 or 3 weeks; they have been known to remain as many months on the voyage: the average voyage is 3 weeks. Freights by sea vary from 12*s.* to 16*s.*:* ——They are likewise sent by Canal, which occupies from 7 to 10 days, sometimes longer. Freights by canal from Bristol to London are 23*s.* per ton, and from Newport to Bristol 2*s.* 6*d.*, which makes it 25*s.* 6*d.*, but it varies according to circumstances. ——Goods sent by sea are always insured, which in time of peace is 3s. per ton, but in time of war or tempestuous weather the charge is very high. An order for tin plates for America is not shipped at London, but Newport. Losses frequently occur to a

Extent of manufac. of Tin Plates in South Wales.
Small portion only goes to London by Sea and Canal.
Uncer. of Carr.
Freights by Sea.
Freights by Canal.
Goods sent by Sea are insured.
Goods for America are ship. at Newport

* This witness afterwards stated the weekly average of Tin Plate to be about 5000 boxes, which does not agree with his former statement—*Editor.*

Irregularity of Carr: considerable amount by sea and canal carriage, owing to their liability to injury from wet; and to their uncertainty, we should be reluctant to undertake the fulfilment of a contract by a given time.——A railway has been projected from the interior of South Wales to Newport, from Merthyr, their conveyance is at present by canal; wheat, barley, and oats would find a ready market with us.——There is also considerable trade in copper plates, the railway would also enable us to compete with Staffordshire, which at present has great advantage over Monmouthshire, in the article of iron.

Opinion upon a Line from Merth: to New:

It would advance the Iron Trade.

Ex. Mr. CHARLES WILKINS, *of Tiverton, near Bath.*

Manufacturer, and employs about 1100 persons.

I have been a Manufacturer in the Clothing business for upwards of thirty years, and employ about 1100 living souls, my present trade is 140 ends per week, or 70 whole pieces of cloth.——There are no manufactories in the West of England larger than mine, but there are some about the same.——The Manufactures depend much on a good supply of Saxony Wool, which is at present imported into the Eastern part of the kingdom, and conveyed principally by waggons, which is rather a dear, but convenient mode of communication.——In a return of the last six months of the past year the carriage amounted to 2,356 cwt. by waggon and barge, of which only 109 was by barge, some of it was by coach.——I prefer the land carriage to the water although more expensive (in consequence of the delays); no manufactured goods are sent to London by water to my knowledge, but by waggon, the carriage to London is 5*s.* per cwt., and from London 4*s.* for wool, and the same for all cloth goods; the rate by coach is 1*d.* per lb.; by barge 2*s.* 9*d.* per cwt. The proposed rate by Railway is rather less than 2*s.* per cwt. According to the return before stated of the last 6 months of last year, the amount of carriage was about £1,150. per year, and at the rate of last month, it would be £1600. per year; the saving in the first case would be about £700. and in the last £900. or £1000.——There is no doubt but the Great Western Line would be the best communication for the clothing trade.——I consider the Gloucestershire clothing districts double those of Wiltshire and Somersetshire, with the exception of Tiverton, and there is much more steam power erected and water power in the former.——The American Market is open to English clothing manufactures. ——There is more bulk of communication required between Bradford and London than between Bradford and Bristol.——The proposed line passes over my premises, and I am a Subscriber to the amount of 54 Shares.——The Basing Line branches from the Southampton at Basingstoke. I attended a meeting at Bath which the Basing gentlemen called, Sir Thomas Fellows took the chair, and the Hon. Mr. Blaquire stated the advantages of the Basing line; but the meeting expressed their approbation of the Great Western.——I believe the Trowbridge manufacture is entirely composed of fancy goods, kerseymeres, and narrow cloths; in Gloucestershire there are generally the fine trade, blacks and blues, except an East Indian manufacture called stripes; at Bradford, fine broad cloths are manufactured: (the Bradford branch of the Great Western starts near Chippenham) at Frome, part kerseymere, part fine cloths, and part coarse cloth.

He requires a good supply of Saxony Wool, which is sent by Land.

The delay attending Water Carriage.

Rates by Waggon, „ Coach, „ Barge, „ Railway.

G. W. best Line for us.

Glouc. Manuf. dble. Wilts and Somerset Tiverton excepted.

American Mar. open

More commu. betw. Lond. and Bradford than betw. Bradford and Bristol.

The G. W. Line appro. at a Pub. Meetg. call. by the Basing Co:

Trowbridge Manuf.

Gloucesters. Manuf.

Bradford Manufac. and Branch.

Frome Manufacture

Ex. Mr. ROBERT CORDWENT, *of Vexford Stockgumber, near Taunton, Somersetshire.*

I am a pretty considerable Farmer in that district, rent and outgoing being about £1400. or £1500. a year.——I am in the habit of sending Cattle to the London market, also seed and corn up to Berkshire, and the country about London.——As the Great Western Railway goes through the principal Towns it would be the means of getting rid of the agricultural produce on its way to London, and we sometimes get a better price for it in those towns than in the former, we therefore endeavour to avail ourselves of the intermediate markets, in order to sell our fat cattle at the best price, and if any thing prevents us we drive on to London.——We send Corn by the Canal, but never Cattle; and we have no other means of sending it to Reading, which is very disadvantageous, as it is of a description that is wanted as soon as we can get it ready after harvest; I have frequently been obliged to make considerable compensation on account of the delay.——We send corn to Weyhill Fair in Hampshire, taking up our teams, and bringing back hops.——The proposed Depot in Bristol is very convenient, because it joins the cattle market, we also keep our stock in the Temple Meadows.

Is a Farmer.

Sends up Cattle to London.

Advantage of the G. W. going through the Principal Towns.

Disadvantage of sending up Corn by Canal.

Convenience of the Bristol Depôt.

Ex. Mr. THOMAS MARLING.

I am engaged in the clothing manufacture, and reside at Stroud, and together with my family employ about 2 or 3000 persons; perhaps 100,000 persons are employed in the Woollen Trade in our district.——The town of Stroud is the centre of the clothing district of Gloucestershire. I should say the manufactures of Gloucestershire are at least twice as much as those of Wiltshire, and the amount of goods much greater.—— It is of great importance that persons engaged in the clothing trade should have ready access to the places where cloths of every description are made.——We get our wool from London, to which we send the greater portion of our goods.——We pay about £1,800. annually for carriage; by means of the Railway we shall save about ⅔ds of it, or about £1,200. per annum.——I am a Subscriber to the amount of 25 Shares in the proposed line.

Manufacturer, emp. 2000 or 3000 perns.

100,000 are emp. in the Trade in his Dis.

Stroud is the centre of the Clo. Trade.

Glou. exceeds Wilt. in Manufactories.

Nec. for the Railw.

This amount of Carriage annually.

Ex. Mr. T. REYNOLDS, *(an Inhabitant of Bristol).*

I was formerly connected with the Iron trade in South Wales as an Iron Master. ——The transport of Iron from South Wales London ways is considerable; also Copper from Swansea.

Great Trade in Iron London ways.

Also Copper.

Ex. Mr. WILLIAM MONTAGUE *(Merchant at Gloucester).*

Direc. of the Glouc. and Berkley Ship Canal.

The feeling towards the Glouces. Branch at Gloucester.

I am a Director of the Gloucester and Berkeley Ship Canal, which carries ships of 600 Tons burthen.——The trade of Gloucester has very much increased of late years. ——Supposing the Great Western Railway should be made, I think there would be a disposition on the part of the Gloucester people to make a Branch to it; which would pass through the middle of the clothing districts of Gloucestershire, namely Stroud and Cirencester.

Ex. Mr. THOMAS BIRCH.

Cheese Factor at Cirencester.

3000 Tons of Cheese sent Ann. from Glo. 2000 from N. Wilts. Land Carr. 4*s.* Cwt.

10,000 Beasts from the above Counties.

He cal. 40 lbs. loss upon a Beast in driving.

12,000 to 15,000 Sheep sent up, which lose from 8 to 12 lbs.

Do. Bacon, &c.

He has known Butter sold as Grease for want of a Conveyance.

Great numbers of Welch Cattle pass through Gloucsster.

Butchers' Meat.

Rem. upon the Basing Line.

I have been a Cheese Factor, &c. at Cirencester for the last 30 years, but I am now retiring from business, and during the whole of the above period no man has had more intercourse with the Agricultural interests of Gloucestershire and North Wilts; about 3,000 Tons of Cheese are sent annually from Gloucester to London, and 2,000 Tons from North Wilts; the amount of my dealings in the above is from 800 to 1000 Tons per year. The Land-carriage is 4*s.* per cwt. which is paid out of the price of the cheese, as we give the dairy man a fixed price, deducting for the carriage; therefore if the cost of carriage was reduced the farmer would derive the benefit.——There is also much Traffic in Cattle and Sheep from the same Counties; I estimate the Beasts at 10,000, which exceeds the amount obtained from the dealers; I calculate the loss upon each Beast drove up to London at about 40 lbs., which at 6*d.* per lb. amounts to £1. but the butchers put it at more; I have sent a few of my own fat cattle up to London, and I think they lost 30*s.*, I estimate it from 20*s.* to 30*s.*; I think from 12,000 to 15,000 Sheep are sent up yearly from the same district, and I consider they lose from 8 to 12 lbs. each, but I hav enot sent any myself.——There are from 400 to 500 Tons of other goods sent up by my brother Tradesmen, such as Bacon, and the carriage is from 4*s.* to 4*s.* 6*d.* per cwt.——Fresh Butter is not sent to London from Gloucester on account of the expence of the carriage, and the want of consumption has reduced it in price very much lately, viz. 5½*d.* and 6*d.* per pound; I have known it sold for grease at 2*d.* or 2½*d.* per lb., both at Gloucester and Cirencester, as they could not dispose of it when it was sweet; as the Railway would be a speedy and certain communication with London, it would be a very great advantage to the Agricultural people; I should have traded in Bacon and Butter also, if there had been a good communication.——There is a great passage of Welsh Cattle through Gloucestershire about 8 or 10 times a year, they come up to be fattened, being in a very poor state; the place of their destination is Essex and Kent, some go to Sussex; I have seen from 1,000 to 1,200 three or four times a year on the Gloucester Road to London.——There is not much Butchers' Meat sent to London, which does not arise from a deficiency in the supply, but the difficulty of conveyance, as they charge more for meat than any other commodity, and an increase in the demand for it would increase the price of Grazing Land in Gloucestershire.—— A Line from London to Bristol passing through North Wilts, with Branches to Gloucester, Cheltenham, and Stroud, would be more desirable to us than a line to Gloucester by way of Bristol, as we should lose a day, and the goods would not be delivered until the second day, which would not suit for Butter and many other commodities. We get from Cirencester to Gloucester, a distance of 20 miles, by land carriage; a Public Meeting was convened at Cirencester, which was in favour of the Great Western Railway coming to Swindon, and the Gloucester Branch, and the Company promised to obtain an Act for the same, if possible, upon the faith of which I became a Subscriber to the amount of 10 Shares; the branch was laid down on the map. The Basing Line would be of no advantage to us, or North Wilts; neither would a line from Tring to Gloucester, across the Cotswold Hills, answer for the Dairy districts.

The proposed Line from Swindon would be very advantageous to North Wilts, &c.

Ex. MR. PHILLIP DAVIS, *Tallow Chandler and Grocer, Reading.*

I consume much Russian Tallow annually, which is sent from London, and it is important that it should travel quickly, as we are frequently short of tallow, and if there is any delay we are prevented executing our orders; the present price of it is about £20. per cask: I sometimes purchase from 20 to 50 casks at a time, (our purchase is greater in Winter than in Summer.) It is conveyed chiefly by water, at 15*s.* a ton, and a cask weighs about 9 cwt.; it is never conveyed by waggon, unless there is a stoppage of the canal: it is 1*s.* 9*d.* per cwt, about 32*s.* per ton, by waggon. If the passage by canal is favorable, we get our tallow in 4 or 5 days from London, and in Winter the average is 10 or 14 days, as the boats have to wait at the wharf before they are loaded; I have known it to be above a month: at one time there was a drought, another time there was a flood and frost together. The Tonnage of the Barges that travel upon the river vary from 25 tons; there are some of 150 tons, and it is their practice to wait in London until they get a full load, which has obliged us to send a waggon as far as Staines, a distance of 23 miles from Reading, to fetch the tallow from the boats, at a great additional expense, as we could wait no longer.——We get our Grocery chiefly from London, although there is much sugar imported into Bristol, but the carriage is objectionable, as it must first go down the Avon, then the Canal, and then the Kennet: I have had some from Bristol, when I suffered from delays caused by floods on the Avon. I carry on a considerable trade in both Dutch and Irish Butter, but although Bristol is more direct than London, yet we get most of it from the latter, principally on account of the carriage. There is much Dutch butter sold in London, and often sold for "fresh," on account of its mildness: Irish butter is not able to compete with it, as it is salter, (but there is a heavy duty on foreign butter, which has its influence upon the market), which I attribute to the time occupied in the voyage; it suffers considerably, particularly in warm weather. Mild butter bears a higher price than the salt, especially in Summer. There is a great quantity sent up from Dorsetshire and Somersetshire by waggon, in firkins, salted, the carriage of which is considerable, and it suffers much from heat.——We have frequently suffered by the men pilfering such goods as soap, tobacco, and sugar, which always increases with the amount of detention.

Neces. of Tallow being conv. quickly.

Price £20. a Cask of 9 Cwt.

Freight by Water 15*s.* a Ton.

By Waggon 32*s.*

4 or 5 Days by Canal

10 or 11 in Winter.

The Tonnage of the Barges vary from 25 to 150 Tons.

Inst. of Detention of Goods.

Obj. to the Comm. with Bristol.

Dutch Butter better than Irish, on acc. of hav. a better convey.

There is also much Dorset and Somers. Butter sent to Lond.

Obj. to the Canals.

Floods are the most prevalent source of delay upon the Thames, but I have not known a flood or a drought at more than a month; I have known a flood and frost together to last longer. I believe the men on the Thames are paid by the Barge Masters at so much a journey, and I have repeatedly known them to press the journey when it was not safe: I have had sugars destroyed by a flood. The distance from London to Reading is 89 miles by the Thames, and 40 by the Turnpike Road. I have known goods, coming from Westward by the Kennet and Avon Canal, to be forwarded to London by land carriage from Reading, avoiding the Thames. The general feeling at Reading is that the trade of the town would be very much increased by the formation of a railway; as they whould be relieved from keeping a large stock, which would benefit the consumers.——I visit London once a month, and although the coaches perform the journey quickly, I should go up much oftener if a railway was established.——Reading is a thriving place, which I attribute to its proximity with the river, and its being on the line to Bristol, (waggons, coaches, and barges, trading to Bristol, pass through Reading.) I know something of the trade at Basingstoke, as I supply them with goods, as shopkeepers with candles and the like: Reading is 15 miles distant from Basing, and is quite a metropolis to it, (I have likewise supplied the following places upon the Basing line, viz. Pamber, Mortimer, Overton, Dean Gate, Whitchurch, and Andover.) The coach traffic through Reading and Basingstoke is nearly equal: a railway direct to Basing would be very likely to interfere with my business, and would be very injurious to Reading, just as a superior line of communication injures an inferior.——We have a silk manufactory at Reading, which has increased lately: we also manufacture pins. There is a great deal of trade in general articles at Reading, and there are more shopkeepers than in most towns: there is also a trade in coals, much of which is sent for the supply of the neighbourhood; also in corn, much of which is sent to London by the Thames, which is generally a very uncertain conveyance, a good deal of it is consequently sent by waggon, particularly when the water has been high, and floods have been expected: there is also a very great traffic in flour; and there are several large corn mills in the neighbourhood of Reading, propelled by the Thames and the Kennet. ——Reading is 17 miles from Newbury, which is also a very large corn market, and is supplied with grocery and tallow by the Thames and Kennet, perhaps the latter is the best communication: I think that upon an average there are two or three stoppages every year upon it for two or three weeks, even since the improvements of cuts, canals, and pounds: I have frequently been kept waiting for tallow: it takes about a day to go from Newbury to Reading, and the country between them is very flat, the Bath Road runs through it; the Kennet is about 80 feet higher at Newbury than at Reading, and I do not think there are more than 10 locks upon it. We also carry on a considerable trade at Wallingford, Thatcham, and Newbury, in the way of candles. We do not go as far North as Oxford, but we go to Andover, which is 34 miles; we send our goods by land, as there is no water conveyance: we also have much trade at Cholsey, Wantage, Maidenhead, and Marlow; the little shopkeepers in the above neighbourhoods are principally supplied from Reading.

Reading is 89 Miles by the Thames and 40 by the Road from London.

Adv. of the propo. Line to Reading.

Reading a very thriving Town.

Reading compared with Basing.

Trade of Reading.

Ditto Newbury.

The Kennett Canal about 80 ft. higher at Newbury than at Reading.

Trade of Reading.

Ex. Mr. THOMAS MORRIS.

I have been an inhabitant of Reading for 30 years, carrying on the trade of a Woollen Draper, and Manufacturer of Smock Frocks and ready made Clothes, with my partners, Letchworth and Co.; our Trade embraces about 80 miles round Reading, to Bath and Bristol in the west, also Bedfordshire, Buckinghamshire, and Northamptonshire, we employ women principally, (about 400 at the present time), in our Manufactory for ready made clothes. Our local trade is considerable; the number of the Inhabitants of Reading by the last census was 16,000, and I believe it is increasing, the town is also increasing in buildings, &c. We get our goods principally from the North of England, viz. Yarns from Barnsly, Cottons from Manchester, and Woollens from Huddersfield and Leeds: the Trade westward is principally in cloth, which would be increased by a better communication.——Reading is in a great measure the point from which the neighbouring country is supplied with goods. I have several customers at Basingstoke, the trade of which I consider about $\frac{1}{10}$th that of Reading, and the same with Newbury; our goods are of light weight but valuable, those from Stroud are sent by land, coming up to Henley, and then across to Reading by the carrier, whose charge is 2*s.* per cwt,, or about the same rate as waggons; both as regards cheapness and rapidity we can generally rely upon them, but it is more expensive than water carriage: (I suppose they may be got up from Oxford by water).——A rapid communication between Gloucestershire in the West, and Huddersfield and Leeds in the North, must be advantageous to all persons engaged in the clothing trade, on and in the vicinity of the line; nearly all the goods I import from Scotland are delivered in London, either by sea or canal, and they are sent to me by water, which is the most economical: I often resort to the waggons in order to prevent detention, (most of our goods from Scotland are dyed in London, which occupies some time as it is dependant on the weather). The water conveyance is 15*s.* per ton, and the land 35*s.*; they average by water from the time they are shipped about 3 days in fair weather, I have had detentions of more than a month in consequence of floods and frost, but those occurrences have not been frequent during the last 2 or 3 years:——I am not acquainted with the Navigation of the Avon, as I have few goods from Bristol.——We are obliged to order our goods before we require them to secure us against delays, which frequently occur, and a large stock on hand adds to the price of an article, as we are obliged to charge the interest upon it for the time; when my sea-borne cloths arrive at London they are not sent direct by the barges, but merely from one warehouse to another; I frequently pay more for the conveyance of 6 cwt. of goods from one wharf to another than the amount of the freight to London: I think their being conveyed at once from the ships by land carriage to Euston Square, and put upon the Railway would be considerably cheaper; if a delay occurred at the London terminus of the Railway, in bringing the goods from the river by the Regent's Canal, equal to the time occupied by the journey it would be objectionable; our clothes from Leeds and Huddersfield generally come by London, either by waggon or water, w e occasionally have them through Oxford by Pickford's Canal, but I prefer the former way.——There is a general depression of Trade at the present time, (but Reading does not suffer more

Woollen Draper, &c. at Reading.

Employs about 400 Women.

Reading con. 16,000 Inhabitants.

Trade of Reading.

Basingstoke about $\frac{1}{10}$th the Trade of Reading.

The want of a rapid Communication.

Carriage from Scotland, &c.

He is obliged to order before he req.

Warehousing Goods in London is very expensive.

Observations upon the Terminus.

than other Towns,) which I attribute to the low price of corn, by which very little money is brought into circulation. We have a large Corn Market, but there is more sold at Newbury.——I attended the first General Meeting at Reading, which was very numerous, it was called in the usual manner by a requisition, thus, " To call a meeting of the Inhabitants to consider the propriety of supporting the Great Western Railway"; the resolutions passed were decidedly in favour of this Line, perhaps there was 1 dissentient in 100, and I have no reason to think that the feeling has changed, some gentlemen connected with the undertaking attended, and explained the line, (I think Mr. Robert Harris, the Banker, spoke against it). I have 5 Shares in the Great Western Railway, which I prefer on account of its passing through the most important Towns, as Maidenhead, Reading, &c.; although ours is not a manufacturing town to any extent, there is some trade in Silk and Pins, but more hands are employed in our Factory than any other. I do not know of any Town on the Line that is a place of manufacture to any extent until it arrives at the Clothing Districts, at Christian Malford, Chippenham, and their vicinities; Stroud and its vicinity is equal in point of manufacture to Bradford and its vicinity. The Reading road and water accommodation is very good, which is the cause of its prosperity.

Newbury is a better Corn Market than Reading.

Popul. of the G.W.

He prefers the G.W. as it pas. thro. the most import. Towns

Reading is not a Manufactur. Town.

The G. W. does not pass any Manufa. Town until it arriv. at the Clothing Districts.

WITNESSES EXAMINED AGAINST THE BILL.

Ex. COLONEL GEO. HENDERSON, *formerly of the Royal Engineers.*

I was elected Chairman of the Committee of Management of the London and Southampton Railway, previous to the passing of the Act on the 25th of July, 1834, (the whole expence of which amounted to £31,000. but it would not have exceeded £15,000. had it not been for the opposition,) which line was first agitated in 1830, the first Prospectus was issued in 1832, a Junction with Bath was also contemplated at the same time, we were consequently in the field one year before the Great Western.—— The first Survey of the Southampton Line was made by Mr. Doswell of Southampton, and the Line crossed the River and terminated at Paddington. I was afterwards instructed by the Committee at Southampton to call in the assistance of some established Engineer. I consulted my brother officers in the Engineering department, who recommended Mr. James Walker and Mr. Giles, and the Committee made choice of the latter gentleman, who had been engaged upon the Newcastle and Carlisle Railway, and was also Engineer of the Basingstoke Canal, and well acquainted with the country. The Committee were very axious for the Northern Side of the Thames, until Mr. Giles represented the difficulty of it to them, upon which they took to the Southern, which afforded an excellent opportunity of reaching the River at Vauxhall Bridge, where it comes up to the wharfs, occupying an unincumbered spot.——I went to Bristol in April 1832, to ascertain the feeling towards this Line, and I mentioned it to several gentlemen, among others to Mr. Robert Bright, the Chairman of the Great Western; I also attended the Exchange, and distributed Prospectus of the Basing Line, I held forth the advantages of connecting Bristol with the British Channel, and the union of both with London, which this combination of lines would effect, and the saving of 45 miles of extra construction, and I was in general well received. I attended a meeting there of the Directors of the Gloucestershire or Coal Pit Heath Railway, by the advice of their Solicitor Mr. Osborne, when I explained the advantages of the Junction to the meeting, the Chairman observed that it was useless to expect a Railway could be carried from Bristol, unless it adopted part of the Coal Pit Heath Line, I replied that the Inclinations of the latter (1 in 70) were considered too steep for the advantageous application of Locomotive power; Mr. Harford, one of the present Directors of the Great Western Railway, attended the meeting, he also suggested to me the appointing of Mr. Brunton as the Engineer, remarking he was much respected in Bristol. The Reform Bill having been thrown out about this period, the Agitation was so great that it was of no use pressing the matter. There are a considerable number of Subscribers at Bath, also at Newbury and Trowbridge, but the Committee of the Basing Line closed their Subscription list, and postponed proceeding with their Bill until the fate of the Great Western was decided. (The

Southampton Rail. Act passed in 1834.

First proceedings on this Line.

Mr. Giles appointed Engineer.

The Depôt.

Early Proc. upon the Basing Line.

Advan. of same.

Coal Pit Heath Rail. has Planes 1 in 70.

Amount of Subscr. Basi. £350,000.

He consid. the Basi. the best Line.

He is General Superintendent of the Southampton.

Southampton Railway and Eton College are at the principal part of the expence of the opposition) at which period the Shares taken amounted to £350,000. ——I do not think any man can look upon a map of England and see these 2 lines that would not *prima facie* determine upon the Bath and Basing.——I am under an agreement to the Southampton Railway as the General Superintendant of the line, to which I was appointed on the 11th of September 1834, and I am bound to devote the whole of my time to that measure; I was a party to the early proceedings of the Basing Line, as it did not then occupy much of my time; during the period of my visit to Glasgow upon the business of a libel upon our line, the Company was remodelled and a fresh Prospectus issued, in which my name was inserted, the parties probably being ignorant of my engagement with the Southampton.——When the Southampton Railway was before Parliament, I was satisfied the Estimates of the Returns were sufficient, and I am still of the same opinion, whether the junction of the Basing is made or not; according to the returns, after deducting about £140,000. a year for the maintenance of the Railway, there would be a clear profit of 20 per cent. to the Shareholders. (The last Dividend that was paid on the Liverpool and Manchester was 9 and 10 per cent.)——The village of Basing is about ½ a mile from the commencement of the proposed Basing Line, Basingstoke is about 2 miles from the line. If Basing was a central Depôt for Troops, they could be sent by these 3 lines in either direction in a very short time; they would reach London (44 miles) in about 2 hours, and Southampton (30 miles) in 1¼ hours, and Bristol (73 miles) in less than 4 hours; the Barracks at Winchester are very excellent, perhaps the best in the kingdom, thus they could be carried the 12 miles down the Railway in ½ an hour, and embark at Southampton at all times of the tide, which would be a great advantage to the Public Service, as the embarkation would not be mixed up with the Naval Arsenal at Portsmouth; in the event of a War it will be most advantageous in the Solent, forming the whole of that vast bay from the Needles up to St. Helens, including Cowes, Portsmouth, and Southampton, the Railway will also be a very rapid conveyance for provisions for Troops embarking and Men of War. ——I have made a comparison of the Population with respect to the Parishes through which these lines pass, the calculation of the London, Basing and Bath Railway amounts to 153,939, and of the Great Western to Bath 158,205, which as it touches the northern point of Hammersmith obtains a Population of 10,202, although they do not pass through the populous parts of the Parish; we touch Lambeth, and if I had included it, we should have had an addition of 80,000. The proposed measure will give a facility of communication to a very large District, according to a calculation which I have it amounts to ⅕th of the Population of all England, as 2,397,900 persons are comprised within the Great Peninsula called the West of England, which includes the Counties of Hants, Wilts, Somerset, Dorset, Devon, Cornwall, and Surrey; and by the Parliamentary Returns of 1831, the whole Population of England amounts to 13,000,000.——In the Evidence upon the Great Western it is stated that a great quantity of Butter and Bacon would travel by their line, I have therefore calculated the average cost of the conveyance of same, and find it will come to 29*s.* 3*d.* per ton, although the freight by sea from the South of Ireland is only 12*s.* to 14*s.*——The Traffic of Passengers in Coaches between

Southa. cal. to pay 20 per cent.

The L. and M. pays 10 per cent.

Basing is about ½ a Mile from the Line.

Adv. of this Line in reference to Troops.

Advantages of the Southampton.

Account of the Population on both Lines.

The pro. Line affor. communicat. to ⅕th the Popu. of Engl. or 13,000,000 Persons.

Exp. of convey. Butter, &c. upon the G. W.

the Southern Coast amounts to nearly ¼ the direct traffic between Bristol and London:

The following is an account of the Coaches that pass through Basingstoke:—

London and Exeter.

2	Herald	14	weekly Journeys.
2	Subscription	14	Ditto.
2	Defiance	14	Ditto.
2	Telegraph	12	Ditto.
2	Traveller	12	Ditto.
2	Mail (to Exeter)	14	Ditto.
12		80	
	Barnstaple and London.		
2	North Devon	12	Ditto.
2	Hero, Bridgewater	12	Ditto.
16		104	
2	Devonport Mail	14	Ditto.
18		118	Ditto.
Also, 2	Old Salisbury	14	Ditto.
2	Light Salisbury	12	Ditto.
22	Total No. of Journeys weekly	144	

11 Coaches pass through Basingstoke, making 144 weekly Journeys.

The amount of Goods sea-borne is also very considerable, as I have ascertained from the Custom Masters at the Ports of Southampton and Portsmouth.——There are but 2 Tunnels on the Basing Line, one of which might very easily be omitted, and the other is not imperatively necessary, by which a great saving will be effected, as well as preventing the cutting up of the country.——There is a considerable Passage by Southampton to Havre, from whence there is a direct road to Paris, a Railroad is also contemplated; the proposed Basing Line will form a convenient communication from Ireland by Bristol to France.——If upon a Comparison of the two lines the Great Western had an advantage of 153 feet over the Basing, but the rises which were afterwards lost by falls were equal to 100, it should be deducted from the above, which leaves 53 feet as the ultimate advantage of the Great Western over the Basing Line upon the whole amount of the risings. The Branch to Bradford and Trowbridge has Planes almost as bad as ours, namely 1 in 264.——The distance from Bath to Willesden is between 103 and 104 miles, and they will have to pay for 6 miles from Willesden to Camden Town, also for another 6 miles from Camden Town to Euston Square, and the Bradford and Trowbridge Branches will be upwards of 12 miles, which makes the Total amount of Mileage on the Great Western 127. The length of the Basing and Bath Line, taking the 44 of the Southampton, is 106 miles, therefore the Passengers of the former must be taxed 20 per cent. more than upon the latter——The Estimate from Willesden to Bath is £2,100,000., and from Willesden to Camden Town has been assumed at £ 100,000., as the works are very extensive, and they have paid large sums for the land, (according to the Evidence they have given £30,000. for 10 acres of land;) the

Tunns. on the Basi.

Faci. of comm. to France by Southa.

Rem. upon compari. of the Gradients.

Bradford and Trow. Branch has Planes 1 in 264.

Comparison of lengths of the G. W. and Basing.

Comparison of the Estimates.

Estimate from Camden Town to Euston Square is £200,000., and adding £750,000. to take them to the Thames, (There is a project for carrying a Sub-way from the end of the Birmingham Line to some point near the Thames, at an expense of £750,000.) will give the total expense of the Great Western £3,150,000. The Estimate of the Basing Line from Basing to Bath is £1,000,000. and from London to Basing £600,000. making a Total of £1,600,000. and a Saving of £1,550,000. capital over the Great Western Railway.——The cost of Fencing upon Commons and upon Lands lately enclosed is 1*s*. a statute rod, or 5½ yards, and we intend passing over the commons with a similar fence; the nature of our soil is such that if Quicks were planted, they would not grow, we shall therefore sow furze seed upon the top of the mound, and by keeping them properly trimmed we shall get a better fence than in the Common Fields, ⅓d of the line will be of this description, which therefore bears out Mr. Giles's Estimate, or very nearly; it costs us 2*s*. 6*d*. the double yard for the other description of Fencing with oak posts and rails, and we can get larch cheaper. I therefore do not consider the Fencing will exceed the Estimate of £12,000. I think Mr. Kyan stated that the expence of applying his patent liquid would be about 2*d*. a sleeper, and he considered it would last 40 or 50 years.——I was present at Meetings at Basingstoke, Devizes, Trowbridge, Bradford, Huntingdon, Newbury, &c. which were carried unanimously in favour of the Basing Line.——Regarding the solvency of the Southampton Company, I can state that there are upwards of £60,000. in Exchequer Bills, bearing interest to the Company, and balances amounting to upwards of £15,000. in the hands of Bankers, we have also purchased and agreed for rather more than half the land between London and Basing, which includes a great quantity of poor land, (perhaps ⅓d) for which we have paid about £10. an acre, and also the expences of their Act, and they are proceeding with their work in a way which will ensure its successful termination, and with the greatest econbmy in every part of it.——We have 44 miles of the Southampton Railway contracted for, but none of our Contractors are bound to time.——About 6,827 running yards of the Railway, or about 4 miles, are nearly completed, *i. e.* the excavations and embankments are made, there may be places were it has to be lowered perhaps 2 or 3 feet, which the contractor has left up for his convenience, and the slopes have to be trimmed, the roads ballasted, and the rails laid.——In the original Prospectus of the Great Western there was one Terminus at Paddington, and another line crossed the Thames at about Kingston, taking the Southern bank of the river, apparently terminating exactly at the same point as the Southampton Railway, and according to a Report on this line in 1833, "the total length of the Railway would be from 115 to 120 miles, depending on its termination, whether at Paddington, or on some part of the southern bank of the Thames; the inducements to adopt the former are the saving of distance, consequently of expense, and of a great part of the expense of the entrance into London; those on the side of the latter are, the advantages of reaching the water side at once, on account of the destination of goods, an equally good if not superior point for the reception and deposition of Passengers, and lastly, a very considerable Revenue derivable from Passengers going short distances on the South, which would be altogether lost by the Northern termination."——We have commenced upon the Southampton, and in 13 places, viz. at Battersea, Wandsworth, Kingston,

G.W. £3,150,000.

Basing £1,600,000.

Cost of Fenc. upon Com. 1*s*. for 5½ yds.

Fencing upon the Southampton.

Oak Fen. upon Do. 2*s*. 6*d*. a double yd.

Memo. Kyan's Pat. 2*d*. per sleeper.

Meetings carr. in favor of the Basing.

State of the Southa. Company.

⅓rd of the Land bet. London and Basing cost only £10. an Ac.

Proceedings upon same, &c.

Original Prospectus of the G. W.

Comp. between the N. and S. Terminus.

Walton (St. George's Hill,) Goldsworth on the East and on the West side of the hill, Frimley, Elvetham, Shapley on the East and on the West side of the hill, Hook, Newnham, Basing on the East and on the West side, and a small portion is done at Purbright. A very considerable delay has been occasioned by the Rails not being supplied, (6000 tons, which were taken at £8. per ton) although ordered the latter end of 1834, (we have since ordered 1000 tons of another house;) we have repeatedly remonstrated about it: I think 1000 tons were to have been delivered in March, and 2000 tons were to have been delivered in May, whereas we only received 700, which are laying down at different parts of the line: we have been obliged to purchase 1¼ and 1½ inch square rails, and work with light waggons, which do not hold above one-third the proper quantity, but we expect several shipments, when the works will proceed upon a greater scale. I consider they are going on honestly and "bona fide," with as much expedition as possible, and that they approximate very closely to the estimates. ——I have spent about 19 days upon the line since the operations commenced, and have watched closely, and ascertained the price which the excavation cost the Contractor, particularly at Shapley Heath, where the larger rails and waggons are used: we have a bridge upon this portion, for which we could not obtain bricks until May or June, it being a thinly populated country, (we have since entered into an agreement for 5,000,000 bricks between London and Basing, all of which are to be delivered this year; we had a ship load of coals from Newcastle to make them) in consequence of which we could not carry the line further. The lead is about one-third of a mile long, and we have at present 4 teaming places, but when we get a little further, and the embankment widens, we shall have 6, by which we shall team more, as we have a great force upon our work at the hills, where it lays in large masses, and presents plenty of space for the men to work. In cutting and filling there has been a low level at Shapley, which the water came into; we therefore left it for a short time, to allow of drainage, and took off the top lift by 2 self-acting inclined planes on each side, and the loaded waggons take up those which have been emptied.——I watched an hour each day on the 23rd and 24th of July, and counted how many waggons came down the inclined planes, and 20 waggons were sent down on the 23rd on both planes, and 22 on the 24th, each containing 3¼ cubic yards. The men are working double shifts, the first commence at 3 in the morning, and are relieved at 12 at noon, the relief going off at 9: (40 minutes are allowed for meals in each shift,) making in the whole 16⅔ hours. Now 21 waggons an hour, at 16 hours only, would team 1092 cubic yards.—I also counted all the men employed during the 2 days, and took down the prices they were paid: there were 2 Getters, to break down and cut the soil, and prepare it for the fillers, who had made a sub-contract with the Contractor to fill this material for 1½*d.* per cubic yard, and carry it to the head of the inclines, and as the work done by each shift is 546 yards, it amounts to £3. 19*s.* 7½*d.*; they are therefore getting 2*s.* 10*d.* each per day;—the Horses, which are the property of the Contractor, cost 2*s.* 6*d.* for food, 1*s.* for diminution, and 6*d.* for shoeing and harness, makes 4*s.*; the Boys are paid 10*d.*; there are 7 horses and boys at 4*s.* 10*d.* per day, which gives £1. 13*s.* 10*d.*;—2 Lads are employed on the incline, who receive 1*s.* 8*d.* each for removing the tongue to allow the empty waggons to go up the hills, making 3*s.* 4*d.*; and there are 9 Teamers at the end of the embankment, and who do the

Acco. of the Works on the Southampton Railway.

The Rails were not deliv. as agreed. (they were taken at £8. per Ton.)

They commen. small Rails and Waggons.

He con. the Works approx. to the Est.

They were also stop. for want of Bricks.

Description of the Works at Shapley Heath.

It is worked by 2 Inclines.

Acco. of his Exp. of the quantity of Excav. working.

He made 1,092 cub. yds. teamed per day.

Details of Expence of same.

work at ½*d.* per yard, which amounts to £1. 2*s.* 9*d.* for 546 yards, (the work done by each shift,) the Teamers are therefore earning 2*s.* 6¼*d.* each per day;—2 men are employed in keeping up the road at 2*s.* 6*d.* each per day, 3*s.* for each shift for the wear of the rope on the inclined planes, making 6*s.* per day; 1*s.* for the wear of the wheels and sheaves for waggons, ¼*d.* per cubic yard, making 11*s.* 4½*d.*; one gallon of oil 2*s.*, and the boy for oiling 10*d.*, amounting altogether to £8. 8*s.* 9*d.* for the half day, which makes the price 3$\frac{7}{10}$*d.* or less than 3¾*d.* per cubic yard, which I consider the true price.—I cannot state the exact number of men constantly employed, as they are perpetually varying: the work is measured up every fortnight, when the payments are made.—The Rope costs 34*s.* per cwt., and amounts to £23. including the carriage: I have calculated it to last 80 days, (to be renewed 4 times a year.)——The large Waggons at Frimley cost £15. 16*s.*, and at Walton £15. 18*s.*; I allow £16. for them: we procure our iron wheels and axles from Newcastle: the frame of the waggon is of oak, the bottom elm, and the sides deal; some of them are sheathed at the bottom to assist the teaming, which is necessary in clay; we contract for the iron, wood, and we pay 18*s.* for the making: I have supposed them to last 1½ years, although I am confident they will last out the work or 3 years: Mr. Stephenson stated in his evidence that our waggons do not hold more than 1 cubic yard, but he must have alluded to the small waggons used at the commencement, as our present waggons hold 3¼ yards. He also calculated the removal of the cutting at 1*s.* 3*d.* per yard, although we have let some of it at 5*d.* and the Opposition generally assumed the Company would never know the expense of the waggons, rails, &c., but I can state that I am acquainted with every expense, and I have given a detail of the materials, &c. which belong to the Company in the first instance, until by degrees the Contractor redeems them.——According to the Evidence given by Mr. Pease in the House of Lords last year, earthwork has been executed at 4*d.* and 5*d.* per yard, and I have no doubt he is correct: he quoted the Stockton and Darlington Railway, which although 40 miles long, cost only £400,000. or £500,000., including all works, branches, shipping staiths, warehouses, and a large establishment of locomotive engines, which is less than £10,000. per mile.——W across Walton Common, also the former site of Walton Race Course, (which is full ¼ of a mile from St. George's Hill;) the depth of the cutting as proposed was 106 feet, but it is now reduced to 80 feet, we having availed ourselves of the fall of the hill within the deviation of 100 yards, by which a great quantity of spoil will be avoided.——684,000 will be conveyed to the Mole Valley, and 760,000 to the Wey Valley, and 370,000 will be taken to ballast the road each way, making a total of 1,814,000; allowing 1000 cubic yards per day, and 250 days to the year, would give 250,000 per annum; the Mole Valley would therefore occupy rather more than 2¾ years to execute, and the Wey Valley about 3 years, but the latter affords greater facilities for teaming, as the descent Westward is very rapid, and admits of being run down upon self-acting planes: I therefore have no reason to think the Southampton Railway will occupy above 3 years in the execution, (without having recourse to work double shifts;) if 800 yards only could be teamed in 1 day, it would not take above 3¾ years, therefore Mr. Locke, and the several gentlemen who have stated it would take 7 years, must be in error; neither has there been any evidence to warrant their supposition that 1,500,000 cubic yards were

He found it came to 3⅞*d.* per cub. yd.

The Rope is renew. 4 times a year.

Allo. £16. for Wag.

Descriptn. of same.

Observ. upon Mr. Stephenson's Evidence.

Mr. Pease stated Earthwork had been done for 4*d.* or 5*d.*

Des. of the Stockton and Darlington.

It cost £10,000. per Mile.

St. Geo. Hill is 80 ft. deep, it was 106 ft.

Amount of the Reduced Quantities.

Time it would take to execute.

He cons. the whole Line will be compl. in 3 years.

Remarks upon the Calcu. of St. Geo. Hill by the Oppo.

required for the Mole Embankment, upon which they have made all their calculations. ——According to Mr. Giles's Section of last year, 1,125,966 cubic yards were to be carried to the Mole Valley, and 1,799,227 for the portion applied to spoil and to the Wey Valley, and 800,000 cubic yards of gravel were to be left in the hill for ballasting. I make the whole of the Mole Valley, according to the original Section, amount to 1,564,883 cubic yards, 510,054 of which come from the East end of the embankment, from the Surbiton Cutting, near Kingston: Mr. Giles's calculation is rather more, he makes it 1,635,133; (we took our dimensions from different drawings, and mine was to a very small scale) he also made 509,000 from Surbiton: even taking this amount it can be completed in 4½ years, at 1000 cubic yards per day, and 250 days to the year, or in 3 years at 1500 cubic yards a day, which was the calculations we assumed. If we can team 1100 per day with 4 teaming places, we can team 1650 a day by increasing them to 6, which quantity I have known done in 1 day at Hartlepool; as 781 waggons were teamed in a day, (we can fill as many waggons as we please) which multiplied by 3½ will give upwards of 2000 cubic yards a day: a large waggon can be teamed as easily as a smaller, as the length of it does not make any difference, although the breadth does.——The whole length of the Mole Embankment, exclusive of the cutting, was 3 miles 886 yards, and I apprehend the extreme length of the lead would have been 5 miles: the average lead of the Mole Valley is upwards of 3¼ miles, and the Wey Valley 1½: the average of the lead of St. George's Hill is therefore 2¼ miles.——Taking into consideration the great extent of the tunneling and bridge-work upon the Great Western, I think that the Bath and Basing, together with the necessary portion of the Southampton line, would take less time to execute than the Great Western.——There is from 800,000 to 1,000,000 cubic yards of cutting at Shapley Heath; the cutting at Hook is nearly on a parallel with St. George's Hill and Shapley Heath.——The method of teaming is attended with some difficulty, on account of the softness of the embankment; when a loaded waggon is brought up to the end of the embankment, and discharged, the soil gives way, and the waggon descends or sinks. I have seen Mr. Grahamsley's Machine at work upon the Newcastle and Carlisle Railway, since the Southampton was in Parliament last year and I consider that it has a tendency to prevent this inconvenience, and to expedite the teaming, and it is very easy to move the machine forward; I observed that the difficulty was more in keeping it back, as the pressure of the earth forced it forward; they were consequently obliged to anchor it back; and the machine is equally applicable for any indefinite length.

Quantities of St. George's Hill originally.

He calculated upon team. 1500 cub. yds. per day.

Instance of 2000 being teamed upon the Hartlepool.

Length of Lead St. Geo. Hill, averag. 2¼ miles.

He con. the Basing will be finis. before the G. W.

Shapley Heath. Hook Hill.

Mr. Grahamsley's Method of Teaming.

Ex. DIONYSIUS LARDNER, LL.D.

I have devoted a considerable time to the study of Practical Science and subjects connected with Civil Engineering, having made upwards of 100 Trips upon the Liverpool and Manchester Railway, and numerous experiments upon the power and speed of Engines of various construction. I have also written upon the same, and have corresponded upon the subject of Steam with Engineers and Scientific Men both on the

His attainments in Practical Science.

His opinion upon Civil Engineers.

He was Professor of Nat. Philos. at the Lond. University.

The Comparison of the Summit Levels of 2 Lines, forms no criterion, it is the Undulations which must be taken.

Gradient Formula. Add all Heights.

Sub. from same all Falls less than 1 in 250, and add all Falls above 1 in 250, which gives the Total Summit.

Resist. on G. W. exclu. the Planes, 16¼ lb. pr. ton going 15¾ do. returning. Do. do. Basi. Line, 18 lb. per ton going, 20 do. returning.

Remar. upon same.

The follow. Table gives a Result of 143,640 lbs. with 1 in 202, and 167,832 lbs. with 1 in 250, in fav. of the Ba. Li.

Continent and in the United States. I have every respect for the opinions of Practical Engineers, but I consider that many are merely judicial men, who have not extensive powers of inference or generalization, which is a matter of Arithmetic, and I do not consider the Practical Engineer to be the originator of the Data, although some may claim the highest situation as Scientific men, but I have not been employed Professionally upon Railways or Locomotive Engines. I was Professor of Natural Philosophy at the London University for 3 years.——I have examined the Sections of the competing lines, having given Evidence upon the subject of the Gradients in the House of Commons, of which I have made a Comparison, confining my observations between London and Bath.——I have drawn two Sections, one of which has a very high summit, and the other a very low one, and yet both require the same total Power to work: Section No. 1 is supposed to represent two Roads, one half consisting of a continuous rising Slope of 1 in 300, and the other a descending slope of 1 in 300. Section No. 2 consists of 6 short ascending sl opes of 1 in 300, interrupted by 6 descending slopes of 1 in 300, (I take 1 in 300, in both cases, by which the necessary power to work them will be alike,) although the latter is a succession of Summits, and does not rise so high as the other, yet it will require the same Power to surmount, as by bringing all the ascending slopes to one end and all the descending to another, they will be found to be equal; therefore the mere summit levels of two Sections does not afford any criterion of the power that will be required, as every thing depends upon the graduation—"If you add all the perpendicular Heights which a Load has to be lifted in ascending, and then subtract from it all the Falls of less acclivity than 1 in 250, or the Angle of Repose, afterwards adding all descending slopes, which are more than 1 in 250, you will get the number of perpendicular Feet which the power has to overcome;" thus, "a descending plane steeper than 1 in 250" will not return a proportionate power, *i.e.* it will not return more power than a plane of 1 in 250, consequently there is a loss of power.——I calculate that in passing from London to Bath, the greatest resistance on the Great Western Line (excluding the Euston Square and Box Planes) would be 16¼ lbs. a ton, and the greatest resistance on the Basing line 18 lbs. a ton, which is the exact proportion; returning from Bath to London by the Great Western, the greatest resistance which an Engine has to overcome would be 15¾ lbs. a ton, and on the Basing it would be 20: if the plane of 1 in 202 was changed to 1 in 250, it would be nearly 18 lbs. a ton, but this does not afford a fair comparison, on account of the 2 planes on the Great Western being omitted. ——— ———I have prepared a Table shewing the comparative Power required, taking all the ascents and descents into consideration.——The quantity of Mechanical Power necessary to draw 1 ton weight from London to Bath, on the Great Western, is therefore the same as would be required to lift 1,799,014 lbs. 1 yard high—and the result of the whole is that 1 ton, carried from London to Bath by the Great Western, and back again to London, would be equivalent to 3,540,965 lbs. raised 1 yard high, and the same upon the Basing, taking the plane of 1 in 202, would be 3,397,320 lbs., giving a difference of 143,640 lbs. in favour of the latter;—but taking the plane of 1 in 250, it would make the amount of the advantage 167,832 lbs., and the effect of this alteration of slopes will not cause any loss of gravity upon the line; therefore, the advantage of the Basing is int he proportion of 339 to 354, with a slope of 1 in 202, and 337 to 354, with a slope of 1 in 250.

Calculation of the Amount of Mechanical Power necessary to draw a Ton from London to Bath, and from Bath to London, on the Great Western and Basing Lines, the Power being expressed in the equivalent Number of Pounds raised Three Feet high.

Calculation of the Amount of Mechanical Power required on both Lines.

GREAT WESTERN RAILWAY.

London to Bath.

	Feet.
Sum of all the Rises . .	383
Sum of all the Falls, not exceeding 1 in 250	243
	140
Fall at Box Hill, estimated at 1 in 250	51·98
To be overcome by Power . .	88·02
	Yards.
Distance from London to Bath .	192,588
Friction at 9 lbs. per Ton in Pounds raised 1 Yard	1,733,292
Power to raise 1 Ton 88.02 Feet .	65,722
Resistance from London to Bath in Pounds raised 1 Yard . .	1,799,014

Bath to London.

	Feet.
Sum of all the Rises	364·5
Sum of all the Falls not exceeding 1 in 250	337
	27·5
Effective Fall of Euston Square Incline	15·91
	11·59
	lbs.
Friction at 9 lbs. per Ton in Pounds raised 1 Yard	1,733,292
Power to raise 1 Ton 11.59 Feet .	8,654
Total Resistance from Bath to London in Pounds raised 1 Yard .	1,741,946

BASING LINE.

London to Bath.

	Feet.
Sum of all the Rises . . .	480
Sum of the Falls not exceeding 1 in 250	181
	299
Effective Fall of Slope 1 in 202 .	141·6
To be overcome by Power . .	157·4
	Yards.
Distance from London to Bath in Yards	187,396
Friction at 9 lbs. per Ton in Pounds raised 1 Yard . . .	1,686,564
Power to raise 1 Ton 157.4 Feet .	117,525
Total Resistance from London to Bath in Pounds raised 1 Yard .	1,804,089

Bath to London.

	Feet
Sum of the Rises	355
Sum of the Falls	480
Total effective Fall	125
	lbs.
Friction at 9 lbs. per Ton in Pounds raised 1 Yard high	1,686,564
Effective Aid derived from Fall of 125 Feet	93,333
Total Resistance from London to Bath in Pounds raised 1 Yard .	1,593,231

The G. W. even tak. as a Dead Level wo. exceed the Basing by 70,274 lbs.

Taking the Great Western, exclusive of the Slopes, and treating it as a Dead Level, the mechanical power required in going would amount to 1,733,792 lbs. raised 1 yard high, and it would be the same in returning, or 3,467,584 lbs. altogether, and even then it would exceed the total mechanical power required on the Basing by 70,274 lbs. raised 1 yard high.

Explanation of the next Tables.

I have likewise made another Calculation of the Power necessary to work both lines with the same results. The Table also includes the Speed subject to two different conditions, in one the maximum of Speed is limited to 30, and in the other to 40 miles an hour, and the "Mechanical Power" necessary to overcome every Slope is expressed in lbs. weight raised 1 yard high, the Resistance of every Slope is expressed in lbs. per Ton, the calculation being made upon the plane of 1 in 202:—

GREAT WESTERN RAILWAY.

	Distance in Chains.	Rise.	Fall.	Gradient One in	Total Height above starting Point.	Speed in Miles per Hour.	Time.	Time from Starting.	Distance from starting Point.	Resistance in Pounds per Ton.	Power expended per Ton in lbs. raised 3 Ft. high	Total Power expended per Ton in lbs. raised 3 Ft. high.	Suppy. Results assuming the greatest allowable Speed to be Thirty Miles an Hour. Speed in Milespr. Hour.	Time.	Time from Starting.
		Ft. In.	Ft. In.		Ft. In.	M. Yds.	H. M. S.	H. M. S.	M. Yds.				M. Yds.	H. M. S.	H. M. S.
London	26	—	5 0	343	5 0	40 0	0 0 29	0 0 29	0 572	2·47	1,415	1,415	30 0	0 0 39	0 0 39
	60	46 0	—	86	41 0	5 0	0 9 0	0 9 9	1 132	35·02	46,227	47,642	—	0 9 0	0 9 39
	25	—	—	L.	41 0	25 0	0 0 45	0 10 14	1 682	9	4,950	52,592	—	0 0 45	0 10 24
	108	—	—	L.	41 0	25 0	0 3 14	0 13 28	2 1,298	9	21,384	73,976	—	0 3 14	0 13 38
	60	—	7 0	566	34 0	40 0	0 1 7	0 14 35	3 858	5·04	6,653	80,629	30 0	0 1 30	0 15 8
	152	13 0	—	772	47 0	18 1,590	0 6 2	0 20 37	5 682	11·90	59,803	120,434	—	0 6 2	0 21 10
	352	—	17 0	1,366	30 0	30 998	0 8 38	0 29 15	9 1,386	7·36	57,003	177,435	30 0	0 8 48	0 29 58
	75	0 6	—	9,900	30 6	24 681	0 2 18	0 31 33	10 1,276	9·23	15,223	192,658	—	0 2 18	0 32 16
	159	11 0	—	954	41 6	19 1,456	0 6 1	0 37 34	12 1,254	11·35	39,695	232,353	—	0 6 1	0 38 17
	41	—	—	L.	41 6	25 0	0 1 14	0 38 48	13 396	9	8,118	240,471	—	0 1 14	0 39 31
	203	—	15 0	893	26 6	34 1,157	0 4 24	0 43 12	15 1,342	6·49	28,994	269,465	30 0	0 5 4	0 44 35
	677	3 6	—	12,766	30 0	24 919	0 20 42	1 3 54	24 396	9·18	136,659	406,124	—	0 20 42	1 5 17
	611	42 0	—	960	72 0	19 1,502	0 23 5	1 26 59	31 1,518	11·33	152,338	558,462	—	0 23 5	1 28 22
	289	4 3	—	4,488	76 3	23 1,208	0 9 9	1 36 8	35 836	9·50	60,395	618,857	—	0 9 9	1 37 31
	149	—	6 9	1,460	69 6	30 265	0 3 42	1 39 50	37 594	7·46	24,462	643,319	30 0	0 3 44	1 41 15
	433	10 9	—	2,660	80 3	22 1,513	0 14 12	1 54 2	42 1,320	9·84	93,761	737,080	—	0 14 12	1 55 27
	82	4 9	—	1,140	85 0	20 912	0 3 0	1 57 2	43 1,364	10·97	19,783	756,863	—	0 3 0	1 58 27
	872	21 0	—	2,740	106 0	22 1,617	0 28 32	2 25 34	54 1,188	9·82	188,336	945,199	—	0 28 32	2 26 59
	324	31 9	—	670	137 9	18 448	0 13 19	2 38 53	58 1,276	12·33	87,859	1,033,058	—	0 13 19	2 40 18
	187	13 3	—	930	151 0	19 1,282	0 7 7	2 46 0	61 110	11·40	46,919	1,079,977	—	0 7 7	2 47 25
	42	1 0	—	2,772	152 0	22 1,655	0 1 22	2 47 22	61 1,034	9·81	9,063	1,089,040	—	0 1 22	2 48 47
	727	68 0	—	706	220 0	18 867	0 29 30	3 16 52	70 1,188	12·17	194,719	1,283,759	—	0 29 30	3 18 17
	707	49 0	—	952	269 0	19 1,443	0 26 45	3 32 37	79 902	11·35	176,573	1,460,332	—	0 26 45	3 45 2
	1,096	—	147 6	490	121 6	40 0	0 20 33	4 4 0	93 374	4·43	106,875	1,567,207	30 0	0 27 24	4 12 26
	48	—	1 3	2,534	120 3	27 1,272	0 1 18	4 5 28	93 1,430	8·12	8,571	1,575,778	—	0 1 18	4 13 44
	195	0 3	—	L.	120 6	25 0	0 5 51	4 11 19	96 440	9·04	38,797	1,614,575	—	0 5 51	4 19 35
	260	36 3	—	473	156 9	16 678	0 11 54	4 23 13	99 880	13·73	78,547	1,693,122	—	0 11 54	4 31 29
	165	22 3	—	490	179 0	16 1,008	0 7 28	4 30 41	101 990	13·58	49,283	1,742,405	—	0 7 28	4 38 57
	196	—	121 6	107	57 6	30 0	0 4 54	4 35 35	104 22	0	—	1,742,405	—	0 4 54	4 43 51
	382	—	43 0	587	14 6	40 0	0 7 10	4 42 45	108 1,386	5·18	43,529	1,785,934	30 0	0 9 33	4 53 24
	30	—	0 6	3,960	14 0	26 1,191	0 0 51	4 43 36	109 286	8·43	5,567	1,791,501	—	0 0 51	4 54 15
Bath -	21	4 6	—	380	18 6	13 1,457	0 1 8	4 44 44	109 748	16·27	7,518	1,799,019	—	0 1 8	4 55 23

	Distance in Chains.	Rise.	Fall.	Gradient One in	Total Height above starting Point.	Speed in Miles per Hour.	Time.	Time from Starting.	Distance from starting Point.	Resistance in Pounds per Ton.	Power expended per Ton in lbs. raised 3 Ft. high	Total Power expended per Ton in lbs. raised 3 Ft. high.	Suppy. Results assuming the greatest allowable Speed to be Thirty Miles an Hour. Speed in Miles pr. Hour.	Time.	Time from Starting.
		Ft. In.	Ft. In.		Ft. In.	M. Yds.	H. M. S.	H. M. S.	M. Yds.				M. Yds.	H. M. S.	H. M. S.
Bath -	21	—	4 6	308	4 6	40 0	0 0 24	0 0 24	0 462	1·73	798	798	30 0	0 0 32	0 0 32
	30	0 6	—	3,960	4 0	23 918	0 0 57	0 1 21	0 1,122	9·57	6,313	7,111	—	0 0 57	0 1 29
	382	43 0	—	587	39 0	17 968	0 16 20	0 17 41	5 726	12·82	107,743	114,854	—	0 16 20	0 17 49
	196	121 6	—	107	160 6	5 0	0 29 24	0 47 5	7 1,518	30·04	129,528	244,382	—	0 29 24	0 47 13
	165	—	22 3	490	138 3	40 0	0 3 6	0 50 11	9 1,628	4·42	16,057	260,439	30 0	0 4 8	0 51 21
	260	—	36 3	473	102 0	40 0	0 4 53	0 55 4	13 308	4·27	24,413	284,852	30 0	0 6 30	0 57 51
	195	—	0 3	L.	101 9	25 0	0 5 49	1 0 5	15 1,078	8·96	38,423	323,275	—	0 5 49	1 3 40
	48	1 3	—	2,534	103 0	22 1,345	0 1 35	1 2 8	16 374	9·88	10,437	333,712	—	0 1 35	1 5 15
	1,096	147 6	—	490	250 6	16 1,027	0 49 35	1 52 3	29 1,606	13·57	327,142	660,854	—	0 49 35	1 54 50
	707	—	49 0	952	201 6	33 1,489	0 15 40	2 7 43	38 1,320	6·65	103,399	764,253	30 0	0 17 41	2 12 51
	727	—	68 0	706	133 6	38 1,097	0 14 7	2 21 50	47 1,474	5·83	93,173	851,426	30 0	0 18 10	2 30 41
	42	—	1 0	2,772	132 6	27 821	0 1 9	2 22 59	48 658	8·19	7,569	864,995	—	0 1 9	2 31 50
	187	—	13 3	930	119 3	34 204	0 4 7	2 27 6	50 1,232	6·60	27,133	892,128	30 0	0 4 40	2 36 30
	324	—	31 9	670	87 6	39 1,150	0 6 8	2 33 14	54 1,320	5·67	40,445	932,573	30 0	0 8 6	2 44 36
	872	—	21 0	2,740	66 6	27 875	0 23 47	2 57 1	65 1,144	8·18	156,976	1,089,549	—	0 23 47	3 8 23
	82	—	4 9	1,140	61 9	31 1,738	0 1 55	2 58 56	66 1,188	7·03	12,689	1,102,238	30 0	0 2 3	3 10 26
	433	—	10 9	2,660	51 0	27 1,025	0 11 46	3 10 42	72 154	8·16	77,707	1,179,945	—	0 11 46	3 22 12
	149	6 9	—	1,460	57 9	21 620	0 5 14	3 15 56	73 1,672	10·54	34,542	1,214,487	—	0 5 14	3 27 26
	289	—	4 3	4,488	53 6	26 823	0 8 11	3 24 7	77 990	8·50	54,049	1,268,536	—	0 8 11	3 35 37
	611	—	42 0	960	11 6	33 1,317	0 13 35	3 37 42	85 352	6·67	89,618	1,358,154	30 0	0 15 17	3 50 54
	677	—	3 6	12,766	8 0	25 875	0 19 55	3 57 37	93 1,166	8·82	131,434	1,489,588	—	0 19 55	4 10 49
	203	15 0	—	893	23 0	19 971	0 7 47	4 5 24	96 352	11·51	51,394	1,540,982	—	0 7 47	4 18 36
	41	—	—	L.	23 0	25 0	0 1 14	4 6 38	96 1,254	9	8,118	1,549,100	—	0 1 14	4 19 50
	159	—	11 0	954	12 0	33 1,451	0 3 31	4 10 9	98 1,232	6·65	23,269	1,572,369	30 0	0 3 59	4 23 49
	75	—	0 6	9,900	11 6	25 1,135	0 2 12	4 12 21	99 1,122	8·77	14,477	1,586,846	—	0 2 12	4 26 1
	352	17 0	—	1,366	28 6	21 261	0 12 29	4 24 50	104 66	10·64	82,389	1,669,235	—	0 12 29	4 38 30
	152	—	13 0	772	15 6	36 1,587	0 3 5	4 27 55	105 1,650	6·10	20,389	1,689,624	30 0	0 3 48	4 42 18
	60	7 0	—	566	22 6	17 636	0 2 36	4 30 31	106 1,210	12·96	17,107	1,706,731	—	0 2 36	4 44 54
	108	—	—	L.	22 6	25 0	0 3 14	4 33 45	103 66	9	21,384	1,728,115	—	0 3 14	4 48 8
	25	—	—	L.	22 6	25 0	0 0 45	4 34 30	108 616	9	4,950	1,733,065	—	0 0 45	4 48 53
	60	—	46 0	86	23 6	10 0	0 4 30	4 39 0	109 176	0	—	1,733,065	—	0 4 30	4 53 23
London	26	5 0	—	343	18 6	14 864	0 1 21	4 40 21	109 748	15·53	8,881	1,741,946	—	0 1 21	4 54 44

LONDON, BASING, AND BATH RAILWAY.

	Distance in Chains.	Rise.	Fall.	Gradient One in	Total Height above starting Point.	Speed in Miles per Hour.		Time.	Time from Starting.	Distance from Starting Point.		Resistance in Pounds per Ton.	Power expended per Ton in lbs. raised 3 Ft. high.	Total Power expended per Ton in lbs. raised 3 Ft. high.	Suppy. Results, assuming the greatest allowable Speed to be 30 Miles an Hour. Speed in Miles pr Hour.	Time.	Time from Starting.
		Ft.	Ft.		Ft.	M.	Yds.	H. M. S.	H. M. S.	M.	Yds.					H. M. S.	H. M. S
London	186	25	—	490	25	16	1,040	0 8 24	0 8 24	2	572	13·56	55,495	55,495	—	0 8 24	0 8 24
	85	17	—	330	42	14	442	0 4 28	0 12 52	3	682	15·79	29,523	85,018	—	0 4 28	0 12 52
	98	—	—	L.	42	25	0	0 2 56	0 15 48	4	1,078	9	19,404	104,422	—	0 2 56	0 15 48
	40	8	—	330	50	14	442	0 2 6	0 17 54	5	198	15·79	13,893	118,315	—	0 2 6	0 17 54
	436	—	—	L.	50	25	0	0 13 4	0 30 58	10	990	9	86,328	204,643	—	0 13 4	0 30 58
	54	—	5	713	45	38	726	0 1 3	0 32 1	11	418	5·86	6,959	211,602	30	0 1 21	0 32 19
	230	—	—	L.	45	25	0	0 6 54	0 38 55	14	198	9	45,540	257,142	—	0 6 54	0 39 13
	74	8	—	610	53	17	1,337	0 3 7	0 42 2	15	66	12·67	20,625	277,767	—	0 3 7	0 42 20
	136	20	—	449	73	16	144	0 6 21	0 48 23	16	1,298	13·99	41,861	319,628	—	0 6 21	0 48 41
	268	—	—	L.	73	25	0	0 8 2	0 56 25	20	154	9	53,064	372,692	—	0 8 2	0 56 43
	113	17	—	440	90	15	1,673	0 5 19	1 1 44	21	880	14·11	35,067	407,759	—	0 5 19	1 2 2
	600	120	—	330	210	14	442	0 31 35	1 33 19	29	0	15·79	208,400	616,159	—	0 31 35	1 33 37
	392	—	—	L.	210	25	0	0 11 46	1 45 5	33	1,584	9	77,616	693,775	—	0 11 46	1 45 23
	40	4	—	660	214	18	271	0 1 39	1 46 44	34	704	12·39	10,907	704,682	—	0 1 39	1 47 2
	152	—	—	L.	214	25	0	0 4 34	1 51 18	36	528	9	30,096	734,778	—	0 4 34	1 51 36
	170	34	—	330	248	14	442	0 8 57	2 0 15	38	748	15·79	59,047	793,825	—	0 8 57	2 0 33
	393		—	L.	248	25	0	0 11 47	2 12 2	43	594	9	77,814	871,639	—	0 11 47	2 12 20
Basing	110	22	—	330	270	14	442	0 5 47	2 17 49	44	1,254	15·79	38,207	909,846	—	0 5 47	2 18 7
	290	—	—	L.	270	25	0	0 8 42	2 26 31	48	594	9	57,420	967,266	—	0 8 42	2 26 49
	61	10	—	400	280	15	791	0 2 58	2 29 29	49	176	14·56	19,515	986,811	—	0 2 58	2 29 47
	503	—	—	L.	280	25	0	0 15 5	2 44 34	55	682	9	99,594	1,086,405	—	0 15 5	2 44 52
	200	40	—	330	320	14	442	0 10 32	2 55 56	57	1,562	15·79	69,467	1,155,872	—	0 10 32	2 55 24
	528	—	—	L.	320	25	0	0 15 50	3 10 56	64	858	9	104,544	1,260,416	—	0 15 50	3 11 14
	588	155	—	250	475	12	945	0 35 10	3 46 6	71	1,474	17·95	232,157	1,492,573	—	0 35 10	3 46 24
	323	—	—	S. L.	475	25	0	0 9 41	3 55 47	75	1,540	9	63,954	1,556,527	—	0 9 41	3 56 5
	260	—	65	264	410	40	0	0 4 52	4 0 39	79	220	0·52	2,947	1,559,474	30	0 6 30	4 2 35
	520	—	65	528	345	40	0	0 9 45	4 10 24	85	1,100	4·76	54,427	1,613,901	30	0 13 0	4 15 35
	534	—	174	202	171	40	0	0 10 1	4 20 25	92	528	0	—	1,613,901	30	0 13 21	4 28 56
	305	—	—	L.	171	25	0	0 9 9	4 29 34	96	198	9	60,390	1,674,291	—	0 9 9	4 38 5
	263	—	31	560	140	40	0	0 4 56	4 34 30	99	704	5·00	28,927	1,703,218	30	0 6 35	4 44 40
	397	—	15	1,746	125	29	271	0 10 13	4 44 43	104	638	7·72	67,406	1,770,624	—	0 10 13	4 54 53
Bath	169	—	—	L.	125	25	0	0 5 4	4 49 47	106	836	9	33,462	1,804,086	—	0 5 4	4 59 57

	Distance in Chains.	Rise.	Fall.	Gradient One in	Total Height above Starting Point.	Speed in Miles per Hour.		Time.	Time from Starting.	Distance from Starting Point.		Resistance in Pounds per Ton.	Power expended per Ton in lbs. raised 3 ft. high.	Total Power expended per Ton in lbs. raised 3 ft. high.	Suppy. Results, assuming the greatest allowable Speed to be 30 Miles an Hour. Speed in Miles pr Hour.	Time.	Time from Starting.
		Ft.	Ft.		Ft.	M.	Yds.	H. M. S.	H. M. S.	M.	Yds.					H. M. S.	H. M. S.
Bath	169	—	—	L.	—	25	0	0 5 4	0 5 4	2	198	9	33,462	33,462	—	0 5 4	0 5 4
	397	15	—	1,746	15	21	1,553	0 13 36	0 18 40	7	132	10·28	89,806	123,268	—	0 13 36	0 18 40
	263	31	—	560	46	17	541	0 11 24	0 30 4	10	638	13·00	75,221	198,489	—	0 11 24	0 30 4
	305	—	—	L.	46	25	0	0 9 9	0 39 13	14	308	9	60,390	258,879	—	0 9 9	0 39 13
	534	174	—	202	220	11	383	0 35 42	1 14 55	20	1,496	20·06	235,652	494,531	—	0 35 42	1 14 55
	520	65	—	528	285	16	1,744	0 22 57	1 37 52	27	616	13·24	151,493	646,024	—	0 22 57	1 37 52
	260	65	—	264	350	12	1,528	0 15 9	1 53 1	30	1,056	17·48	100.013	746,037	—	0 15 9	1 53 1
	323	—	—	S. L.	350	25	0	0 9 41	2 2 42	34	1,122	9	63,954	809,991	—	0 9 41	2 2 42
	588	—	155	250	195	40	0	0 11 1	2 13 43	41	1,738	0·05	691	810,682	30	0 14 42	2 17 24
	528	—	—	L.	195	25	0	0 15 50	2 29 33	48	1,034	9	104,544	915,226	—	0 15 50	2 33 14
	200	—	40	330	155	40	0	0 3 45	2 33 18	51	154	2·21	9,733	924,959	30	0 5 0	2 38 14
	503	—	—	L.	155	25	0	0 15 5	2 48 23	57	660	9	99,594	1,024,553	—	0 15 5	2 53 19
	61	—	10	400	145	40	0	0 1 9	2 49 32	58	242	3·44	4,611	1,029,164	30	0 1 32	2 54 51
Basing	290	—	—	L.	145	25	0	0 8 42	2 58 14	61	1,342	9	57,420	1,086,584	—	0 8 42	3 3 33
	110	—	22	330	123	40	0	0 2 4	3 0 18	63	242	2·21	5,353	1,091,937	30	0 2 46	3 6 19
	393	—	—	L.	123	25	0	0 11 47	3 12 5	68	88	9	77,814	1,169,751	—	0 11 47	3 18 6
	170	—	34	330	89	40	0	0 3 11	3 15 16	70	308	2·21	8,273	1,178,024	30	0 4 15	3 22 21
	152	—	—	L.	89	25	0	0 4 34	3 19 50	72	132	9	30,096	1,208,120	—	0 4 34	3 26 55
	40	—	4	660	85	40	0	0 0 45	3 20 35	72	1,012	5·61	4,933	1,213,053	30	0 1 0	3 27 55
	392	—	—	L.	85	25	0	0 11 46	3 32 21	77	836	9	77,616	1,290,669	—	0 11 46	3 39 41
	600	—	120	330	35	40	0	0 11 15	3 43 36	84	1,716	2·21	29,200	1,319,869	30	0 15 0	3 54 41
	113	—	17	440	52	40	0	0 2 7	3 45 43	86	682	3·89	9,681	1,329,550	30	0 2 50	3 57 31
	268	—	—	L.	52	25	0	0 8 2	3 53 45	89	1,298	9	53,064	1,382,614	—	0 8 2	4 5 33
	136	—	20	449	72	40	0	0 2 33	3 56 18	91	770	4·01	11,995	1,394,609	30	0 3 24	4 8 57
	74	—	8	610	80	40	0	0 1 23	3 57 41	92	638	5·33	8,679	1,403,288	30	0 1 51	4 10 48
	230	—	—	L.	80	25	0	0 6 54	4 4 35	95	418	9	45,540	1,448,828	—	0 6 54	4 17 42
	54	5	—	713	75	18	933	0 2 11	4 6 46	95	1,606	12·14	14,425	1,463,253	—	0 2 11	4 19 53
	436	—	—	L.	75	25	0	0 13 4	4 19 50	101	638	9	86,328	1,549.581	—	0 13 4	4 32 57
	40	—	8	330	83	40	0	0 0 45	4 20 35	101	1,518	2·21	1,947	1,551,528	30	0 1 0	4 33 57
	98	—	—	L.	83	25	0	0 2 56	4 23 31	103	154	9	19,404	1,570,932	—	0 2 56	4 36 53
	85	—	17	330	100	40	0	0 1 36	4 25 7	104	264	2·21	4,137	1,575,069	30	0 2 8	4 39 1
London	186	—	25	490	125	40	0	0 3 29	4 28 36	106	836	4·44	18,161	1 593,230	30	0 4 39	4 43 40

[Time of Transit by both Lines which is in favor of the Basing.] The journey from London to Bath on the Great Western, allowing 30 miles an hour as the maximum speed, and 25 miles upon a Level, would occupy 4 hours, 55 minutes, 23 seconds, and on the Basing 4h. 59m. 57s.; from Bath to London on the Great Western 4h. 54m. 44s., and on the Basing 4h. 43m. 40s.; therefore both together would make 9h. 50m. 7s. on the Great Western, and the same on the Basing 9h. 43m. 37s., the difference being 6m. 30s. in favor of the Basing. The journey from London to Bath on the Great Western, allowing 40 miles an hour as the maximum speed and 25 miles upon a Level, would be 4h. 44m. 44s., and on the Basing 4h. 49m. 47s.; from Bath to London on the Great Western, 4h. 40m. 21s., and on the Basing 4h. 28m. 36s., therefore both together would make 9h. 25m. 5s. on the Great Western, and the same on the Basing 9h. 18m. 23s., leaving a difference of 6m. 42s. in favor of the Basing.——— [Explanation of the Mechanical Power.] I consider this a fair mode of comparing the practical working of the two lines, (but I did not imagine my calculations would have proved so disadvantageous to the Great Western at the time of my commencing,) as the Mechanical Power is a good general Index of the general power required, it is the first approximation which an Engineer or Scientific Man would think of making, although there are other circumstances to be taken into consideration, as "the effect of Graduation upon the particular nature of Steam Power," which although a consideration of some moment is not the principal, neither can it be expressed exactly by numerical calculation, for which reason I did not take it into account, but I have included every thing that could be expressed numerically; I have also taken the Mechanical Power in order to make it general, it may be applied to Steam, Water, Man, or Horse Power. If a spring Steelyard was attached to a trace and pulled by a horse, it would indicate the power he exerted, thus if it was 10 lbs. the drawing would be strained in the same way as a rope with a weight of 10 lbs. suspended to it, and if the horse exerted this strength for the distance of 10 miles, the mechanical Power expended would amount to 100 lbs. The cost of Fuel is in precisely the same proportion, although it varies according to the price but the proportion remains the same, [Mecha. Power only $\frac{1}{5}$th the importance, the rem. $\frac{4}{5}$ths is ex. upon Wear & Tear, &c. &c.] and is about $\frac{1}{5}$th of the importance, the remaining $\frac{4}{5}$ths being expended upon the Wear and Tear of the Engines, (which alone bear almost as great a proportion as the Fuel,) also the Maintenance of the men, attendance upon the engines, &c.; the extra size of the same also forms part of the calculation, although a small addition in the power does not increase it, but a heavy Engine wears out the Road quicker and breaks the Rails [Origi. Rails L. & M. tota. insufficient.] oftener.——— Engineers had no experience whatever in Rails at the time the Liverpool and Manchester was constructed, as they laid them down 35 lbs. to the yard, which was altogether insufficient, and accordingly as they renew the rails they are laying them at 50 lbs. a yard, and I believe even 60 lbs. are contemplated on some railways, which is quite sufficient to bear Engines equal to a plane of 1 in 250; and the only difference is [1 in 264 does not req. an Ass. Eng.] the cost of the iron.——— An Inclination of 20 feet in 1 mile, or 1 in 264, does not require an Assistant Engine, but it must have an Engine made expressly for it, and of double the power of one upon a Level; if it can be avoided it would be better, it would [Remarks on same.] be worth going "any distance of less than double the length" to avoid it.

I have likewise made another Comparison of the two Lines with respect to their average Power and greatest resistance, in which the results of the foregoing Tables are brought together, as follows:—

COMPARATIVE VIEW of the GREAT WESTERN and BASING LINES.

	Great Western.	Basing.
Total Mechanical Power necessary to work the Line both ways, calculated by estimating the Resistance upon each successive Slope from the Table of Gradients, expressed in Pounds Weight lifted Three Feet high	3,540,965	3,397,316
Difference of Total Mechanical Power in favour of the Basing Line	—	143,649
Total Mechanical Power necessary to work the Line both ways, calculated by allowing Nine Pounds per Ton for Friction throughout the whole Distance, and then estimating the Power necessary to lift the Load through the Sum of all the Rises, and the Quantity of this Power restored by the Sum of all the Falls .	3,540,960	3,397,320
Difference in favour of the Basing Line .	—	143,640
Total Length of the Line in Yards	192,588	187,396
Difference in favour of the Basing Line .	—	5,192
Average Resistance of the Line, worked both Ways, in Pounds per Ton	9·1879	9·0645
Difference in favour of the Basing Line .	—	0·1234
Maximum Resistance on ascending Slopes from London to Bath, in Pounds per Ton	35·05	17·96
Difference in favour of the Basing Line .	—	17·09
Maximum Resistance on ascending Slopes from Bath to London, in Pounds per Ton	29·93	20·09
Difference in favour of the Basing Line .	—	9.84
Time of Transit from London to Bath, and from Bath to London, Thirty Miles an Hour being taken as the greatest allowable Speed	H. M. S. 9 50 7	H. M. S. 9 43 37
Difference in favour of the Basing Line .	—	0 6 30
Time of Transit from London to Bath, and from Bath to London, Forty Miles an Hour being taken as the greatest allowable Speed	9 30 5	9 18 23
Difference in favour of the Basing Line .	—	0 11 42
Length of an absolutely Level Line requiring the same Quantity of Mechanical Power	Yards. 196,721	Yards. 188,739
Difference in favour of the Basing Line .	—	7,982
Effect of the Gradients expressed in equivalent Increase of Length	4,133	1,343
Difference in favour of the Basing Graduation .	—	2,790
Comparative Amount to which the Power necessary to work the Line both ways would be reduced if the Box Hill and Euston Square Planes on the Great Western were converted into absolute Levels, expressed in Pounds raised One Yard . .	3,466,586	3,397,318
Difference in favour of the Basing Line .	—	69,268
Greatest Resistance from London to Bath, exclusive of Euston Square Slope, in Pounds per Ton	16·27	17·95
Greatest Resistance from Bath to London, exclusive of the Box Hill Slope, in Pounds per Ton	15·53	20·06

Summary Comparison of the Power required on both Lines (with the Plane of 1 in 202 upon the Basing)

COMPARATIVE VIEW of the GREAT WESTERN and BASING LINES, with the Gradient of 1 in 202 changed to 1 in 250.

Summary Comparison of the Power required on both Lines. (with the Plane of 1 in 250 upon the Basing.)

	Great Western.	Basing.
Total Mechanical Power necessary to work the Line both ways	3,540,965	3,373,128
Difference of Total Mechanical Power in favour of the Basing Line	—	167,832
Total Length of the Line, in Yards . . .	192,588	187,396
Difference in favour of the Basing Line .	—	5,192
Average Resistance of the Line worked both ways in Pounds per Ton	9·1879	9·0000
Difference in favour of the Basing Line .	—	0·1879
Maximum Resistance on ascending Slopes from London to Bath, in Pounds per Ton	35·05	17·96
Difference in favour of the Basing Line .	—	17·09
Maximum Resistance on ascending Slopes from Bath to London, in Pounds per Ton	29·93	17·96
Difference in favour of the Basing Line .	—	11·97
Time of Transit from London to Bath, and from Bath to London, Thirty Miles an Hour being taken as the greatest allowable Speed	H. M. S. 9 50 7	H. M. S. 9 43 37
Difference in favour of the Basing Line	—	0 6 30
Time of Transit from London to Bath, and from Bath to London, Forty Miles an Hour being taken as the greatest allowable Speed	9 30 5	9 18 23
Difference in favour of the Basing Line .	—	0 11 42
Length of an absolutely Level Line requiring the same Quantity of Mechanical Power	Yards. 196,721	Yards. 187,396
Difference in favour of the Basing Line .	—	9,325
Effect of the Gradients expressed in equivalent Increase of Length	4,133	—
Difference in favour of the Basing Graduation .	—	4,133
Comparative Amount to which the Power necessary to work the Lines both ways would be reduced if the Box Hill and Euston Square Planes on the Great Western were converted into absolute Levels, expressed in Pounds raised One Yard .	3,466,586	3,373,128
Difference in favour of the Basing Line .	—	93,458
Greatest Resistance from London to Bath, exclusive of the Euston Square Slope, in Pounds per Ton	16·27	17·95
Greatest Resistance from Bath to London, exclusive of the Box Hill Slope, in Pounds per Ton	15·53	17·95
Average Resistance from London to Bath, and from Bath to London, the Box Hill and Euston Square Slopes being supposed to be reduced to Levels, in Pounds per Ton . . .	9	9

Application of the above Tables.

The length of a Line absolutely Level, requiring the same Mechanical Power as the Great Western, would be 196,721 yards, and in the case of the Basing would be 188,739 yards, including the Plane of 1 in 202, which would give a difference of 7982 yards in favor of the latter, and is partly caused by the Basing Line being absolutely shorter. ———The average Resistance on the Great Western worked both ways would be 9·1879 lbs. per ton, and upon the Basing 9·0645 lbs. per ton, taking all the Slopes in both cases.

———I have made all my Calculations by two different Formulas; in one—I considered the resistance which the power had to overcome, from one end of the line to the other, by Friction, which I then combined with the Mechanical Power by the before stated process; I added or subtracted them, accordingly as the gravity assisted or opposed the friction, which gave me the Mechanical Power required to transfer the load from one end to the other, and I then took it in the other direction, and added them together, which gave the Total Mechanical Power required.———In the other—I took all the Slopes from one end of the line to the other, taking the common method of expressing the resistance to the drawing power on each, viz. in pounds weight per ton; from which Resistance, and the length of the Slope, I obtained by a simple arithmetical process the total Power required to draw a load from one end of it to the other; having ascertained the amount of every Slope I added them together, by which I obtained the Total Mechanical Power in both directions—and the results were the same by each method; in some cases they agreed to the unit, where they did not it probably arose from a few decimals being taken in one case and omitted in the other.———I have included the Power absolutely expended in working the Euston Square and Box slopes, but I have not made any allowance for the "Waste of power," which will be incurred whichever way they are worked; viz. the power necessary to pull the rope back, or to work the rope, in the event of a single rope being used, but I have merely taken the power necessary to draw the rope on the slope.—I included in my Calculations of the Great Western, the first Gradient from Bath to London at 1 in 308 of 21 chains, and the first Gradient from London at 1 in 343, both of which fall into the Stations, which may partly account for their introduction. If an engine and train were to arrive at the foot of an Inclination of 1 in 308 with a speed of 25 miles an hour, and at the moment of its arrival the impelling power was suspended, the engine would stop at a distance of 2280 feet up the plane; but if the impelling power was continued instead of suspended, and the same pressure exerted as upon a level, viz. 9 lbs. per ton, it would be brought to a state of rest at 6468 feet, as the power is less than the resistance of the plane, and if the same should occur in an inclination of 1 in 343, the number of feet would be 3024 and 7224 respectively; therefore, if the "*vis inertiæ*" would take them up those planes, lighter engines might certainly be used; I believe the next in steepness (upon the Great Western) are 1 in 473 and 490: and the Power required upon 473 is 13¼ lbs. per ton, and upon 490 it is $13\frac{6}{10}$, the difference between 1 in 202 and 1 in 473 being nearly in the proportion of 2 to 1, or 50 per cent., as stated by Mr. Locke.———An ordinary Engine similar to those used upon the Liverpool and Manchester, is able to surmount 1 in 250, (there are two descriptions of engines used upon this railway, one class weighing 8 and the other 10 tons, and coupled, *i. e.* both wheels are worked:) Engines always go slower up the Slopes on all lines, and generally work below their power on a Level, keeping a quantity in store to help them up the slopes. I think that an engine running 25 miles an hour on a level, would by its store of power continue at the same rate up the slopes, provided they were not too long, but it is desirable to have engines with as little "extra power" as possible. If there was a Slope of 1 in 250 upon a railway, and the remainder was nearly Level, it might be desirable to work it with an Assistant engine, but in the case of the Basing there are so many slopes of 1 in 205, that it would be better to build

He made his calcu. by 2 Formulas.

Description of the First.

Description of the Second.

Expla. of the way he took the 2 Inc. Planes on the G. W.

Do the Gradients of the G. W.

Calcula. how far the "*visinertiæ*" would take an Engine up 1 in 308.

Do. 1 in 343.

Power req. on the 5th and 6th Plane of the G. W.

The Eng. upon the L. and M. can rise on 1 in 250.

Des. of Eng. upon same.

All Eng. go slower up Slopes.

Observ. upon Work. of Engines upon 1 in 250.

Expla. of same 1 in 250 taken both ways is equal to a Level as reg. the Mechan. Power and Fuel.

an Engine expressly to surmount them.—In ascending a Slope of 1 in 250, the Engine is obliged to pull at the rate of 12 lbs. per ton, *i. e.* 9 lbs. to cover the Friction of the road, and 3 lbs. for Gravity; but in going back the pull would only be 6 lbs. per ton, as the Gravity would return the same relief to the moving power as it took away from it in ascending, and therefore, considered in reference to "Mechanical Power," is the same as a level; thus there is nothing lost by the Rises and Falls upon the Great Western, except upon the Box and Euston Square planes, where there is a waste of power in this way, "the Engine must be built of a certain strength, and therefore of a certain weight, "in order to be able to exert the maximum pull; and although it does not exercise it upon "the other parts, it nevertheless has to carry the increased weight," this Waste of Power applies to the Fuel, but not in the same degree as it does to the Wear and Tear of the Engine and Rails.———Suppose the Basing ascended at an inclination of 1 in 250, or 21 feet per mile, for one-half the distance viz. 53 miles, passing over a summit level of 1124 feet, and descended at the same inclination, it would be similar to a Level in reference to "Mechanical power," *i. e.* the inclination would be neutralized; but it would be different with "Steam power," because "you must have a heavier engine to ascend, upon which there would be a waste in descending:" therefore, the above line would be better than the Great Western as regards the application of "Mechanical power in general," (or Horse power) but not in the application of "Steam power," the more we can equalize the resistance in the latter, "*cæteris paribus,*" the better. I therefore certainly consider "the most Level line is the best."———

Application of the above Theory.

He pref. the most level Line.

Remarks upon Inclines, and Power of Breaks.

A light weight may be stopped but not a heavy one.

Mr. Brunel's experiments upon the Canterbury incline, viz. of passing down at full velocity in a carriage without a rope, and stopping it by the break, within 60 yards forms no criterion of its safety, because the power of the Break is inversely to the load, and although he may stop a waggon containing 5 persons, it would require a greater power, also more in proportion, to stop a greater load, one is a light feather in a single carriage, or 750 lbs., (averaging 150 lbs. each person) and the other a heavy Train of carriages weighing 50 tons, going down the Slope.———A Train descending a slope of 1 in 107 has a downward tendency requiring 12 lbs. per ton to balance; and a carriage descending from the top of the Box Plane, from a state of absolute Repose would arrive at a velocity of 48 miles an hour by Gravity only, (allowing 9 lbs. per ton for friction) occupying about $6\frac{1}{2}$ minutes in the descent, and whatever velocity it possessed at starting must be added, thus if it was 20 miles an hour, the result would give 66, (the engines generally arrive at the top of a slope at a speed of 20 miles an hour.) Upon starting from a state of rest, and proceeding by Gravity only without any power, it would acquire a velocity of 30 miles an hour after passing over 5612 feet, if the Break was then applied to check any further increase of velocity, it must exert a force of 12 lbs. per ton throughout the remainder of the slope, a distance of 7588 feet, the total force required would therefore be equivalent to 4,500,000 lbs. raised 1 foot high.* But the pressure of the Break upon the wheel must be 5 or 6 times the amount of the resistance required, or the proportion which the actual pressure of the former bears to the resistance,

A Train descen. the Box has a downwd. tendency of 12 lbs. per Ton.

Calculation of the Velocity in descen. the same.

The press of the Break must be 5 or 6 times the Gravity of the Load to stop it.

* The learned Doctor must here have alluded to the total force required "supposing the Break to be of a description capable of exerting a Power equal to the Gravity of the Load," but none of our present Breaks resist in that proportion or act upon more than $\frac{1}{5}$ or $\frac{1}{6}$ of the total amount of Gravity; the following lines explain the real power required.—*Editor.*

(supposing it to be Elm); thus to produce a resistance of 12 lbs. a ton we must press the Break upon the wheel with a force amounting to 60 or 70 lbs. per ton.——Upon the Euston Square Plane of 1 in 86, it must be 17 lbs. per ton, or 85 or 90 lbs. in the whole; the pressure must be applied upon the velocity becoming rapid, which depends upon the speed with which it starts.———I have seen the Break totally fail on the incline of 1 in 96 upon the Liverpool and Manchester, (as stated by Mr. Stephenson in his evidence,) the Engineer having let the train down at considerable velocity, upon applying the break it completely failed and was burnt, and we did not stop until at a considerable distance from the bottom, when it was found that the wheels of one of the waggons were broken, and had been dragging along the rails during the whole descent, which must have formed a more powerful check than the break, notwithstanding we went down at a most furious velocity, perhaps 40 or 50 miles an hour, we probably started at about 15 or 20 miles.———There is a Curve ¼ of a mile from the end of the Box plane of ¾ths of a mile radius, which is objectionable, as it has a tendency to throw the waggons on the outside of the rail, and a very slight pressure added to which, would turn them off, particularly if the former should occur at the upper corner of a joint rail, which after they have been used a short time are seldom flush; this tendency increases in the same proportion as the square of the speed and the smallness of the radius. With a load of 100 tons travelling 50 miles an hour it would amount to 93 lbs. per ton, or 9,300 lbs. altogether, upon a curve of ¾ths of a mile radius, which would apply according to the weight; if the waggon was loaded with 5 tons, then the outward tendency would be 5 times 93 lbs. At 40 miles an hour it would be 57 lbs. per ton, or 5,700 lbs. altogether; at 35 miles an hour it would be 47 lbs. per ton, or 4,700 lbs.; at 30 miles an hour 33 lbs. or 3,300 lbs., the pressure would be upon the flanches of the two outward wheels, upon which it would be divided.———I therefore consider the Curve at the end of the Box plane to be highly dangerous to the passenger train descending from Bath, and I think it would be necessary to lower the speed from 23 miles, (which is the average) to 20 miles an hour, in order to descend with any degree of safety.———I also recollect an Accident arising on account of a Curve on the Liverpool and Manchester line, which prevented the Engineer seeing a train of stone waggons occupying the road in advance, although a signal was made to cut off the steam and put on the breaks, (which he alleged he did) yet the velocity was so great that we dashed the former to pieces, but as there is provision made in the passengers' carriages to prevent the effects of a collision, and being likewise protected by the Engine, we escaped, although some of the passengers were much bruised;—Accidents are not likely to occur on Curves if proper precautions are taken, unless the Break should fail. If it was worked by a single rope, and it was to break, I do not think any resistance would stop the train dead, although it would gradually effect it, if applied the moment it broke it would prevent their descending, but the *vis inertiæ* will not extend far;—A Break is not likely to act as well upon a long as upon a short plane, as the train may get beyond its power.—The space of ½ of a mile would not be sufficient to stop it upon the Box Plane, neither do I think they would have much effect upon a great velocity.———The circumstance of Mr. Locke's having descended the Tunnel for goods on the Liverpool and Manchester of 1 in 48 from a state of rest, without applying the break, does not prove that it is not

The downw. tend. on the Euston would be 17 lbs. per Ton.

Insta. of the Break failing upon 1 in 96 L. and M.

The Curve at the bottom of the Box Inc. is very dang.

Remar. upon same.

100 Tons have a force acting upon the outer Rail of 93 lbs. per Ton, at 50 Mil. an hour, upon ¾ Mile Curve, 57 lbs. at 40, 47 lbs. at 35, and 33 lbs. at 30.

Instance of an Accident arising from a Curve.

Remarks upon Power of Breaks upon Inc. Planes.

dangerous, as by the time it arrived at the bottom the velocity must have been 54 miles an hour; he must have taken especial precaution; this Tunnel is used for Goods exclusively, there being another for Passengers.——Mr. Brunel has referred to the

A very dangerous Curve is not appar. to the eye.

inclination of the Lowther Arcade, about 1 in 107, in illustration of the little danger to be anticipated, which cannot apply to any more than the appearance: and an inclination quite inadmissible upon a Railway would be scarcely perceived by the eye.——

The Box Plane 2¼ Miles long.

The Tunn. 1¾ of it.

The Box Plane of 2½ miles is also objectionable, on account of 1¾ of it being in a Tunnel: I am not aware of any parallel, and nothing but an overruling necessity can justify it, as the Power requisite to pull a load up a plane of 1 in 107 is greater than

The Power requir. upon 107 is 3½ to 1 com. with a Level.

Comparison of the Destr. of Air in this Tunnel with one upon a Level.

is required upon a Level, in the proportion of 30 to 9, or nearly 3½ to 1, and requiring a proportionate increase in the consumption of fuel, which will produce an increased destruction of atmospheric air: compared with a Tunnel of similar length, but upon a Level, it would be equal "to one nearly 6 miles long and 9 feet high, (as the proposed height is 30 feet) the breadth being the same in both cases," and I cannot conceive such a Tunnel in the middle of a line exactly practicable.——

Do. with Coals.

Do. with Coke.

I consider Mr. Brunel in error when he says the Tunnel will not affect the atmosphere; the use of coal would be intolerable, and if coke was used it would not be merely the gas which would escape from the combustion, but the gas decomposed in the atmosphere, which mixed together are similar to that which is found in wells, or carbonic acid

Obsv. upon Shafts.

Their insufficiency.

gas, which in large quantities is fatal to life.——I have assumed in my Calculations, that the Shafts will not produce any effect upon the passing train, although they will probably ventilate it for the next, (which I consider the extent of their utility) and I consider that some other means must be resorted to carry it off: it is impossible to state what degree of foul air would remain in the tunnel, and affect the next train; but part of it is sulphuric acid gas, a small portion of which will produce serious inconvenience, as 1 gas pipe in 10,000 left open in the streets frequently does; it is

Effect of an Engine in the Tunnel.

only to imagine a street contracted into a tunnel to judge of it: This air is lighter than the atmosphere when it comes out of the chimney, being hot, but it becomes heavier when cold: all the high pressure steam that works the engine ascends the chimney, not with the natural force of draft, but blown with prodigious violence, and forming a jet, carrying along with it the noxious air, and striking the roof with such force that it rebounds on the first carriage like a ball, and when the steam is condensed it issues in the form of a white cloud of particles of water, and goes off with the wind, but this is

½ lb. of Coke per Ton per Mile Lev.

The quant. of noxi. Gases form. by the Fuel in the Tunnel would be 3090 lbs.

not the case with the foul gases, which are incondensable: The passage of a load of 100 tons through the Box Tunnel, allowing ½ a lb. of coke per ton per mile to draw a train upon a level, would deposit about 3090 lbs. of noxious gases incapable of supporting life, viz. carbonic acid 1077 lbs., azote 1077 lbs., and an uncertain quantity of sulphureous acid, (perhaps 150 lbs.) which varies according to the different kinds of coke.

The Mecha. Power regulates the speed.

——It is the Mechanical or horse power (horse power is equal to 33,000 lbs. raised 1 foot high per minute) which would regulate the time occupied in passing through the tunnel, and not the actual consumption of fuel, although the rate of consumption does;

To conti. the same Speed the Fuel must be incr. in the pro. of 30 to 9 comp. with a Lev.

thus, "the more Fuel consumed per minute, or per hour, the greater will be the Horse power," and although the Smoke consumed by the engines would be in the proportion of 30 to 9, it does not follow that an engine in the proportion of 30 to 9, as compared with

Rem. upon wrk. the Box by an Ass. Eng.

those used upon a level, would pass it at the same speed as upon a level, unless the "Fuel consumed" was in the same proportion, a weak as well as a strong power may be produced at a rapid rate.

The Fuel consumed upon the Box must exceed that used upon the latter in the pro. of 30 to 9.

If 30 is consumed in a tunnel 30 feet high, 9 will be consumed in a tunnel 9 feet high, and it is not essential to compare the lengths, because the Consumption of Air throughout will depend upon the Total quantity of Mechanical Power expended; the effect produced per yard, or per mile, will be the same in both cases.—— ——

Expl. of same.

There is not as great an objection to a Tunnel of $1\frac{1}{4}$ miles upon a Level; it is the slope which forms the difficulty: the Passengers passing up the Box Tunnel would breathe the same quantity of vitiated air, for the same length of time, as they would in a Tunnel of 9 feet high and nearly 6 miles long upon a Level, the breadth being the same in both cases: there is no consumption of vital air, practically speaking, in passing down, although they would vitiate the air in a small degree:

If 2 Trains meet it will be 18 to 30.

in the event of 2 trains meeting in the tunnel, those descending would receive the ill effects as well as those ascending, and the consumption of air under these circumstances, in comparison with a Tunnel of the same length upon a Level, would be in the proportion of 18 to 30, but it is not very likely to occur.———

Remar. upon same.

A fair Comparison between two Tunnels, one upon a Slope of 1 in 107, and the other upon a level, the necessary power required to work each being in the ratio of 30 to 9, would be by supposing 3 Level Tunnels, each 9 feet high: and the passengers would breathe the same quantity of vitiated air in each case, the length not affecting the question.———It depends upon circumstances whether an Assistant Engine working from behind would affect the passengers; it would not if the air was in a state of repose, but they would be involved in all the foul air of the leading engine:

14 or 15 Miles an Hour the most upon the L. and M. Inc.

much depends upon the relative velocity of the current, which might sometimes be equal to 10 miles an hour, but 20 would be a very high wind; if less rapid than the velocity of the train it would not blow any bad air upon them, possibly it might overtake them, (they do not exceed 14 or 15 miles an hour on the slopes of the Liverpool and Manchester) but I think it very unlikely: (short tunnels are more likely to be ventilated from the ends.)———Assuming an engine upon a Level has the power of drawing 9, there is no difficulty in conceiving that an Assistant Engine may be made capable of drawing 21 at the same velocity, which together with the other would make 30, by which they would continue up the plane at the same velocity as upon a level, by consuming fuel and atmospheric air in the same proportion.

Plane of 1 in 96 on the L. and M. work. by an Assis. Eng.

Light loads go up alone.

The Plane of 1 in 96 upon the Liverpool and Manchester is generally worked by an Assistant Engine, (but in cases where the train is underloaded they go up alone,) which frequently delays them, as it may be employed upon another train, sometimes it is not lighted, (there is the extra cost of keeping an engine always ready for use, whether required or not,) and I have known them come to a dead stop upon this plane.———

Obs. upon work. the Box by a Stati. Eng.

Calculati. of same.

The Gravity upon the Box Tunnel would amount to $\frac{1}{107}$th part of the load, thus if the load was 52 Tons it would amount to 1,088 lbs. and taking the Friction at 9 lbs. per Ton gives 468 lbs. upon 52 Tons, which added to the Gravity would give 1,556 lbs. as the power of Draught necessary to overcome the load, exclusive of the weight of the rope, which would require to be 7 in. in circumference, and $6\frac{1}{2}$ lbs. per yard, at which rate an endless rope 5 miles long would weigh 57,000 lbs., and according to Messrs. Stephenson and Locke's Experiments, which appear very satisfactory, the resistance may be taken at $\frac{1}{12}$th its weight.

Friction of the $\frac{1}{12}$th of Load.

Therefore $\frac{1}{12}$th of 57,000 lbs. will

4752 lbs. the Power req. for the Rope, or 3 times more than the Load.
25 lb. pull at 15 mil. an hour 1 Hor. pow.
It would req. an En. of 252 horse powr.
140 hor. pow. is int. for the L. & M. Pla.

give the force necessary to pull the rope up, which alone amounts to 4752 lbs., making a Total of 6308 lbs. (the Power required for the rope compared with the load, is therefore in the proportion of 3 to 1).———Now 25 lbs. pulled at the rate of 15 miles an hour is equal to 1 Horse power, therefore 6308 lbs. pulled at that rate would be equal to 252 Horse power, and two Engines would also be required, one to work while the other was being repaired, (Mr. Stephenson has stated the Engine at 70 Horse power,) an Engine of 140 Horse power is intended to be used upon the Tunnel now making upon the Liverpool and Manchester for Passengers and Light Goods, which is 1 mile long 1 in 100, upon which 50 tons will be an ordinary load, including carriages, which therefore bears out my calculation, (I do not apprehend that there will be any foul air generated in the latter tunnel, except by respiration, if it is worked by Stationary Engines).

Des. of working the Plane of 1 in 48 upon the L. and M.

———It is incorrect to suppose that a $5\frac{1}{2}$ inch rope will be sufficient on account of the Inclination being less than that in the tunnel upon the Liverpool and Manchester (1 in 48 and 2,000 yards long, upon which they use a 6 inch rope) in which case the load brought up by the Locomotive is divided into 5 or 6 different portions, and drawn up separately in loads of 25 tons each, (this system would not do in the middle of a line, on account of the time it occupies,) at about 10 miles an hour,

Observat. upon the size of an Endless Rope for an Incline.

but the rate at which they travel does not affect the strength of the rope, the Length must be taken into consideration but not the Acclivity, as far as the Rope is concerned, unless there is an increase of load, it balances itself, thus, " if you put an Endless rope over a pulley actually perpendicular, and apply a Power to put it into motion round the pulley, you require the same force as you would with the same rope upon a level, as it balances itself," (the steepness would affect a single rope).

He consid. a 5 Mi. Rope impractic.

I therefore believe a 5 miles Rope is impracticable in the common sense of the word, and that no one would attempt it, yet it is within the bounds of Mechanical possibility.

Do. single Ropes.

———A Single Rope would be only half the weight and half the resistance, yet the Power required to draw it back and unroll it would be considerable, however a small Locomotive would be equal to it, but as the Fuel would consume the vital air in the Tunnel it is objectionable, and it would also be an extra expence in the consumption of Fuel; the cost, also wear and tear of a rope is considerable; the rope upon the Liverpool Tunnel is spliced twice a week, and the Sheaves and Pulleys are constantly being repaired, there are 220 pulleys in a mile, which cost from 15*s.* to 20*s.* each.

Obj. to the Basing.
The Eng. must be heav. than the G. W. in the pro. of 20 to $15\frac{1}{2}$
Obs upon same.

———To sum up, the Engines of the Basing must be built capable of exerting a greater pull than those of the Great Western in the proportion of 20 to $15\frac{1}{2}$, they must be stronger and heavier, and the power necessary to carry this increase of weight is so much power to their comparative disadvantage, but it is erroneous to suppose that as the Engines are built capable of exerting this greater Pull, that they will use it throughout the line, they will only bring it into action upon the Planes, by which there will be no waste of power. There is also the increased Wear and Tear of the Railroad if the Rails are not of equal weight to resist the above, or the interest of the capital necessary to make them so. I believe these are all the disadvantages upon the Basing.

Objec. to the G. W. Euston Plane, Box Plane, Tunnel and Curve.

—The principal objections to the Great Western is the waste of Power upon the 2 Inclined Planes at Euston Square and Box Hill, also on account of the latter being partly in a Tunnel, and situated in the middle of the line, with a Curve at the foot of it; (I do not object to the remaining Tunnels on the Great Western Line): I am quite sure that the idea of working the same by a Rope, endless or single, must be abandoned, and recourse must be had to an Assistant Engine, however objectionable; a Stationary Engine is likewise a waste of power, as the fire must be kept up all day, with the steam roaring out of the safety valve, always ready for action; but I am not sure whether an Assistant Locomotive could work up a Plane of 1 in 86. The remaining Gradients being better, lighter Engines may be used upon them, which would consequently diminish the Wear and Tear upon the Rails and Engines, and the cost of the latter would also be rather less.——I am therefore of opinion, after considering the several advantages and disadvantages of both lines, that the Basing Line is the most preferable, particularly for Passengers, independent of the communication with the Thames.

He cons. the Box must be worked by an Ass. Eng.

He doubts whether an Ass. Eng. could work upon the Eust.

He pref. the Basing and Bath Line.

Ex. Mr. JOHN URPETH RASTRICK, C.E.

I have practised as an Engineer 40 years, and have been consulted upon nearly all the several railways laid down since 1824, and I have travelled upon every railway in the kingdom. I executed the Kenyon and Leigh Railway, which forms the junction between the Bolton and Leigh and the Liverpool and Manchester, which is 2½ miles long, and cost £13,500.: I also executed the King's Swinford, which is a line connected with some iron works, and 5 miles long including its branches; it commences with an Inclined plane, the loaded carriages bringing up the empty; a Locomotive then takes them 2 miles, when another Plane takes them to the bottom: the engine runs at a rate of 7 or 8 miles an hour, which is above its power, but it is seldom out of order. I also executed the Stratford and Moreton, which is 16 miles long, and worked by Horses: the only circumstance attending the working of a Locomotive engine in place of the latter is that the chairs and rails must be heavier, and the tunnels a little larger: (the fish-bellied rails are used upon the three last mentioned railways).——I was the Engineer under the Act for the Grand Junction, (I laid out the line from Birmingham to Basford) which is also intended for Locomotives: I was directed to take the shortest possible line, without considering difficulties: the Company afterwards put it into the hands of Mr. Locke, under whose direction it is executing, Mr. George Stephenson being the Consulting Engineer, and they have rejected 42 out of the 52 miles which I laid down, and I will stake my respectability as a professional man that they will regret it.——I was also Consulting Engineer on the Dublin and Kingstown Railway, which is opened, and I believe answers.——I have been largely engaged in the Manufacture of Steam Engines all my life; I was apprenticed to my father, and employed in the construction of them for the first 5 years, but I have resigned manufacturing for the last 8 or 10 years. I have also given much attention to the subject of Locomotive Engines for the last 11 or 12 years, (manufacturers of steam engines can at any time turn their attention to them, and it has been the case with many others besides myself): I also combine with it every information. I communicate with Mr. Dixon, of Manchester, the

His Ex. and Works.

He was Engineer of the Kenyon & Leigh and of 2½ mi.

Ditto the King's Swinford of 5 mil.

Des. of same.

Ditto the Stratford and Moreton of 16 m.

Des. of same.

Do. the Gra. Junc. of 52 miles.

Des. of same.

Remarks upon his Experience connected with Locomotive Engin.

Resident Engineer of the Liverpool and Manchester, upon the subject, by which I can draw as good conclusions upon the several engines as though I had made them; but I do not consider that an individual who has not made an engine can have the experience of a manufacturer, or be as capable of judging of their economical working.——— I was also engaged, in conjunction with Professor Barlow and Mr. Nicholas Wood, to select the pattern entitled to the Premium which the Directors of the London and Birmingham offered for the best Rail and Chair, none of which were of a description that we could recommend to be employed. I also gave Evidence in Parliament upon their Bill; I considered Mr. Robert Stephenson's Estimate (£1,875,527.) too high, and when first before Parliament I objected to support it on that account, (I believe Mr. S. is in the habit of making his estimates high in order to prevent opposition in Parliament, and to cover any unforeseen contingencies): we apportioned it differently; he estimated the whole at 1*s*. 6*d*. per cubic yard; I left a larger amount than usual for contingencies, £374,473., (17 per cent.) in order to cover the difference between us, by which I brought it to his estimate, (his price for the same was £195,496.) I calculated upon the Contractor using the Permanent Rails and Chairs belonging to the Company, but he would find all other materials, including sleepers, &c. My estimate for Clay was 1½*d*. for getting, 2*d*. for filling, 1*d*. for spreading, 3*d*. for the first half mile, and 3*d*. for every successive mile of lead; in Chalk, ½*d*. for getting, 2½*d*. for filling, 1*d*. for spreading, and the same ratio for leading; in Sand, ½*d*. getting, and the remainder similar to clay; Stone and Marl, 2*d*. getting, 2*d*. filling, 1½*d*. spreading, and lead the same as before; Red Sandstone in the neighbourhood of Meridon was 6*d*. getting, 2½*d*. filling, 1½*d*. spreading, and the lead was the same as above: these calculations were made upon the supposition of Horses being used upon the teaming, (at least I do not think I intended Locomotives, as I have allowed a very high price). If an Engine properly constructed for the purpose was used, it might be done for 1*d*. or 1½*d*. per cubic yard, or about half the price of Horses, but much will depend upon the length of the road. Some soils do not require spreading, as sand, which teams out of the waggons at once, and forms itself into an embankment; clay, on the contrary, comes out in very large pieces, perhaps 2 feet square, in the Summer; I always describe them in my Specifications to be cut into pieces of not more than 6 inches. I think I averaged the whole of my Excavations at 13½*d*., allowing 3*s*. 6*d*. a day for the navigator's wages, which I did not consider too high.———Nothing is more fallacious than making a calculation with particular reference to one railroad, and applying it to another.———I was also called as a witness to prove the applicability of Locomotive power on the Liverpool and Manchester, and I established the principle on the first application to Parliament for a Bill. ———I have looked over the Gradients of the Basing Line, and I do not find any Plane that cannot be overcome by Locomotive Power (without Assistant Power) at a rate of 20 miles an hour, which is about the general speed, and by generating or reducing the quantity of steam it can be regulated to suit every inclination upon the line: I find it best to have Engines above their work rather than below it: and it was by my advice that the Directors of the Dublin and Kingstown used heavy Engines on their line in preference to light ones.———I have always used the Permanent Rails in the construction of the line, which is a great saving; they sometimes became bent, but we easily straightened

Lond. and Birming.

He was one of the judges when they gave a Prem. for best Rail and Chair.

Mr. R. Stephenson's first Estimate.

Aver. of same. 1*s*. 6*d*.

Mr. Rastrick's do.

The Perm. Rails are always used in the forma. of the Line.

Details of his Estim.

For Clay, do.
Chalk,
Sand,
Stone,
Marl,
Red Sandstone

Teaming with Loco. about half the price of Horses, or 1*d*. to 1½*d*. c. yd.

Avera. of his Price for L. and B. 13½*d*.

The Basing Line can be worked by Loco. Power.

He prefers heavy Engines.

He has alw. used the Perm. Rails in the Works.

them, and the injury was very trifling.——The commencement of a great work always proceeds slowly, the materials not being provided causes considerable delay.—— ——

The comm. of great works always slow.

I have made a Comparison of the Gradients on the competing lines from Calculation, taking all their several undulations, as I should any other mechanical subject, commencing from the level of the Depôts in each case, (but I am not aware of their difference in height.)——

His Comparison of the Gradients.

The Sum of all the rises from London to the Summit on the Basing Line is 480 feet, and the same upon the Great Western is 319 feet 9 inches, leaving a difference of 160 feet 3 inches in favor of the former. From the Summit to Bath by the Basing the rise remains 480 feet, and from the Summit of the Great Western at Swindon they fall and rise up again to the Tunnel; and the Sum of all these rises from London to Bath amounts to 383 feet, therefore the Rise upon the whole distance is only 97 feet in favor of the Great Western.——

The Rises from Lon. to Bath 97 ft in favor of the G. W.

The Sum of all the rises from Bath to London by the Great Western is 364 feet 6 inches, and by the Bath and Basing 355 feet, which leaves a balance of 9 feet 6 inches in favor of the Basing; the rises from the Summit of Burbage by the Basing Line towards London is 350 feet, and the Sum of all the rises from the Summit at Swindon towards London by the Great Western is 313 feet 9 inches; this leaves a Balance of 36 feet 3 inches in favor of the Great Western.——

Ditto from Bath to London 36 ft. 3 in. in fav. of the G. W.

——I have also made Calculations of the Power required on each line. It is a fundamental principle in all calculations to determine exactly the force exerted by a certain power. It is generally understood that 33,000 lbs. raised 1 foot high is equal to 1 Horse Power, which is my basis; having ascertained the Friction (which I have taken at 10 lbs. per ton) and the Resistance, I multiplied them into the length of the Railway: the Gravity depending entirely upon the inclination, I take the difference between the Sum of all the falls and rises, and then ascertain what Power would be required to raise 1 ton up that height, which I reduce to the height of 1 foot, as the standard upon which all Mechanical calculations are compared, *i. e.* as so many lbs. raised 1 foot high.——

His Calculat. of the Pow. req. on both li.

33,000 lbs. rais. 1 ft. hig. eq. to 1 hor. pow.

He all. 10 lb. per Ton for Friction.

His Formula for calcula. the above.

Applicati. of same.

As the Sum of all the Rises on the Great Western from Euston Square to the Summit at Swindon is 319 feet 9 inches, and the Sum of all the Falls within the same distance is 50 feet 9 inches, there is consequently a total rise of 269 feet; now the distance being 79 miles 41 chains, which if reduced will make 419,826 feet, which multiplied by 10 lbs. per ton, gives the Friction over this distance equal to 4,198,260 lbs., and the Altitude to be surmounted, or the Gravity to be overcome, is 269 feet, which multiplied by 2240 (the number of lbs. in a ton) gives 602,560 lbs. as the Gravity to be overcome, whichadded to the amount of Friction gives 4,800,820 lbs. raised 1 foot high; which is the Power required to take 1 ton from London to the Summit of the Great Western:—

Rise upon the G. W. from Lon. to Summit 269 feet.

Length of same 79 in. 41 ch.

Friction on same 4,198,260 lbs.

Gravity of the above 602,560 lbs. giving a Total Resistance 4,800,820 lbs. raisd. 1 ft. high to draw 1 Ton. to the Summit of the G. W.

the Sum of all the rises from London to the Summit of the Basing is 480 feet, and there is a fall of 5 feet between those points at the Mole embankment, which being subtracted leaves a total rise of 475 feet to be surmounted; the length being 71 miles 67 chains, and reduced to feet as before, makes 379,302 feet, and multiplied by 10 lbs. gives 3,793,020 feet as the amount of Friction: the total Altitude or Gravity to be surmounted is 475 feet, which multiplied by 2240 gives 1,064,000 lbs., and added to the amount of Friction, makes a total of 4,857,020 lbs., leaving a difference of 56,200 lbs. raised 1 foot high in favor of the Great Western:—

Do. on the Basing,

Friction	3,793,020
Gravity	1,064,000
Total	4,857,020

the Power required to take 1 ton from the Summit at Swindon down to Bath is 1,161,060 lbs., and from Burbage upon the Basing Line

To the Basing Sum.

1,082,201 lbs., leaving a difference of 78,859 lbs. in favor of the Basing, after deducting from this amount the advantage of the Great Western up to the summit, or 56,200 lbs., a balance will be left of 22,659 lbs. in favor of the Basing, upon the whole distance from London to Bath.——The result of the Calculation from Bath to London is as follows:—the Basing Line from Bath to the summit at Burbage, including the level plane at the Eastern end, will require a power of 2,612,860 lbs. raised 1 foot high:—the Great Western from Bath to the summit at Swindon, including the tunnel, will require a power of 2,140,500 lbs., leaving a balance of 472,630 lbs. in favor of the Great Western; and from the summit of the same down to Euston Square requires a power of 3,659,167 lbs., and 2,729,020 lbs. upon the Basing, leaving a balance in favor of the latter of 930,147 lbs., from which must be deducted the balance in favor of the Great Western, from Bath up to the summit, which leaves a balance upon the whole distance of 457,787 lbs. in favor of the Basing, which added to the amount of its advantage from London to Bath, makes the total advantage of one complete trip from London to Bath and back 480,464 lbs., thus the average each way is 240,223 lbs. raised 1 foot high in favor of the Basing Line.——The average Power to take a ton 1 mile upon the Liverpool and Manchester is equal to 56,415 lbs. raised 1 foot high: upon the Great Western it is 53,742 lbs., and upon the Basing 52,975 lbs.: the difference as compared with the Liverpool and Manchester is about $\frac{1}{16}$th, and as compared with the Great Western is $\frac{1}{70}$th of the whole power required upon these lines in favor of the Basing Line.——This power is the Power required to take 1 ton of goods only, but as it is always accompanied by a certain quantity of "Dead Weight," (the tender, coals, water, &c.,) it is necessary to increase this power in the "same ratio which the dead weight bears to the weight of the goods," in order to arrive at the real degree of power required.——An Engine of 10 tons weight, and 11 inch Cylinders, would be capable of taking 110 tons upon a Level, including the weight of the Engine and tender, fuel and water (4 tons): the Carriages for the Goods being $27\frac{42}{100}$ tons, makes the dead weight $41\frac{42}{100}$ tons, the weight of the goods is therefore $68\frac{58}{100}$ tons, or in the proportion of $1\frac{6}{10}$ to 1 compared with the whole Train, (the weight of the goods generally bear a proportion of $2\frac{1}{2}$ to 1 to the weight of the carriages). The Power to take 1 ton over a mile upon a Level being 52,800 lbs. raised 1 foot high, allowing 10 lbs. per ton for Friction and the Dead Weight, would increase the Power required for the same to 84,691 lbs. raised 1 foot high, and the Ratio of the dead weight to the general weight of a train would be increased by the use of Assistant Engines.——The steepest Gradient upon the Great Western, excluding the Box Plane and the fall at Euston Square, also a short rise of 1 in 308, is 1 in 473, upon which an Engine of 10 Tons and 11 inch Cylinders would be required for a load of $64\frac{64}{100}$ Tons, the gross weight of the whole train would therefore be $74\frac{63}{100}$ Tons, which is the utmost the Engine would draw up this inclination, much less would it proceed up the Box Plane, but an Assistant Engine would be required of $12\frac{1}{2}$ tons, and $13\frac{7}{8}$ths inch cylinders; the Plane at Euston Square of 1 in 86 will require a Fixed Engine to assist in taking up the train, with the Locomotive Engine and Tender, and if they travel at the rate of 20 miles an hour, it must be of 102 Horse power, with ropes, chains, pulleys, drums, barrels, &c.; if a Locomotive Engine was descending at the same time that

Differen. from Lon. to Bath 22,659 lbs. in favor of the Basi.

Diff. from Bath to London 457,787 lbs. in fav. of the Basi.

Ave. Power to take 1 Ton 1 Mi. on the L. and M. is equal to 56,415 lbs. raised 1 ft. high.

Ditto on the G. W. 53,742 lbs.

Ditto on the Basing 52,975 lbs.

10 Ton Eng. 11 in. Cyl. would take 110 Tons upon a level.

Total Dead Weight $41\frac{42}{100}$.

Total Weight of Goods $68\frac{58}{100}$.

Pow. to take 1 Ton 1 Mile upon a Lev. 52,800 lbs. raised 1 ft. high.

Do. incl. dead wht. 84,691 lbs.

The G. W. would req. a 10 ton. Eng. with 11 in. cylind allo. $74\frac{63}{100}$ Tons gross Load.

The Box wo. req. a $12\frac{1}{2}$ tons Ass. Eng. with $13\frac{7}{8}$ cyl.

Euston Plane would req. a 102 hor. pow. Fixed Eng.

another was ascending, a considerable degree of power might be saved, but I apprehend that they would always wind up the moment they are ready, without waiting for another Engine; therefore the above stated Locomotive would work throughout the whole distance, at least I consider that would be the most practical mode of working the line.——The greatest Inclination on the Basing from Bath to the foot of the incline of 1 in 202; of 14 miles 14 chains is 1 in 560, which an Engine of 10 Tons and 11 inch Cylinders will work backwards and forwards, and if the line was continued on to Bristol it would run about 11 miles further at nearly the same speed; the Engines with Passengers upon the Liverpool and Manchester run through the whole distance of about 30 miles without taking in fuel or water, which is the utmost that I should recommend an Engine to go, it would be highly injurious to continue beyond that distance, as the fire bars become clogged by clinkers, a certain portion of the coke also runs into glass or cinders, the Engine therefore requires to be opened and cleansed, oiled, &c.; a saving of time would be effected by taking a fresh Engine, and proceeding at once with the train, leaving the old one.——The next Plane is 1 in 202 for 6 miles 54 chains, after which there is another ascent of 1 in 528 for 6 miles 40 chains; the next is 1 in 264 for 8 miles 20 chains: the first of the above planes is near Bulkington Mill, 14 miles from Bath, upon the summit of which there is a Level of 4 miles 3 chains, when there is an inclination of 7 miles 28 chains of 1 in 250, making the distance from the foot of the Incline of 1 in 202 to the foot of the Incline upon the Eastern side of the summit at Burbage, 27 miles 65 chains long and 1 in 250, upon which I propose to work an Engine of 12 Tons with $13\frac{1}{3}$ inch Cylinders; the same Engine would return back, and as the highest inclination is only 1 in 250, it would be comparatively easy work.——I propose dividing the remaining distance to London (64 miles 39 chains) into two Trips, the first Engine would run 32 miles 19 chains, and the second 32 miles 20 chains, each being 11 Tons weight and 12 inch Cylinders; as the greatest inclination upon this distance is 1 in 330, (they would run at a greater velocity than 20 miles an hour upon the level portions of the Line):—If the Inclination of 1 in 202 upon the Western side of the Burbage Summit is altered to 1 in 250, the size of the Engine would be reduced, the cylinders would be from $13\frac{1}{3}$ to $12\frac{5}{8}$ inches, and the weight from $12\frac{1}{2}$ tons to 12 tons.——If the Locomotive Engine is not run to the bottom of the Plane at Euston Square, but left at the top, and a fixed Engine employed to wind up the train ($60\frac{63}{100}$ tons) without assistance of any kind, an Engine of 132 Horse power would be necessary,—and a Horse or small Engine would also be required to drag the carriages wherever they were wanted upon the small Plane at the bottom in Euston Square, (of about 572 yards long.)——I have also Calculated the Power by another mode, and with as nearly as possible the same results, I have taken each inclination separately, taking the Friction at 10 lbs. and the Gravity equal to the inclination: (if an ascending line, the Gravity must be added, if descending, it aids the Friction;) by which I find the Mechanical Power required over that distance, this method of taking the inclinations leads into decimals, which requires a long process to arrive at the conclusion accurately, but it is not the case with the former mode, which I consider the most correct.——I have likewise made a Calculation of it another way, by which the Power required to take 1 ton of Goods upon an average of both ways over 1 mile of the Great Western came to 93,485 lbs.

Remar. upon same.

The Basing Line would req. a 16 Ton Eng. 11 in. Cylin. next Bath.

Obs. upon the emp. of Locomo.

Eng. should not run above 30 Miles at a time.

Remarks on same.

Then An Eng. of 12 Tons and $13\frac{1}{3}$d Cyl. upon the 1 in 202 & 250.

And 2 more Eng. 11 Tons 12 in. Cylin. (next Lond.)

If the Plane of 1 in 202 is alt. to 1 in 250 a less Powerf. Eng. will do.

Rem. upon Euston Square Plane.

It would req. a F. Eng. of 132 H. Po.

He calculated each Pla. separ. and with the same results.

Expla. of Gravity.

He proved it by another process.

Wh. gave a Balance of 777 lbs. per Mile raised 1 ft. high in fav. of the Basing.

raised 1 foot high, and upon the Basing, taking into account the weights of the different sized engines, dead weight, &c., to 92,708 lbs., leaving a balance of 777 lbs. raised 1 foot high, upon every mile in favor of the Basing, which also confirms my former statements, and I am confident my calculations are right, as I have had them checked two or three times.——I find the average Expence of 1 Ton of Goods conveyed each way would be 5*s.* $0\frac{95}{100}$*d.* upon the Great Western, and 4*s.* $10\frac{81}{100}$*d.* upon the Basing, although I have taken the most powerful Engines upon the latter, a difference of about $1\frac{47}{100}$*d.* per ton, which multiplied by their weight, or $43\frac{1}{3}$ tons would make it 5*s.* $3\frac{65}{100}$*d.* in favor of the Basing Line.——I consider 60 lb. Rails are sufficiently strong to carry these Engines, although very heavy: (Mr. Robert Stephenson states in his evidence they are capable of carrying 16 tons): the great Wear and Tear of the Engines hitherto has been through the weakness of the Rails, but if they are made of sufficient weight I consider Heavy Engines will be found the most advantageous, although they should cause a little more Wear and Tear of the rails; I have always recommended the use of them, (the best Engine that I am acquainted with, is that which I recommended to the Dublin and Kingstown Railway, which weighs $11\frac{1}{2}$ tons with the water, and has 11 inch cylinders). The Rails originally laid down upon the Liverpool and Manchester were only 35 lbs. to the yard, which occasioned great expence with the Engines, and they cannot be made of a suitable weight on account of it; if they make the Engines sufficiently strong they destroy the Rails, therefore they are working under great disadvantage; I have always found that Engines which are tied down to their power are generally out of order, but it is seldom the case when they are sufficiently powerful; I prefer the Parallel Rails, as it allows of the blocks being placed closer together, thus extra strength can be obtained to carry heavier Engines should they be required, an arrangement which cannot be made with the Fish-bellied: (I consider Professor Barlow, in his pamphlet upon Rails, demonstrates the superiority of the Parallel Rails in the most satisfactory manner): the heaviest Rails which are at present used are 45 lbs., but I have always considered it would be more economical to increase them.——I made a Report (7th March 1829) previous to the opening of the Liverpool and Manchester, upon the comparative merits of Locomotive and Fixed Engines, in which I took the annual expence of a Locomotive at £367. 4*s.* 4*d.* which calculation was made upon Engines exclusively connected with Coal Mines, but running upon a good length of Railway at a speed of 5 or 6 miles an hour. (There is no difficulty in adjusting the machinery of an Engine to make it run 20 miles an hour, the piston must merely be made to work a shorter stroke, and the wheels increased in diameter). Messrs. Robert Stephenson and Locke published their "Statement of the Expence of working a Locomotive," about 12 months afterwards, in which they stated it would amount to £324. 12*s.* 10*d.*, and insisted that I had considerably over-rated it; notwithstanding their then recent experience of the Rocket Engine, (which obtained the prize of £500. at the opening, I was appointed one of the Judges upon the occasion, in conjunction with Mr. Nicholas Wood, and Mr. Kennedy,) which had then been running 6 months on this line, and it was considerably improved in construction compared with the Engines upon which I formed my data, but experience has shewn that neither of our Estimates were sufficient, (we assumed a speed of 5 or 6 miles an hour, which forms no comparison

Aver. Exp. on the G. W. of 1 Ton carr. bo. ways is 5*s.* $\frac{95}{100}$*d.*

Do. Bas. 4*s.* $10\frac{81}{100}$*d.*

Dif. in fa. Ba. $1\frac{47}{100}$*d.*

60 lb. Rails quite suff. to carry the Engines of Basing.

His reasons for preferring Heavy Engines.

Eng. should alw. be above their work.

His reas. for prefer. Parallel Rails to the Fish Bellied.

He does not consi. the Rails at present used heavy enough.

His Report to the L. and M. in 1829.

He took Locom. at £367. 4*s.* 4*d.* pr An.

Messrs. S. and L. ans. it and made it £324. 12*s.* 10*d.*

The Rocket Eng. obt. the Prize at the opening L. and M.

Account of same.

with our present rate of 20).——Mr. Robert Stephenson states in his evidence that the Engine at Willesden cost £1500. a year, or £5. 2*s*. a day for 300 days; and referring to his details of same, I cannot understand why two people should be employed to pump the water, (I certainly do not know the depth of the well which communicates with the tank,) the Fireman, or the person that attends the engine should do it when the latter was at rest, and I think 24*s*. or 27*s*. a week is quite enough for the Engine-man for 10 hours work, (the wages on the London and Birmingham are £2. 2*s*. in consideration of their working as long as may be required, without extra charge), he should also attend to the fire without a Fireman at 18*s*. (my late partner's (Mr. Foster) engine man is paid 27*s*. 6*d*. and he attends to the fire); I am sure the Repairs cannot amount to 15*s*. per day, a good engine will not require any for the first six months, those on the Dublin and Kingstown have not yet been repaired, with the exception of a few brasses, (I san state this positively, being a Shareholder); he has also allowed 10*s*. a day for the Depreciation of the Engine, arising from its getting old, they are of comparative less value every day; but it is not always the case, as many of the Engines on the Liverpool and Manchester after running a number of years and being repaired, have become better than they were originally, (the Engine which has run the greatest distance upon the Liverpool and Manchester, was made by Mr. Berry of Liverpool). I therefore consider his calculations are made upon Engines running 20 miles an hour, working to their full power, and under totally different circumstances which forms no criterion of the expence of an engine for leading earth at about 4 or 5 miles an hour, which is quite sufficient speed, (but 8 or 10 would not affect the amount of Repairs or Depreciation); an Engine of such expence is quite unnecessary, although the cost of Fuel in the neighbourhood of London may add considerably to it: all the Engines in the North burn Coal, (at about 5*s*. 6*d*. per ton), the expence of which I took at £111., if Coke is used it will make an addition of £40. upon an Engine of that size; this brings my Estimate to £407. 4*s*. 4*d*., which divided into the number of working days would give £1. 7*s*. 4*d*. per day, and I am quite satisfied that £2. 2*s*. is sufficient for a suitable Engine; if I were a manufacturer, I would engage to furnish one for that price, as it is not working above ½ of the day. I calculate the cost at about £550., which would be worth about 12½ years purchase, supposing it to last 20 years, I make it about £160, 4*s*. per annum, including Repairs and Depreciation (I took the latter at £55. 16*s*.) The Engine before mentioned which I made for my partner, (the only Locomotive that I have constructed), has been at work upon the King's Swinford ever since the opening in 1829, and the expence of working it amounts to £360. or £370. a year; (both Mr. George Stephenson and Mr. Booth came down to see it before they commenced their own Road).——I used Locomotives upon the Kenyon and Leigh for the removal of earth, as it was desirable to get it finished by the opening of the Liverpool and Manchester; they went a certain distance on the former line to bring materials, and then passed about 1 mile upon the Kenyon, part of which was laid down permanently upon stone blocks, and about 100 yards was temporary and upon sleepers, but quite solid, (if the Railway had been well laid over them they would have lasted as long as stone blocks): there was merely two ends of an embankment to join together, and to the best of my recollection it did not take more than one month, the Temporary Railway preceded the Permanent one but a

Mr. R. Stephenson states the Engine at Willesden £5. 2*s*. per Day.

Mr. Rastrick con. £2. 2*s*. sufficient.

His reasons for same.

Many Engines on the L. and M. are better now than when originally made.

Mr. S.'s Calcu. are made upon a too expensive Eng.

Details of the Expense of Working an Engine considered.

Mr. R.'s Engine on the King's Swinford cost £360. per An.

Locomotives were used on Kenyon and Leigh for the removal of Earth.

Account of same.

Loco. very advan. in wet weather.

very short time; our reason for applying the Locomotive was on account of the uneven surface of the road, the season being exceedingly wet the horses were working up to their knees in clay, which prevented their proceeding sufficiently rapid; the weight of the Locomotive did not cause any settlement of the embankment, (but it could not well take place during the short period of the works, being only 6 or 7 feet high);

Sand is the best Mat. for form. Emb.

I also consider that they may be used with safety in the formation of Embankments 30 or 40 feet high in Sand, which is the best material for forming embankments, as it soon comes to a settlement:

Observ. upon same.

I remember a case in which I ordered a line to be kept 2 feet higher upon a Sand embankment, to allow for settling previous to the rails being permanently laid, and I was obliged to take 18 inches off again; Mr. Treadwell* executed the work under me, who I believe to be a honest and skilful Contractor, the men did not work in double shifts, neither was the locomotive used until the last fortnight, (for which he paid £2. 2*s*. a day).———

There were Planes of 1 in 211 and 214 orig. on the Grand Junction Railway.

I had a Plane of 1 in 211 upon the Grand Junction, (in consequence of being obliged to rise over one of the streets next Birmingham) which Mr. George Stephenson considered extremely dangerous, and objected to, on account of its being next the Station, (there was also a 7 miles plane upon the same line at 1 in 214) but Euston Square Plane would be considerably more dangerous, as it descends immediately into the Station.———

Objections to the Box Plane.

The Box Plane is also very objectionable, it will neutralize much of the most perfect part of the line, and prevent their availing themselves of gravitation; on the contrary, they will be obliged to make use o breaks to check it: the Inclination of a Railroad should be arranged in such a manner as to render breaks unnecessary, as they destroy part of the power; although I do not consider there is any danger in descending a plane of 1 in 107 if the breaks are properly attended to, but you dare not let the engines run down amain, as they would arrive at a very great velocity by the time they reached the bottom.—

Calculation of the Velocity in descen. the same.

I have calculated that a carriage started at the top from a state of rest, would acquire a velocity of about $43\frac{1}{2}$ miles an hour by the time it reached the bottom, occupying about 5 minutes 46 seconds in the transit, and if started with a velocity of 20 miles an hour, it would attain $47\frac{90}{100}$ miles, which would be exceedingly dangerous.

Advantage of a Plane 1 in 250.

I should therefore recommend that all Rises should be brought within the Angle of Repose, or 1 in 250, which requires very little assistance from the engine, and by shutting off the steam in descending, the piston would be acting against a vacuum, which alone offers considerable resistance; it is therefore better both for the economy and safety of the concern.

Cost of Carriage upon the Basing.

———The Cost per mile of 1 ton of goods, drawn upon the plane of 1 in 250 by the heaviest engine required upon the Basing, (an engine of $12\frac{1}{2}$ tons weight, and $13\frac{1}{2}$ inch cylinders) would be 1·05498*d*. in ascending, and ·05866 of 1*d*. in descending; but as the above engine possesses more power than is required for 1 in 250, as it is calculated for 1 in 202, which plane is intended to be altered to 1 in 250, the weight of the engine will consequently be reduced to 12 tons, with $12\frac{5}{8}$ inch cylinders, which would make the cost 1·04782*d*. in ascending, and ·05826 of 1*d*. in descending.———I have likewise made a Table of the Power required, which includes every thing, as follows:

* The party that is now employed upon the Southampton Railway under Mr. Giles.

Comparison of the Power and Expence of Working the Great Western Line of Railway, and the London, Basing, and Bath Line of Railway.

The Great Western Line of Railway to be worked by Locomotive Engines of 10 Tons Weight, with Two 11 Inch Cylinders throughout, having a fixed Engine of 102 Horse Power to assist in taking up the Train at Euston Grove Incline (an Ascent of 1 in 86), and an assistant locomotive Engine at 12½ Tons Weight, with Two 13¾ Inch Cylinders at the Box Hill Incline (an Ascent of 1 in 106). Total Length, 109 Miles 34 Chains. Plan of Working the G. W.

The London, Basing, and Bath Line to be worked by locomotive Engines of 10 Tons Weight, with Two 11 Inch Cylinders, from Bath to the Foot of the Incline of 1 in 202, (14 Miles 14 Chains). Locomotive Engines of 12 Tons Weight, with Two 13½ Inch Cylinders from the Foot of the Incline of 1 in 202 over the Summit at Burbage, and to the Foot of the Incline of 1 in 250 (27 Miles 65 Chains), and locomotive Engines of 11 Tons Weight, with Two 12 Inch Cylinders over the remainder of the Length (64 Miles 39 Chains) to London. Total Length, 106 Miles 38 Chains. Do. Do. the Basing & Bath.

The Rate of Travelling 20 Miles per Hour, and the Weight of the Goods conveyed being 43$\frac{3}{10}$ Tons. Table of Power and Expence.

—	Power required to take One Ton of Goods from London to Bath.	Power required to take One Ton of Goods from Bath to London.	Power required to take One Ton of Goods from London to Bath or from Bath to London on the Average of both Ways.	Power required to take One Ton of Goods One Mile, on the Average of both Ways.	Cost of taking One Ton of Goods One Mile, on the Average of both Ways, in Decimals of a Penny.	Cost of taking One Ton of Goods from London to Bath or from Bath to London on the Average of both Ways. £. s. d.	Total Cost on 43$\frac{3}{10}$ Tons of Goods to be conveyed each Trip to or from Bath. £. s. d.	Average Ratios of the Gross Weight of the Trains to the Weight of the Goods conveyed, on the average of both ways.
Great Western Line	10,307,536	10,151,605	10,229,571	93,485	·55699	0 5 0$\frac{45}{100}$	10 19 11	1·7396
First Plan recommended, with different sized Engines, as above; London, Basing, and Bath Line	10,387,933	9,354,133	9,8 71,033	92,708	·55236	0 4 10$\frac{81}{100}$	10 12 2½	1·7500
Differences in favour of the London, Basing, and Bath Line	—	797,472	358,538	777	·00463	0 0 2$\frac{14}{100}$	0 7 8½	—
Differences in favour of the Great Western Line	80,397	—	—	—	—	—	—	·0104
Second Plan of working Great Western Line as above.	10,307,536	10,151,605	10,229,571	93,485	·55699	0 5 0$\frac{45}{100}$	10 19 11	1·7396
London, Basing, and Bath Line, supposing the Incline of 1 in 202 to be reduced to 1 in 250, and locomotive Engines of 11½ Tons Weight with Cylinders 12⅝ Inch. Diameter to work over the whole Line	10,441,151	9,391,025	9,916,088	93,131	·55488	0 4 11$\frac{8}{100}$	10 13 2	1·7580
Differences in favour of the London, Basing, and Bath Line	—	760,580	313,483	354	·00211	0 0 1$\frac{37}{100}$	0 6 9	—
Differences in favour of the Great Western Line	133,615	—	—	—	—	—	—	·0184
Third Plan of working Great Western Line as above	10,307,536	10,151,605	10,229,571	93,485	·55699	0 5 0$\frac{45}{100}$	10 19 11	1·7396
London, Basing, and Bath Line, if the Incline 1 in 202 is to be retained, and locomotive Engines of 12 Tons Weight with Cylinders 13½ Inches diameter to work over the whole Line	10,512,421	9,455,128	9,983,775	93,766	·55866	0 4 11$\frac{48}{100}$	10 14 7½	1·7700
Differences in favour of the London, Basing, and Bath Line	—	696,477	245,796	—	—	0 0 1$\frac{42}{100}$	0 5 3½	—
Differences in favour of the Great Western Line	204,885	—	—	281	·00167	—	—	·0304

Object. to working the Box Tunn. by a Station. Eng.

The only advantage of working the Box Plane by a Stationary Engine would be the clearing it of smoke; but in the event of an accident, the ropes and pullies will be very much in the way, and when one train ascended, another could not be passed until the former was wound up, and the rope sent down again; if they pass up alternately, or first one side and then on the other, they must cross over at the bottom

Do. Assistant Eng.

which will occasion considerable danger and delay:—I should prefer the use of Assistant

Insuff. of Ventil. a Tunn. by Shafts.

Engines if the tunnel could be ventilated.———I consider the more Shafts used the greater will be the confusion in the draughts, as they will counteract each other. The

Sugges. for Ventil. Tunnels.

best way of ventilating a tunnel is by one Main Shaft in the centre, which should be proportioned to the draught required to clear the tunnel, and as it should be well supported at the bottom, and carried up exceedingly strong, it may become expensive.

Rem. on the draught of Chimneys, &c.

I should prefer placing them upon one side, similar to the chimneys of Iron Works, (of which I have built many) which are frequently 2 or 300 yards from the works, into which all the smoke is led, and the higher the chimney the more powerful the draught: we are also obliged to keep them at a considerable distance off in coal pits.—

Pow. req. to work the Box Tun. must be 3 times that used upon a Level.

If the Box is to be worked by an Assistant Locomotive, it must be twice the power of the engine that takes the train, and both engines must work up to their full power; consequently,

Comp. of the Box with the Claverton.

there will be three-times the smoke of 1 locomotive with 11 inch cylinders: but the Claverton Tunnel being upon a Level, the engine need not exert more than ⅔ds its power

Smoke in the formr. in the pro. of 4½ to 1 comp. with latter.

in passing through; the Smoke in the Box Tunnel would consequently exceed that of the latter, in the proportion of 4½ to 1.—The respective lengths of these tunnels bear a proportion of 7 to 5, the Box being 1¾, and the Claverton 1¼ miles, and there is no difficulty in making Shafts in either case, it being merely a matter of expense, and the

Difficulty of Ventil. the Box.

higher the shaft the more powerful the draught. We have no experience in the ventilation of a tunnel connected with such powerful engines, it is a problem yet to be solved. I cannot concur with Mr. George Stephenson if he said "he had no objection to a tunnel 20 miles long," as it would be impossible to ventilate it, and therefore would be

The Tunn. of 1 Mi. on the Leicester and Swan. very badly ventilated.

dangerous. I think a Tunnel of 1 Mile is objectionable, there is one of that length upon the Leicester and Swannington, and although it has a number of shafts, it is extremely badly ventilated:

Description of same.

I went expressly to inspect it during the progress of the London and Birmingham through the House of Lords, which I was supporting; I arrived at the mouth of the tunnel the moment an engine was passing, and I attempted once or twice to enter but failed; it was full 10 minutes before the smoke cleared out, although there was a strong gale blowing into it; they certainly use coal upon this line, but I do not think the use of coke upon the Box Tunnel would make much dif-

Injurious Effects of using Coke.

ference, the difficulty would be nearly the same.—When Coke is used a quantity of carbonic acid is created, which is very noxious, being what is termed choke damp, into which if a man puts his head he instantly drops; it is similar to the air in the Grotto del Cane in Italy. I should not have any objection to pass through the Tunnel, but if it could be avoided it would be much better, considering it is a public

The Line proposed inst. of the Box the most preferable.

passenger railway.———I have seen the Plane suggested instead of the Box, and I certainly prefer it; it is of much greater length, but less slope, and would have saved

His Report reg. the exec. of the L. & B.

an amount of power equal to 246,187 lbs. raised 1 foot high.———I made a very full and voluminous Report to the Directors of the London and Birmingham Railway immediately upon their obtaining the Act, (Messrs. Robert Stephenson and Palmer also made Reports,) but as it was a private communication, I do not think I am justified in going into the details of it; however, I recommended the works to be advertized, and let by Contract, also Security required to the amount of 10 per cent., and I divided the line into 8 Contracts, of which I made estimates.——I was the first that practised the system of letting works by Contract, having introduced it upon the Kenyon and Leigh: the Directors at first insisted it should be executed similar to the Liverpool and Manchester, viz. apportioning the work out to different men, and employing a number of Superintendents, as they considered it the most economical; (they also thought that I should not get any person to undertake it my way); however, I refused to do it by any other mode than contract, as it would have occupied so much more of my time. I was in the habit of going over the works on ce a month, to see that they were going on satisfactorily, having a Resident Engineer, who informed the Directors the amount of money to be paid.—The system of letting the Work in several "Small Contracts" may be executed as cheaply as "Contracts by the Lump" under good management, as the Superintendent may of course make the same saving as the Contractor, provided he has equal experience, but it is not likely; although an Engineer is supposed to be as well acquainted with and capable of superintending this department as any other, yet not having devoted his attention exclusively to it, he cannot be considered as competent as a Contractor. I recommended the system of Contract with a view of getting people of capital to undertake the work, who could find Security; by which the Engineer would be relieved of all trouble and anxiety attending the minutiæ of the same, and it would also be equally beneficial to the Company. I would not undertake the management or execution of any works by the system of "Small Contracts," as I should always be responsible to the Directors for the execution of it, which I never intend to be.——The Cost of Fencing upon the Stratford and Moreton was about £220. per mile, (or 2*s*. 6*d*. a running yard for both sides) which includes posts and rails, quicks and side drains.——The expense of Ballasting, laying of the blocks, and finishing the surface, depends upon the situation, also upon the quantity used; I have used it under the blocks only upon some lines, where horses were not employed.—I have been along the Southampton Line nearly as far as Basing, and I found the principal part of the excavation at Shapley Heath in sand, which is very favorable, as upon their digging it out from the bottom, it falls over at the top; they have been obliged in some places to drive a few piles to keep it up. I have also been over the country from Bradford to Bath, and I found plenty of material for ballasting, as the Oolite stone ranges all round Bath: (I think a Line between these places is practicable, but I cannot say positively without making a survey: the Kennet and Avon Canal passes between them, and is level all the way from Bradford to the Dundas Aqueduct, which forms some criterion). Ballasting is very scarce upon the London and Birmingham Railway: I have been over every portion of the line, and I observed at Willesden that they were actually carrying oyster shells to ballast part of the line; there are some situations were a little may be procured: between Watford and Tring it is principally in chalk, with but few flints: at Blisworth they would be well provided, as the greater part of the cutting is in stone: some of the Contractors told me that they had priced it as much as 3*s*. in

He recom. them to exc. the Works by Contract.

Was the first Engi. who intro. Contract, which was upon the Kenyon and Leigh.

Descrip. of same.

Comparison of Con. with Job work.

Fencing Stratford and Moret. 2*s*. 6*d*. a yard for both sides.

Obs. upon Ballast.

Do. on the Basing.

Do. Bradf. & Bath.

Remarks upon a Line between Do.

Ballast very scarce upon the L. and B.

Account of same.

He allow. £880. per Mile for Ballast on the L. and B.

their estimates: I estimated the surface material in my estimate of this line at £880. per mile.——I consider the average quantity of earth that can be removed in 1 day depends entirely upon the means and skill of the party executing it; I am quite sure that I could succeed in teaming 2000 cubic yards per day, but it would be more expensive than the usual method, although much depends upon circumstances, as whether the cutting was very deep, or the embankment was raised in 1 or 2 lifts: I have never had occasion to team more than 5 or 600 a day, in which case the embankment was 30 feet wide, and had 2 teaming places upon it, (4 places might have been obtained by placing the rails close together) and was executed by the usual method. I consider 800 cubic yards a day the average throughout the year with 4 teaming places, which (allowing 250 days) would give 200,000 cubic yards per annum.—I reckon there are about 250 absolute working days in the year, and as you can work much longer than 12 hours upon some of them, 300 makes a fair average throughout the year. About 20 waggons, containing 3 cubic yards each, is the average work of 1 hour. An Embankment of 30 feet high should be made in 2 lifts, but it depends upon the nature of the soil; I would endeavour to make it in one if it were entirely in sand; a strong clay is a long time consolidating, and is therefore best carried up in small lifts: there cannot be a better mode of forming embankments than the method adopted on Turnpike Roads with three-wheel carriages; I have known an embankment upon the same, of only 20 feet, executed in 5 lifts, in which case the frequent passing of the waggons tended much to consolidate it. The most advantageous plan is to widen the top of the embankment, in order to get in 6 teaming places, and as you go on dress the slopes down towards the bottom, so as to preserve that number, (which is the most I have ever known used); it can be arranged with a slope of 1½ to 1, but it can be done better at 2 to 1.——The works can be executed with much less comparative difficulty in Winter, if the material is favorable: rain would also pass off a high and dry road, but working in clay causes considerable delay.—The effect of Frost upon Railroads is very trifling, and the Liverpool and Manchester has never been stopped by snow; it is the custom under such circumstances to place 1 carriage before the engine, which presses the snow down on the rails, and clears the road: we remedy it on the King's Swinford by fastening 2 besoms upon the engine, which sweep the rails as they pass along.——I have been in London the last 6 or 7 weeks, during the whole of which period I have been engaged upon this business, and in making the several calculations, &c., with the exception of a short time occupied in attending the Assizes at Stafford.

He thinks 2000 cu. yds. may be teamed per day.

Although he has not teamed above 600.

He consid. 800 the aver. thro. the year of 250 days.

Aver. num. of Days in the year 300, but there are only 250 possi. work. Days.

Method of forming a 30 ft. Embankmt.

Do. on Turnpike Roads, which are car. up in many Lifts.

Method. of work. 6 Team. Places.

Frost affects Railw. very trifling deg.

Precautions against Snow.

Ex. Mr. GEORGE LEATHER, C.E.

His Experience.

I have been actively engaged during the last 30 years in various Engineering Works, previous to which I superintended the execution of the Surrey Iron Railway from Wandsworth to Croydon. I am Engineer of the Aire and Calder Navigation, which Company have found it necessary to lower their dues very much, in consequence of the opening of the Leeds and Selby Railway for the conveyance of Goods in November

He is Engineer of the Aire and Calder Navigation.

Obser. upon same.

1834, (it opened for Passengers in the September preceding,) but the quantity of general merchandize carried by the Canal has increased notwithstanding; we have a Railway about 1 mile long connected with the Canal, and another of 1½ miles is going to be made.——I have been employed in the superintendance of 7 Railways, amongst others the Clarence Railway, having been called in to revise the Plan, I also assisted in the procuring of their 2d Act: and I then laid out the mode of executing the works: they subsequently got into difficulties, and called me in a second time.—— ——I have been engaged upon the Gradients of the Competing Lines with Mr. Rastrick during the last 5 or 6 weeks, and prepared the Table given in by him from our calculations; I also perfectly agree with the several statements, having gone minutely into and proved them to be correct.——I have been over the London and Southampton Line and inspected the works nearly as fa ras Basing.——The extent to which Locomotives can be used upon Embankments depends upon circumstances, as the nature of the soil and whether it takes long in subsiding, which regulates the laying down of the rails and sleepers; London Clay requires some time, but Sand and Gravel, of which there is a preponderance upon the Southampton Line, consolidates almost immediately.—— Sand is likewise the most favorable material to team: (a large waggon will hold about 3 cubic yards of Sand, but not above 2 of Clay if it is very strong:) I should think about 900 yards of it might be teamed in a day of 10 hours; 250 is about the average of positive working days in the year, but as we are able to work double shifts during a considerable number, the fair average number amounts to about 300.——I think wooden Sleepers may be advantageously employed during the formation of a Railway, but I do not approve of them for a permanent support, if stone can be obtained at any reasonable rate.—The cost of Earthwork varies according to the country and situation. ——I prefer Letting the works in large Contracts if I can get respectable people to undertake them, as a large Capitalist has great advantages over petty Contractors, and consequently can execute them more economically, but if a Company could find a trustworthy Contractor, and were to supply him with Capital, they would be enabled to make the same saving.—I executed many years back some works of small amount by the system of small Contracts, and by bestowing great attention they did not amount to any more than they would have cost a Contractor, whereby I saved his per centage, but it is difficult to find Overlookers sufficiently competent and trustworthy: small light works are also much easier let by this system than works containing large cuttings. ——My Contracts state generally that the Contractor shall be paid "such sums on account as may be due, reserving a certain per centage," but I frequently exceed the proper balance when I see a Contractor exert himself, especially if he is a man of small Capital.——When I lodge a Section I generally consider myself bound to within 100 yards of it, and it is quite customary to remove the line within this ditsance, if any advantagecan be gained by the same.——I consider it much better for a Railway to cross Public Roads "by Bridges," in preference to "Crossings upon a Level," in order to prevent the possibility of accidents, and to avoid the expence of Gate-keepers.

Do. Clarence Rail.

He made the Cal. of the Gradients of comp. Lines wi. Mr. Rastrick, and agrees wi. him in the resul.

Obs. upon the use of Loco. on Emban.

Sand is the most fav. Material for Team. 900 Yds. of it may be Teamed in 1 day.

A Wagg. will hold 2 Yds. of Clay, but 3 of Sand.

250 Work. Days in the yr. but 300 av. d.

Memo: Blocks and Sleepers.

Comparison between Large and Small Contracts.

Remar. upon same, &c.

Terms of his Contracts.

It is custom. to alter a Section within the 100 yds. Deviation.

He objects to Level Crossings.

Ex. Mr. FRANCIS GILES, C.E.

He was Art. in 1803. Mr. Rennie ranked first in the Profess. Mr. Telford next.

I was Articled to my Brother for 6 years in 1803, at the age of 16, and I afterwards became his partner, we were much employed by the late Mr. Rennie, who ranked first in his profession, (next to whom was Mr. Telford,) having the advantage of his advice and assistance; we were not in his office, but entirely independent of him, and our business was the planning and surveying of Public Works of all descriptions, but we did not execute any of great extent until after the decease of Mr. Rennie in 1821.

His Experience. The Boards who have employed him. The Canals he has constructed. Do. Surveyed.

———I have been employed by the Navy Board, the Admiralty, the Trinity House, the Corporations of London, Liverpool, Dublin, Newcastle, and Bristol; I have executed 4 Canals, the Joal in Bedfordshire, the Sankey Navigation near Liverpool, the reconstruction of the Basingstoke Canal, and the Lea Union near London; I have surveyed many, as the London and Portsmouth, the Military Canal, Romney Marsh, Lock Erne in Perth, the Birmingham, the Weald in Kent, the Cambridge, the Bedford, the Stockton and Darlington, (which was proposed before the railway,) the Portsmouth and Arundel,

Do. Harbours. Do. Ports, &c. Do. Rivers. Do. Bridges.

the Aire and Calder, the Berks, the Hants, and the Grand Trunk; I have also surveyed many Harbours, as the Rye, the Dover, the Margate, the Boston, the Ramsgate, the Port Patrick, Ports in the Irish Channel, the River Thames, the Blythe, the Tyne, Shields, the Pembroke, the Dublin Bay, the Wexford, the River Mersey, and the Port of Liverpool, the Preston and Chester, the Sunderland, the new Quay in Cardigan Bay, the River Conway, and the River Medway. I executed the Harbour at Courtown, and renewed the Harbour at Bradford; I have surveyed all the new Bridges in London, and many other smaller works; I executed the Hayward Bridge in Dorsetshire for the County,

The Eden Bridge is the highest in the Kingd. being 100 ft.

the Warwick in Cumberland, the Eden, which is 100 feet high, and I believe the highest bridge in the kingdom, the Corbey and Beck Bridge, and the Gelt, the three last bridges being upon the Newcastle and Carlisle Railway, and of great magnitude.

Do. Turn. Roads. Do. WaterWorks. Do. Drainages. Do. SeaWalls, &c.

I have also had occasion to cut down hills in order to improve Turnpike Roads; and I have erected several Water Works, Drainages, as Romney Marsh, and Water Beach Fen which is below Cambridge, and Sea Embankments, as the old Roman Wall at Dunchurch, and Freestone in Lincolnshire, and Wallasea on the Dee; and I have always been in the constant pursuit of my profession.———Many of the above works were surveyed for Mr. Rennie, who was responsible for their accuracy; but they were all signed by my name, and I also considered myself responsible (in my department) to the several Boards.

He was employed to make Surveys for the Southampt. in 1833.

———I was employed in 1830 by a Committee of Gentlemen to make surveys for the Southampton Railway, and I was formally appointed their Engineer at a meeting at Winchester in 1831, at which Sir Thomas Baring presided; I deposited the Plans that year, and I resurveyed it in 1833.

His reas. for aband. the Northern Line. Difficulties of the same.

———My original instructions were, to see if a line could be advantageously laid down on the North side of the Thames, but I considered it impracticable, on account of the difficulty of obtaining the consent of the owners of property in that direction.

Memo. Exten. Kennet and Avon.

(I surveyed a plan for the extension of the Kennet and Avon Canal to London, " which was to join the Thames at Isleworth, and the Grand Junction Canal at Cowley, and although it did not approach near London, yet the objections of landed Proprietors were so great that it was abandoned, although the Plans were deposited.—A plan was also agitated for bringing the waters of the Colne to London, which passed through Lady Berkeley's park in embankment, by whom I was consulted regarding the possibility of its removing, and the Opposition to the Bill in the Committee of the House of Commons was immense; I therefore thought it quite impracticable to carry any public works through Middlesex,) which I represented to the Committee, but they repeatedly urged it, and I as frequently objected and finally declined, from a conviction that I should have led the Company into difficulties.——— I made the Estimate, and gave Evidence upon the Bill as their Engineer when it was before Parliament, and I believe the Works, "quasi works," (or such as come under calculable matter,) will not exceed my amount; we have purchased 30 miles of the Land at £36,000. (which includes Compensation to several owners for dispensing with occupation Bridges,) which is very near my estimate; 8 or 9 miles of the line is Common land, for which we have paid double its value, or £10. per acre, one-half to the Lord of the Manor, and the other half to the Commoners.———The Act of Parliament allows 7 years for the execution of the works, but I have no doubt we shall do it in 3 years from the time of commencing the heavy work, or about the present time. We have been much delayed by the British Iron Company failing to deliver the Rails according to their contract, owing to a strike among the Colliers, as detailed by Colonel George Henderson in his evidence: Our Deed not having been signed until January was also another obstacle, as many of the Land Owners objected to sell until they had legal authority, but I am not aware of any Railway that has progressed more quickly, with the exception of the Liverpool and Manchester.———I have passed several Residences on the Wandsworth Road without interfering with them, although it was stated that I could not in the Evidence of the Opposition to the Bill; I am also effecting a passage under the Wandsworth Road, the possibility of which was also denied.——— We do not cross any Public Road upon a Level, except one in connection with our Depôt, which the Locomotives will not cross, nor any Occupation Roads, except one upon Lord Calthorpe's Property, (I passed several roads upon a level last year).——— I allow 14 feet Headway under all arches over the Railway, and 16 feet Headway over Turnpike roads: I have allowed 18 feet under the bridge over the Ditton and Leatherhead road to oblige the Trustees.———The Acclivity for Turnpike Roads is fixed at 1 in 30, and Parish Roads at 1 in 13, which is insufficient, we shall make them 1 in 20.——— We have sunk 6 Shafts to ascertain the nature of the Soil, the first at Sir Richard Frederick's Park has confirmed the Boring as far as the depth of Gravel went, but it was found to exceed the Boring by 4 feet in Clay. (The Borers stated last year that when they got through 20 feet of Gravel they found Stone, and could not get any further, the fact was, that some stones had fallen into the hole and stopped their augur, which deceived them.) The second Shaft is by the palings of Oatland Park, and is in 16 feet of Gravel and 11 feet of Sand, which no doubt continues as far as we go, or 27 feet, the Gravel is of the description required for ballasting, and the Sand is good and easily cut through, and I think it will settle down again very well, (but it did not stand perpendicular in the shaft,) some water was found at 16 feet below the gravel, which I pumped out in order to protect the men. The next Shaft is in Sand, and some Blue Silt was found bordering upon Clay at about the bottom of our cutting, a patch of it was also found

Memo. Exten. River Colne.

He consi. his Estim. of the South. suffic.

Cheap. of the Land.

8 or 9 Mi. purchas. for £10. per Acre.

The Act allows 7 yrs. to exec. the Works.

Cause of the Delay of the Works.

Works effected, the possibility of which were denied last yr.

Southamp. does not cross any Publ. Rd. upon a Level.

14 ft. Headway for Rail. & 16 for T. Rd.

Accli. for T. Roads . . . . 1 in 30, Do. P. Rds. 1 in 20.

Description of the Soil at the several Shafts.

between our borings. There is nothing peculiar in the other Shafts, as they are all in Sand, from which I conclude that we shall get rid of the Clay before we reach Oatland palings, which are about ¼ of a mile from the line; the above Soils are the best for getting, moving, and placing, also for settling down upon embankment.——We commenced our Work with light rails and waggons, not being able to procure heavier; I was at first fearful of making the waggons too large, but they are now sufficiently large to carry 3 or 4 yards, to which they are often loaded, (the weight of a cubic yard varies from $1\frac{1}{10}$ to 1¼ Tons, which my present Rails are quite able to crary, as our Engine weighs 8 tons. I consider it is desirable to keep the Waggons low, as it assists the filling, the bodies of mine are placed upon a bed, without any intermediate fabric as those upon the London and Birmingham, (by which a constant leverage is kept up in running along the rails), they are also much higher than mine, but do not appear to hold more than 2 cubic yards; mine cost about £16. each, and are made of the best materials, they take 2 cwt. 3 qrs. 9 lbs. of Iron, which amounts to £5. independent of the wheels and axles, we have some lined with Iron, which were the contrivance of a Contractor. ——I have raised the Level at the Battersea Road about 6 inches, but I have sunk it 2 feetat the Station, and start from Trinity High Water Mark, (the heighth of the wharf from which I started last year is 2 feet above it); instead of being 1 in 528 up to the Wandsworth Road, it is now 1 in 500 until it arrives at Wimbledon, at 50 feet above Trinity High Water Mark: passing through Kingston Hill at this level, I then sink to 45 feet across the Mole Valley, and again resume the 5 feet: the embankment in the valley was made 25 feet high, in order to admit of a bridge under it; I have altered it to 20 feet, by which it will be less prejudicial to the country: I was enabled to reduce the height by introducing iron beams at the bridges instead of arches. The Gradients I delivered last year represented certain lengths, at a general inclination of 1 in 330, which are now divided into 3 steps, in order to give a relief to the engine, which is the same thing, as I have the same total height.—The first Mile from the Depôt is all under Trinity High Water Mark, but I have the means of lowering the road down to the present drainage level, the top of which is 10 feet below Trinity High Water Mark, (the tide being prevented passing up into the sewer): my original section was below it, which I have now increased by 2 feet more.—New Street is about ¼ a mile from the depôt, and has a clear height of 11 feet between the railway and the ground; if I lower it 5 feet below Trinity High Water Mark, it will give me 14 feet head-way, (as $11+5-2=14$ feet clear) and I can sink it 8 feet if I please.—There is only one occupation road between the latter and the depôt, where there is 6 feet between the rails and the ground, and 9 feet to the bottom of the marsh, but I have made an arrangement by which it is to be confined to a cattle arch. The Mole Embankment passes 11 feet above the tide valley, which agrees with the gradient I gave in the Commons.——I stated in my evidence of last year that the line was the best and cheapest that could be obtained, by which I referred generally, *i. e.* within the 100 yards, which I have not exceeded, (I contemplated some of the improvements at the time, and mentioned them): I consider that I have a right to make any alterations, provided they are not prejudicial to the public; Latitude is always allowed in Public Works.——We are at work at 12 places on the line, which comprehends all the heavy and some of the smaller parts, which were

Favorable nature of the Soil.

Method used in commencing the Southampton Line.

A Cub. Yd. weighs from $1\frac{1}{10}$ to 1¼ Tons

Descripti. and Cost of the Waggons.

Comparison of same with the Waggons on the L. and B.

Description of the Deviations made in the Gradients upon the Southampton.

The 1st Mile is under Trinity High Water Mark.

Height under the Bridge at New Street.

Explanation of the Deviations.

Latitude alwa. allo. in Public Works.

commenced on account of their connection with other parts, and as we obtained early possession of the land.——I have reduced the Cutting at Wandsworth about £180,000. out of £700,000., and I have got lower down Kingston Hill, and reduced it £250,000., but my greatest reduction is at St. George's Hill, which I have reduced from a depth of 116 to 80 feet, viz. 20 by raising the level, and 16 by passing lower down the hill. I have also raised the Embankment over the Wey, to suit the improved level, and instead of throwing the soil to spoil I shall now employ it to form this embankment, (the spoil banks of last year amounted to 1,750,000 cubic yards, which is about the amount of my present saving).—St. George's Hill last year amounted to 3,725,000 cubic yards, 800,000 of which were required for ballasting. The quantities are now 684,000 for the Mole, (it is ridiculous to suppose 1,500,000 are required for the Mole) 760,000 for the Wey, and 370,000 as a reserve for gravel, which is sufficient for 37 miles, at 10,000 per mile; therefore, the time occupied in executing the work will be considerably lessened, (the cutting at Frimley will now occupy a longer space of time to execute than St. George's Hill).—The length of the Lead at St. George's Hill is nearly 3 miles, and at the Wey 1½, therefore the average is nearly 2¼ miles.—I am negotiating with some land-owners with the view of making further reductions in this work, which is partly my reason for not commencing the Western side.—St. George's Hill is in clay, (silty clay) as I expected, but we have at present only cut into the crest: we first came to gravel, then sand, which is most valuable for brick making; as we do not require all of it for that purpose, we therefore throw part to embankment, for which it is excellent, but I do not throw the gravel to embankment. The portion which is finished stands very well. I am working it with Inclines, at merely sufficient inclination to run the waggons down with safety, and I do not intend using any ropes: the side roads enable us to make the cuttings more extensive, and employ more men.—There is a Hill at Woking of some extent, but trifling compared with the other at St. George's Hill, which I am prepared to work with heavier rails and full strength.——The soil at Frimley is sand and loam; I also apprehend there is a vein of clay, which I am prepared for.——There is an embankment at Elvetham of considerable extent, which is in sand, and stands well, and we have a bridge built there.——It is also in sand at Shapley, which joins the latter, and I expect to find clay in the Hill, for which I am prepared.——Hook Hill is in gravel and clay; there is about 1,000,000 cubic yards of the latter on the top of it, and there is but one more considerable work upon this side of Basing.——If I should not find sufficient ballasting at Hook Hill, plenty may be had from St. George's Hill, as I have reserved 370,000 cubic yards, which will not be required there, and I do not know how much more may be got out of the sides. ——I have let all the Earth-work between London and Basing in 4 Contracts, and I do not let it in small Contracts, (as stated in the evidence of the Opposition) as the whole of the earth-work and bridges from the River Wey to Basing are in 1 contract, upon which there are 5,000,000 cubic yards, and taken at 6*d.* would amount to £125,000., (the whole contract amounts to £170,000. or £180,000.)——We do not require any Sureties from Contractors, and there is sufficient evidence upon the London and Birmingham to shew its inefficiency, as it is well known that the party who took the first contract from London failed, and left the Company £10,000. in debt: I believe

Reductions made in the Wrks. at Wands.
Do. Kingston Hill.
Do. St. Geor. Hill, 116 ft. to 80 ft. being about 1,700,000.
Acct. of the Altera.
Original quantity of the Hill 3,725,000.
Reduced quantity of same.
Frimley will now occupy a longer time to ex. than St. G. Hill
Aver. Lead of Do. 2¼ Miles.
Description of the Soil and Method of Working St. Geor. Hill.
It is worked by Inclines without Ropes, &c.
Works at Woking Hill.
Do. at Frimley.
Do. at Shapley.
Do. at Hook Hill.
Work bet. Lon. and Basing let in 4 Con.
He does not let the Work in small Cont.
One of his Contracts amou. to £170,000. or £180,000.
He does not require Sureties.
Observ. upon same on the L. and B.

the latter have claimed the forfeiture in one case of failure, but the amount of same is very trifling compared with the contract, and I should be very sorry to take any thing from a Contractor if he had acted honestly and done his best.——I am bound to say the Directors have placed full confidence in me, and done me justice, and I believe my men will do the work at the prices which they have agreed upon.——We have some Engine Houses and Shops erecting upon our Line for temporary purposes, which are built in the cheapest manner, being made of the tops of the fir purchased for the sleepers, walled and plastered with clay and lime whitened, also floored and covered with common pantiles, which is the best sort of covering: large buildings for these purposes are any thing but profitable to a Company.—Mr. Bainbridge has also built some Cottages on his land, which we have rented.——I have heard the evidence which has been given upon the Cost of Earth-work, and I am still of opinion that so far from costing 1*s*. it can be done for half that amount, exclusive of Rails and Sleepers: I did not originally contemplate supplying the Contractors with sleepers, but having reflected that the stability of the rails and excellency of the road depended upon them, and as I had also found a cheap way of obtaining the same, I considered it would be economical to do so: I presume the first sleepers will be sacrificed in the execution of the works, but the expense of them will not exceed £150. a mile.—A saving is made by using the large Permanent Rails in the formation of railway, instead of having temporary rails; the former have merely to be ordered somewhat earlier, and if some should become bent, they only require hammering straight again.—I include the temporary rails in my estimate, as far as we have used them, also all waggons, tools, inclines, ropes, oil, locomotive power, wear and tear, &c.: I made the Expense of my Locomotive £2. 2*s*., instead of £5. 5*s*. per day, as stated by Mr. Robert Stephenson: I did not go into the details of the locomotive engines, as my business is the applying and not the making of them: I took the cost from the best authorities, viz. from engines used in Lancashire, and Mr. Rastrick has fully confirmed it, which I was not aware of until a day or two before he gave his evidence. Much depends upon the price of fuel, and we may be worse situated in this respect than the London and Birmingham; but I am cutting through some Peat of the finest quality, which I find was used in the Smelting Works at Durham; I have also tried it at the smith's forge, when it produced a fine flame, which is the source of heat, (perhaps it may injure the fire box by producing too much sulphur, although it has not been proved): I have therefore no doubt peat will supply us with fuel, allowing for a little extra expense in repairing the coal box, although I have allowed quite sufficient for fuel without having recourse to it.——I have been obliged to increase the Price of Earth-work since I gave Evidence in the House of Commons, although not from absolute compulsion; but I have reduced the quantity of work in a much greater ratio, and as I proceed I shall find greater facilities, and the men will understand their work better: the Harvest no doubt has tended to increase the rate of wages, but they will fall as soon as it is over.—I shall carry the work through at a fraction above 6*d*. at the London end, and less than that price at the other end, making about 6*d*. from end to end: I am not paying above 5*d*. for excavating, except in one case, which is at Kingston, where I pay 5½*d*, the upper part of this hill being stiff clay, but the lower

Description of the Engine Houses on Southampton.

Works of the above description should not be Expensive.

He is still of opinion the Works will be execu. for 6*d*. exclu. of Rails & Sleepers.

He uses the Perma. Rails in the Works.

Remarks on same.

He estim. the Loco. at £2. 2*s*. per Day. Ditto Mr. R. Stephenson, £5. 5*s*.

Observations upon the use of Peat as Fuel.

Peat destroys the Coal Box more than Coke.

He has increased the Prices since he gave evid. in the Comms.

But has reduced the Works in a greater ratio.

Account of same.

He pays 5*d*. for Ex. exc. one case of 5½*d*.

part is milder.—The Company provide all materials and waggons to start the Contractors, also planks and barrows if required, for which 1*d.* per cubic yard, or 20 per cent., is reserved from their payments; we also find Mr. Tredwell and the person at this end with horses, although the former is a man of some capital; they lay their road and keep it in repair. The materials and tools, also horses, locomotives, waggons, &c., become the property of the Contractor at the conclusion of the works, upon his redeeming them, (no extra charge is made for interest, but they have the several materials at prime cost) otherwise they remain the property of the Company; perhaps it will take Mr. Tredwell 2 years to redeem his materials.——Our Rails are being supplied by the British Iron Company; the Specification describes them to be hammered and rolled, but I cannot say whether they are so manufactured; there is very little difference between one and the other: those which have been delivered bear chipping and cutting very well, which is a proof of the good quality of the iron, which is made from pure mine and not cinder mixture; after the first blasting it is received as puddled iron, and it is then hammered or should be hammered into bars; it then goes through another process of smelting and refining, after which it goes into a state of rolling, which is repeated before it comes to its final size.——I intended them to be 50 lbs. to the yard, but they are 57, some are nearly 60 lbs. (I am not aware of their being used heavier upon any railway): they are Parallel Rails, which I prefer to the Fish-bellied, as they afford greater facilities of fixing the chair, you are not confined to one particular spot, as in the case of the former. The Rails of the Liverpool and Manchester were 35 and 36 lbs. only, and upon the Newcastle and Carlisle 42 lbs. We shall adopt the use of Sleepers, which are better upon embankments than stone, on account of their weight causing them to sink; (they are also a considerable expense at first, on account of the carriage,) particularly as our line passes through a country possessing very little of the latter, but plenty of the former: I prefer Larch, and shall use all I can get, but it is a very scarce wood; the remainder will be of Scotch Fir, prepared with Kyan's liquid, in which I place great confidence: I have used Scotch Fir sleepers upon the Newcastle and Carlisle Railway; (I also know places where it has been used as fencing); I consider they will last from 5 to 7 years, when stone blocks may be substituted, if considered more desirable, which will last until they are broken: it depends upon circumstances how long they will remain level, but wet weather affects them considerably; they sometimes require re-setting the week after they have been laid. Both stone and wood cause dislodgements of the rail, but they do not require levelling as soon upon a sand embankment as upon one of clay.——350 cubic yards have been teamed at St. George's Hill upon an average every day during the last 2 weeks. It will not be necessary under the present reduced height of embankment to team from more than 1 head, but I shall have 2 at Shapley Heath, where the embankment is 40 feet high, and contains from 860,000 to 1,000,000 cubic yards, 30,000 of which are done, although Mr. Locke in his evidence stated it at 2 or 3,000, but it has not been touched since he was there, and the work is open to any person to measure and see who is right.—I calculate that their present average is from 9 to 1200 per day, I certainly have no doubt about their having teamed 1100 cubic yards per day, as there are about 280 waggons teamed daily, each of which contain about 3¼ cubic yards,

The Compy. retain 20 per Cent. for Mat.

Account of same.

Description of the Manufacture of the Rails. upon the Southampton.

They are Parallel Rails, 57 lbs. to the Yard.

Comparison of same with the Fish-belli.

L. and M. 35 lbs.

N. and C. 42 lbs.

Remar. on Sleepers.

He intends using Scot. Fir Kyanized.

He considers they will last 6 Years.

Comparison betwn. Blocks & Sleepers.

350 cu. yds. Team. per d. at St. Geo. Hl.

Description of same

Do. Shapley Heath.

900 to 1200 per Day are Teamed at same.

Work Dble. Shifts.

He considers Col. Henderson's Calcu. correct, and that the Wk. will not ex. 4*d*.

each shift working 7 hours, (they used to work 8,) I therefore consider that Colonel Henderson's calculation of the cost of the works is quite correct, although it may not always average the amount he stated, but there are more men now upon the works than when he was there, therefore I consider it will not cost above 4*d*., including filling, teaming, and the draught by horses, also ropes, waggons, keeping the roads up, &c.—Our Ropes frequently break, we have had 3 new sets of Ropes at Shapley, the first is gone to Newnham, and the second was very bad, and lasted only one month.—I allow the Contractor 5*d*. per cubic yard, which he has not overdrawn: he has to take into consideration every Contingency, thus, the men are sometimes stopped while the rails are being relaid.—There have been some persons from the Great Western Railway spying upon our works, and I have to complain of their having made mischief among my men, saying that they were not paid sufficient and the like*: and in consequence of the information I gave in my evidence of yesterday, the men working in the double shifts struck, although they were getting 2*s*. 2*d*. per day, but they were willing to work in single shifts.——— I have drawn out a Statement shewing the Basis upon which all the Earthwork upon this line is to be carried on, which is conformable to the system I adopted upon the Newcastle Railway, and it may serve as a foundation upon which to calculate all Earthwork, as follows:—

Wear of Ropes.

The Contract of the above is 5*d*. includ. Contingencies.

Mr. Giles com. of the Agents of the G.W.

The Men at the above Works have struck, in consequ. of Mr. G.'s evidence.

Earth Work.

1 Waggon carrying 3 Yards, 7 Journeys per Day of 1½ Mile each, 10½ Miles out and back, will give 21 Yards a Day for 250 Days, or 5,250 Yards; but say 5,000 Yards per Year for each Waggon, which, at ½*d*. per Yard, is 10*l*. 8*s*. per Year.

Mr. Giles's Table for Calculating Earthwork.

The Cost of a Waggon being . .	£16	For 3 Years.
Repairs	16	
	£32	
Deduct Value of Iron-Work at the end .	£ 4 or £5.	
	£28 for 3 Years.	

So that 10*l*. 8*s*. a Year for 3 Years, will more than cover the 28*l*., *i. e.* Cost and Wear of Waggon. 100 Waggons, at 5,000 Yards a Year each, will move 500,000 Yards, or 500 Waggons (which Number will be provided to Tredwell) will move 2,500,000 Yards per Year, or 5,000,000 in 2 Years, which is about the Amount of Tredwell's Contract.

Horse Draft.

1 Horse and Boy 5*s*. a Day for 250 Days per Annum, 7 Journeys 1½ Mile out and back, carrying 12 Yards each Journey, in 4 large Waggons, upon a good Railroad, level or slightly descending, will equal 84 Yards a Day, or say ¾*d*. a Yard.

* The subject of this difference was gone into, and supported by evidence on both sides, but nothing very important having been elicited, it is omitted.—*Editor.*

Prices of Earth Work.

Mr. Giles's Table for Calculating Earthwork.

	Clay and Gravel, 1½ Mile Lead.		Sand, 1½ Mile Lead.
Waggons	0½*d.*	.	0½*d.*
Horse Draft and Boy	0¾*d.*	.	0¾*d.*
Men, Horses, &c. getting in and out of Work	0¼*d.*	.	0¼*d.*
Getting and filling	2½*d.*	.	1½*d.*
Teaming	0½*d.*	.	0½*d.*
Trimming Banks	0½*d.*	.	0½*d.*
Laying and keeping up Roads, Oil, and Boy, and Drains, on a large Scale of Work	0½*d.*	.	0½*d.*
	5½*d.*	.	4½*d.*

	Surrey and Hants.	Surrey, near London.
Say Clay and Gravel, Half the Quantity to be moved by Tredwells	5½*d.*	Add to 5½*d.* 1*d.* for extra Wages, making 6½*d.*
Sand—other Half to be used	4½*d.*	
	10*d.*	
Rope and Inclines only on about 1,000,000 of Yards	5*d.* Mean.	

Engine Power.

If 4*l.* a Day—say 1,000*d*, or 1*d.* per Yard, to carry 1,000 Yards Two Miles a Day, or Half a Yard at 2*l.* a Day, for the Engine.

In order to work an Engine effectually the Work of Excavation must first be opened to the Extent of 1,000 a Day; this Quantity can be teamed over One Embankment with Four Roads in One Day, or 2,000 Yards a Day over Two Embankments, each having Four Roads.

Specification.

Mr. Giles's Specifications upon the Lond. and Southn. Railway.

We hereby undertake and agree to Excavate all the Hills and Cuttings upon the London and Southampton Railway, from Nine Elms in the Parish of Battersea to the Western Side of the Parish of Pirford, a Distance of about Twenty-one Miles and a Half, and to convey the Soil of the same to and to form all the Embankments in this District with it, including the necessary cutting and Trimming the said Soil in forming the said embankments, including also the Trimming the Slopes of the above Excavations and Embankments, and the temporary Drainage necessary for carrying on the above works, at the price of —— per Cubic Yard, to be measured in the Excavations only, and not to include any Measurements of the Embankments, or temporary Drains.

We also agree to dig the Foundations for all Bridges, Culverts, and other Buildings at the Price of —— per Cubic Yard, exclusive of the Expence of pumping Water. Also, to fill and deposit and pound the Soil, at the backing of all the Bridges, Culverts, &c. at —— per Cubic Yard. We also agree to remove all Top Soil from the Excavations, also from the Seat of the Embankments, and to lay it on each side of the Railway, also to relay the same and trim it upon the Slopes of the Excavations and Embankments as a coating to receive Grass Seeds, at —— per Cubic Yard, to be measured only in the First Excavations of the Top Soil. We also agree to Ballast the Permanent Way Six Cubic Yards in every Lineal Yard, at —— for every such Lineal Yard. We also agree to find all Labour, Waggons, Barrows, Horses, Carts, locomotive Engines, and Tools of every description necessary for laying the Roads, and carrying on and completing the above Works, except that the Railway Company agree to find all Rails, Pedestals, Sleepers, Nails, Points, Crossings, and Turn Plates requisite for forming the Temporary Railways for executing the said works; the Railway Company also agreeing to provide the necessary Money for purchasing all the above Implements and

Mr. Giles's Specifications upon the Lond. and South. Railway.

Tools, but which are to be the Property of the Company until the whole Value of them at Cost Price is redeemed by us the Contractors, which Redemption shall be made progressively at such Rate per Cent. at the Cost Price of the said Implements and Tools may bear to the Value of the Work done by us from time to time under this Contract; we agreeing also to bear all Expences of keeping all the above Implements and Tools in repair during the progress of our Works; and as soon as the said Cost Price of the said Implements and Tools is repaid by us to the Railway Company they will be our Property and at our entire Disposal. Payments are to be made to us every Fortnight within Five per Cent. of the Value of our Work done, after deducting the Per Centage as above for the Implements and Tools provided by the Company. All Payments for Wages and Labour shall accordingly be made to the Workmen in Money every Fortnight. The final Balance due upon the Completion of each Part of our Work shall be paid to us when certified by the Engineer Mr. Francis Giles, or other principal Engineer to the London and Southampton Company for the time being. The whole of these Works are to be performed under the entire Direction and to the Satisfaction of the said Mr. Francis Giles, who shall arbitrate all Differences between us and the Railway Company, and whose decision shall be final. If in case of Death, or any circumstance of Negligence, or other causes, we should fail to carry on the above Works in such order and time as the said Engineer may require, he shall have full Power in the name of the Railway Company to take possession of the Works under this Contract, at any time, upon paying to us the Value of our Work done, according to such Valuation as he may make of our Works upon the Basis of the Prices in this Contract.

The Ave. Lead upon the Southampton is 1½ Miles.

I stated the Lead of the Southampton last year at 2 miles, but I have since balanced the work better, and adjusted to 1½ miles; the greatest lead was 5 miles, the average at St. George's Hill 3 or 4, and at Frimley it was about 3 or 3½.——I have been applied to by many Contractors, but I am quite satisfied with my present men; amongst the former were Messrs. Brown and Bewick, who executed the Willesden Contract, for which they had 9¼*d.* (but they were charged £5. a day for the Locomotive,) upon my representing the nature of the work they made me an offer for the Wandsworth Contract, and I have recommended that it should be accepted, and they will commence as soon as I shall have got through with the Bridge.*——I have advantageously employed the local Poor upon my Works, as upon the Newcastle and Carlisle, my plan is to head them by a few experienced Gangers; I have some now that earn 2*s.* to 3*s.* per day at piece work, and I will match them with any Navigators.—— ——I was not concerned in getting the Bill of the Newcastle and Carlisle, but I was called upon to make a Report upon the line in 1829, when the Bill was in Parliament, and I made an Estimate for the same, which amounted to £300,000. for a single line, (but sufficiently wide to take 2,) with cast iron rails, after which I was appointed to execute it, and commenced in 1830, but I did not reside at the works; I am now only their consulting Engineer; I also have been over the line occasionally for the Loan Fund, and any remarks or suggestions which I then gave have always been attended to, and the same Resident Engineer whom I appointed is still retained.—A part of the line which I superintended, from Bleadon to Hexham, a distance of 17 miles, is opened for a double line, (but only 4 miles are absolutely laid down with a double line of rails,) and 23 miles at the Western end are nearly finished, but they are very short of Funds; there are 6 more miles in hand from Hexham to Haydon, and 11 from Haydon Bridge to Brenkensop which are

The Contractors of the Willesden Cont. applied to him.

He has employ. the Local Poor advanta.

Account of his connexion with the Newcas. & Carlisle.

Est. of sa. £300,000.

It was comm. 1830.

He has ceased to be their Cons. Engin.

Account of the Progress of the Works.

* This Party declined the Contract the day after Mr. Giles made this statement.—*Editor.*

not touched.—Wages in the North are quite as dear as in Surrey and Hampshire, and as this line proceeded circumstances required continual improvements, and the several sources of Trade developed themselves in such a manner that the Directors required alterations in the character of the Works, the difference was principally in the levels. The Line had consequently cost £300,000. at the time of my last survey for the Loan Board, but it was not for works only, as the Directors had paid interest upon the shares for a certain time, and the expences of management were heavy, owing to the great length of time they had occupied, (the Bridges were also increased in width for the double line,) perhaps the sum expended upon the works amounted to £ 280,000. : I consider a double line may be executed for £450,000., or about £8,000. per mile.— They wished me to reside at the Works, which I declined, and retired from the superintendance, as I could not in consequence of my engagement with the Southampton Railway, having pledged myself to reside upon that line, (I have accordingly given up my residence at Reigate). And three gentlemen of the Committee who I differed with eventually carried on the works, regularly receiving Salaries, which would not have been the case had I remained.—The works have since been proceeding very sluggishly, owing to the want of funds, and works once began, if not followed up will very soon get into decay, and require constant expences; they have followed my plans as closely as they could, which they have admitted, and so far from any dissatisfaction having been evinced towards me by the Company, the following is part of a Report upon the subject, "The office of Operative Engineer having been found to be incompatible with the "numerous engagements of Mr. Giles, the able and eminent Engineer by whom the "works have been some time carried on, the Directors found that there were persons "among their own body, Messrs. Benjamin Thompson, George Johnson, and Nicholas "Wood, of whom, if they could be prevailed upon to appropriate their time, it would be "highly desirable to have their services, as well on account of their experience as of "their extensive knowledge in the making and laying of roads of the description "required for this undertaking, and they were glad to ascertain that those Gentlemen "were willing to act as a Managing Committee, the valuable assistance of Mr. Giles "being retained as Consulting Engineer to the Company;" and I believe the feeling displayed in this Report was really felt by the majority of the Company.—Cowran Hill had a lead of full 1 mile, and the extent of the Embankment was 1 or 1¼ miles, ¾ths of the Cutting went to Spoil, and was conveyed by gravitating planes, the waggons being drawn up again by horses: They have had occasion to increase the length of this embankment since I left them, and I have a letter stating that the price has not increased although the bottom of the cutting is very hard; but the Contractor does not find Rails and Sleepers in the present case as he did in the former.—I therefore had experience upon this Railway before I began the Southampton, the materials of which are superior to any at present constructing. I have let the Mole, which has an average lead of 3½ miles, for 6*d*., but they have the best description of waggons and rails, and a Locomotive Engine to lead it with after a horse leaves it, it therefore possesses advantages over the Newcastle and Carlisle, and will consequently be executed cheaper. I can further state that I do not know any Railway that is not objected to by one party or another.

Reason for the Works exceeding the Estimates.

It was only laid out for a single Line, and Cast Iron Rails.

He believes a double Line will not exceed £800. per Mile.

Observations upon Mr. G. declining the superintendance of this Railway.

Further Account of the Progress of the Works.

Mr. N. Wood and 2 other Gentlemen were afterw. appoin. to superint. the Wks.

Description of Cowran Hill Cutt.

The Work was exec. for the same Price since Mr. G. left.

He has let the Mole under more advan. circumst. for 6*d*.

The Works upon the Carlisle was let by Public Advertis.

——The work upon the Newcastle and Carlisle was let by Public Advertisement, and some of the Tenders were so low that it was improper to receive them, however as the parties appeared to understand the nature of the work, I suggested that the Company should try them, and that a trifling advance should be made, perhaps ½*d*.——We are enabled to execute the works cheaper than Contractors, on account of our large capital. I do not think 10 per cent. too great a profit for a Contractor if he has to lay out much Capital; 15 per cent. is reasonable, and large Contractors seldom undertake it at less, and generally more; but my men are in totally different circumstances, and profess themselves satisfied with 5, but I think they will clear 10 per cent. and they have no other risk than that of losing their profit, which is quite sufficient stimulant, (they would be glad of 3 upon a contract of £100,000.)——

LargeContr.seldom undert. for less than 15 per Cent. Profit.

His Men satisfied with 5, but he thinks they will get 10 p.c.

He is joint Engineer of the Basing with Mr. Brunton.

It is 2 Miles shorter that the G. W.

Accou. of the Route of the same.

——I am joint Engineer with Mr. Brunton for the Basing and Bath Line, and I made a survey from London to Bath, and examined the intermediate country last year: the distance is 2 miles shorter by the Basing than by the Great Western; 43¼ miles of the former pass along the Southampton Railway, and is already in a forward state, it will form a very good junction with the Bath line at Basing, which passes through Shirbourne Banghurst to King's Clear, then over Cookham Hill into the Valley of the Kennet, opposite Newbury, up to which place it may be called back country, and is very favorable: it passes over the top of Greenham Hill (Sandleford) through the common in front of the park, thence through the Village of Enborne, and at the back of Hampstead; we interfere with some of Lord Craven's lands, but not with the park; thence we descend into the Valley of the Kennet above the right bank; from thence we go the South side at Kimpbury, and to the South of Hungerford, which is in the valley; we then pass on to Shalbourne, thence to Burbidge, passing within 1½ miles of Savernake Forest, Lord Aylesbury's property, who resides in the middle of it, (I shall be happy to remove the line ½ a mile further off if it will satisfy his Lordship, or we could pass still more Southerly through Burbidge, which I have had surveyed, and proceeding from near Worting on to Laverstock to Amesbury, and so on to Ludgershall, thence down to Market Lavington, when it would again unite with the original line near Trowbridge, but I prefer the line I have laid down, although there is not much difference: the steepest inclination upon this Deviation would be 1 in 200, and it would make the distance 2 miles shorter); thence I proceed on to the North side of Pusey, crossing the river or rivulets which are the contributory streams to the Hind at Pusey, passing below Devizes nearly 3 miles south of it, near a hill called Lydmay; thence it passes into the Valley of the Avon to Weston, from thence at the South of Bulkinton to the North of Steeple Ashton and Trowbridge; it then goes into the Valley of the Kennet at Bradford, within ¼ of a mile of the town, thence down the valley, crossing the river and Avon Cliff; it then keeps the North side of the Kennet, and re-crosses at the Dundas Aqueduct; we then pass under Claverton Hill by a Tunnel; we could curve round the hill if there was any necessity for it, but I see no physical difficulty in the tunnel, except the extreme height, which would also affect the ventilation more than usual.——We then pass on to the Depôt at Bath, which is 70 feet above the level of the river, and of little local value; the Western end will come exactly opposite the old

Passes nr. Newbury

Near Hungerford.

Near Savernake Forest.

Proposed Deviation to avoid Lord Aylesbury's Property.

Mr. G. prefers the Original Line.

It pass. nr. Devizes.

Tunnel through Claverton Hill.

Description of the Bath Depôt. 70 ft. above the River.

bridge, with which we intend having a Passenger communication by steps, as at Manchester: the Carriages will get up by an inclined plane, which for about 100 yards will be 1 in 9; I cannot say it will be very easy for them, but it is quite practicable; there is a road already (the Whitcombe Road) part of which I intend retaining; great weights frequently pass up and down it: the communication for goods will be at the East end.—I could descend to the Level of Mr. Brunel's Depôt, but I object to crossing the river at that part if I could possibly avoid it, as it is low: I keep clear of the river and of the neighbourhood of the visitors, as there are merely a few stationary residences of Whitcombe.—The Line could not affect Prior Park, as we pass at a considerable distance from the New Lodge, although we are not far from the Old Lodge (which is not used) and the Lake, but we should not touch the Terrace or Park. ———We have an Embankment about ¾ of a mile long near Lord Craven's, (who dissents) part of which is 60 feet high, and part 22 or 23, but the greater part is 40; we have another within a ¼ of a mile of his Lordship's Park, ½ a mile long, which is 37 feet in the highest part; also an embankment near Trowbridge, which (with the exception of a small portion in the middle) is 36 or 37 feet in the highest part; the length is about 7 miles, being similar to that West of Grittenham Wood on the Great Western; it contains 2,435,000 cubic yards, and will be in clay with slopes 2 to 1, and some portion of the lead will be 6 miles long.———The total amount of Excavation upon the Basing line is 10,338,000.———I delivered Lists of the Gradients to Dr. Lardner, Mr. Rastrick, and Mr. Leather, and they suggested to me the adoption of an Inclination of 1 in 250 upon the Plane 7 miles long, instead of 1 in 202, and I have agreed to the same, and consider my estimate will cover any extra expenses, although the work is of considerable magnitude.———I have made a Calculation of the Power requisite to overcome the resistance on both lines, as follows: on the Basing line, from London to Bath the resistance will be 89,641 lbs., from Bath to London 81,163 lbs.; on the Great Western, from London to Bath the resistance will be 90,327 lbs., and from Bath to London 87,897 lbs.; therefore, from London to Bath it is 686 lbs. and from Bath to London 6,734 in favor of the Basing line. There are also 2 planes on the Great Western which require assistant power in ascending, and a powerful check in descending, viz. the Euston Square and the Box Plane: (100 is about the average number of Passengers in 1 train, and the danger attending these planes is connected with the whole of them at once,) but there are none upon the Basing line: the descent of 1 in 86 to the Depôt at Euston Square is very objectionable, but it would have been less bad the plane been upon an ascent.———The flatter the levels can be got upon a railway the better. The Stockton and Darlington Railway, also the Seaham and the Clarence, are Colliery Railways, and are not used with a view of going 20 miles an hour, although they carry passengers; the Newcastle and Carlisle Railway is fitted for passengers as well as goods.———Our Line would be very convenient for the transport of Passengers, Cattle, and Provisions from Ireland to Southampton, and there is a considerable quantity of Irish Provisions remitted for the use and consumption of the Naval Arsenal at Gosport; a communication may be made from Southampton to the latter; I have not examined the country minutely, but I have gone over it, and

Passeng. Approach.

Carriage ditto.

Goods ditto.

He could descend to Mr. Brunel's Level.

Remarks on same.

Property intercep. by the Bas. & Bath.

Description of the Heavy Works upon the latter.

Total Exc. upon the Bas. 10,338,000 c. y.

Dr. Lardner, Mr. Rastrick and Mr. Leather alt. the Pla. 1 in 202 to 1 in 250.

Mr. G.'s Calculat. of the Power required:

London to Bath,
G. W. 90,327
Basing 89,641
Balance 686 lbs. in favor of the latter.

Bath to London,
G. W. 87,897
Basing 81,163
Balance 6,734 lbs. in favor of the latter.

Fur. Obs. upon same

Memo. Gradients.

Observations on the conv. upon the latter in refer. to Ireland.

Line from Southam. to Gosport

Communic. between Southampt. & Paris.

have had it levelled.——Packets also sail weekly from Southampton to Havre, from whence there is a direct road to Paris; a railway has also been agitated: this line would also form a communication between Vauxhall Bridge, London, and Redcliff Wharf, Bristol, which is very desirable.——The line might likewise include Reading, by being carried round from Frimley down the Blackwater Valley, which I have examined: Frimley is 30 miles from London, and from thence to Reading would make 47, being 9 miles longer than upon the Great Western, making a difference in the journey of about 26 minutes. It would pass through most excellent country, and would afford the means of reaching the Thames direct, avoiding Eton College; and the section would be tolerably easy, the steepest gradient being 1 in 330, with not above 1½ miles of tunneling, which would be in 6 tunnels: it would also avoid the crossing of the Thames at Maidenhead, the valley of which is one of the richest in England, and the tunneling under Mr. Palmer's and Mr. Wood's property, also the Euston Square plane.

Observations upon a Branch from the Basing to Reading.

Advant. of a Line from Gloucester by Cheltenham, Oxfd. and Tring.

—— ——A Line may be taken to Gloucester through Cheltenham and Oxford to Tring (I called upon Mr. Creed respecting it, and understood from him that the London and Birmingham intended making it), and the following is a List of the Gradients:

Gradients of the same.

List of Gradients on the Gloucester, Cheltenham, Oxford, and Tring Railway.

	Miles	Chains	Gradient
Tring	11	34	1 in 431
	8	0	1 in 1,760
	10	0	1 in 6,600
	3	0	Level.
	3	0	1 in 880
	18	0	1 in 528
	3	0	1 in 311
	3	23	1 in 250
	1	37	Summit Level.
	9	20	1 in 176
	3	60	1 in 320
	5	20	1 in 283
	2	0	1 in 528
Gloucester Depôt	0	0	
	81	34	

Observations upon same.

The latter part can be easily effected, and the line between Oxford and Cheltenham would pass along the Valley of the Winrush, which, although rather narrow, contains sufficient space; there is a slight inclination previous to arriving at the summit, and we should have some difficulty in passing Hailes' Hill, although there is a gap in it, yet we should require both cutting and tunneling to a great extent; thence to Cheltenham would be easy: we should require an extra or side cutting after passing the valley, as there would not be sufficient excavation: we should have 1 in 176 for 9 miles, with a tunnel 2 miles long upon the same, and another Plane of 3¼ miles 1 in 250.—I have no doubt the Line from Gloucester and Cheltenham can be carried on by Winchcombe (which is 14 miles from Gloucester) to Birmingham and Coventry, and I am satisfied it will be done eventually; therefore, although the communication between London and Cheltenham is rather circuitous, it would compensate by affording part of a communi-

Planes upon same.

The above Line may be carr. on to Birm. and Coventry.

cation to Birmingham, and it is also the best for all trade directly Northward of Gloucester, but not Southward; $\frac{9}{10}$ths of the traffic from Gloucester is from the North and North-east.—This Line might also be carried on from Tring to Cambridge and Lynn; I know the country well, having surveyed it 25 years back: Oxford is a nucleus at which many roads concentrate, and Cheltenham is another, and all the traffic from South Wales and Worcester, and the North-west from Tewkesbury, passes through it.———

$\frac{9}{10}$ths of the Traffic from Gloucester is from the N. & N.E.

This same Line may also be car. on from Tring to Cambridge

A Line from Swindon to Gloucester by Stroud would have to pass down the narrow and much occupied Valley of Stroud, which is completely intersected by mill streams, roads, factories, canals, houses, &c.; there is no passage whatever left for a railway without removing some of the present works.—I presume the commencement from the Great Western up to Cricklade would be easy, but there must be a tunnel through Sapperton Hill, 1½ miles long, at an inclination of 1 in 176, (which I assume is the general inclination of the Stroud Valley) which will run for that 9 miles; but this part of the country is not shewn on Mr. Brunel's Plan, neither is the Section sufficient to judge of the merits of the line, as the country between Stroud and Gloucester, or 20 miles out of the 88, is not expressed: the Branch to Cheltenham, at about 2 miles from Gloucester, would be inconvenient and circuitous; it would also have a tendency to remove the traffic which at present passes through it, to Stroud: the Population of the former, which are of a travelling description, amounts to about 30,000: but the inhabitants of Stroud are of a manufacturing and stationary description, receiving their raw material from London, the principal part of which returns, but it is of a very light description, and cannot be of sufficient amount to require a railway. I think that a railway would increase the traffic more between Cheltenham than between Stroud and London, but a Branch from our Line to Stroud could be made if considered necessary.—A person travelling from Cheltenham to London by this line must first go to Gloucester, and again return on his road to Stroud, (making altogether 18miles) when he will be further from London than when he started.——

Remarks on the Gloucester Branch from the G. W.

Difficulty of same.

Sapperton Tunnel of 1½ Miles long, at 1 in 176, which conti. for 9 Miles.

Disadvant. of the Cheltenham Branch

30,000 Inhabitants in Cheltenham.

Comparisn. between same and Stroud.

——I have made a Survey of the Great Western, and I consider that it passes through a Country possessing very little Trade.———

Acct. of his Survey of the Grt. Western.

Bridges over the Thames are very expensive on account of the great width of the River, and there should not be any more made than is absolutely necessary, as they must form additional impediments to the Flood Waters: the whole drainage of the River passes through the Valley of the Thames, which is already very much obstructed, and any increase would be a great public evil: The crossing of the Thames at Maidenhead is therefore a serious objection, also the Kennet and Reading, the crossing over Pangbourne, and the crossing of the River Colne will also be objectionable, as the work must be considerable, consisting of several arches, which will narrow the flood way: the crossing of the Loddon is also objectionable: The embankment crossing the Valley of the Brent at Hanwell is objectionable, and the covered way through Sydney Gardens, being only 16 feet from the rails to the crown, although the distance is short.———

Bridges over the Thames are very objectionable.

His objections to same.

The Box Plane and Tunnel are exceedingly objectionable, and ought not to be permitted, (nothing can justify them in a country in which they can be avoided,) I am sure they will prove dangerous to the public, which will increase proportionate to the traffic; the smoke, gas and vapour will also be considerable, on account of the number of Engines and their diminished speed.—This Hill

Objections to the Box Plane and Tunnel.

may be completely avoided, as there is a naturally low level from Chippenham down to the Vale of the Avon at Bradford, although it is rather circuitous. I examined this Valley at the time of the opposition of the Great Western to the Southampton Railway Bill and observing the line was circuitous I inspected Box Hill, and immediately perceived that a passage could not be made from the White Horse Valley to Bath without a Tunnel of such immense difficulty that I considered it impracticable, (the Box Tunnel had not then been mentioned,) I therefore concluded the Great Western must pass down the Valley of the Avon, upon which I am prepared to shew its excess of distance over ours.——There is no direct communication by the Great Western from Bradford and Trowbridge to Bath, which are the principal commercial towns connected with it.—I therefore cannot understand how any Engineer can support a line from Bath through Box Hill and onwards, with a Branch to Bradford and another to Trowbridge, in preference to a line running along the level of the Canal from Chippenham to Bath, and passing through Bradford and Trowbridge, I would not do it for any employment.——I have many heavy works between Bath and Bradford, as the Valley of the Avon, which I propose passing by Viaducts, (I have 3 Viaducts on the line, one 640 feet long and 62 or 63 feet high, another 840 feet, and the other 640 feet, and from 50 to 60 feet high,) by which the floods will not be intercepted, and we have plenty of stone upon the spot; there was a short tunnel here originally, but having a great objection to them I omitted it, by which the distance is increased ¼ of a mile; the ground on this side of the hill is of a slipping nature, but we can carry a line through it notwithstanding, as the Kennet and Avon Canal passes through it, (it is more likely to affect a canal than a railway): I therefore do not see any difficulty in executing the line.—The Claverton, which is our only Tunnel, is 1 mile 1¼ furlongs long, and 30 feet high, and is upon a level, (but if the line is carried down to the Angle at Bath in order to intersect the Termination of the Great Western, it would be 1 in 330, as Mr. Brunton's Bristol line crosses the road higher up), I stated in the Commons that it was to be executed without Shafts, which do not assist the ventilation, however serviceable they may be in the formation.—The Ventilation of the Tunnel upon the Leicester and Swannington by Shafts is a complete failure, I sent my son to investigate it, and he was nearly smothered. (It is not a Passenger but a Colliery Railway.)——I shall use Shafts in the execution of the Tunnel upon the Southampton, (which is 18 feet high and 22 feet wide,) as the depth is not objectionable, but I have not had experience in them except upon Canals, neither am I aware of any Engineer that has had much experience upon the subject.——There are 4 Tunnels upon the Great Western between Bath and Bristol, which may certainly be shortened, the low position of the line in the valley of the Avon is also very objectionable; the termination of the line at Bristol is at Temple Meads, and the principal part of the basin where the larger vessels come is at the other side of Bristol.——I have seen Mr. Brunton's Plan between Bath and Bristol, (it was arranged that he should undertake the examination of this portion of the line, and I the remainder,) which has but 1 Tunnel upon it, which is for goods and at the termination at Redcliff Wharf, where there is plenty of room for 3 vessels to lie; and it affords immediate communication with sea-going vessels, but the upper part of the river is accessible to lighter vessels only, as the swing Bridge below Redcliff Wharf will not admit Steam-boats,

Marginal notes:

Remarks on Box Hill.

Comparison of the G. W. and Basing in refer. to Bradford and Trowbridge.

Remarks upon the Basing Line betwn. Bradford and Bath.

Des. of the 3 exten. Viaducts upon the same.

The Claverton Tun. 1 Mi. 1¼ Furl. long and 30 Feet high, and upon a Level.

Mem. regarding the taking his line to Mr. Brunel's Depôt.

Formation of the above Tunnel.

His objections to the Great Western between Bristol and Bath.

Mr. Brunton's Line between same.

Comparison of the Depôts at Bristol.

but it might be made sufficiently large by removing a portion of it, Irish cattle and provisions could then be landed at the Wharf, but it is not of much consequence as the distance is very short; the Passengers' Depôt was at Somerset Square, and a Bridge across the river would have increased the convenience of the traffic; I have not seen Mr. Brunton's Section, but I have had the distance levelled, and am satisfied it can be made.——I have not had any communication with the Bristol people upon this business, (I remember their advertising for Engineers to make Surveys of their line to London, and a letter was sent to me suggesting that it was worth attending to, and as I was passing through Bristol at the time, I wrote a letter offering to make a preliminary Survey for £1500)——I have observed that the earth at the mouth of the Tunnel under the Harrow Road at Willesden is not removed, which obliges them to wheel the Soil over it, in order to convey it to the lead, which is therefore a waste of power. ——The Great Western is 115 miles from Willesden to Bristol, supposing a Branch to Oxford practicable it would be 12 miles long, the Gloucester Branch would be 35 miles, and 7 more to Cheltenham, and the Bradford and Trowbridge 11 miles, making in the whole 180 miles; from London to Swindon is 80 miles, thence to Gloucester 35, making together 115, and to Cheltenham 122 miles; abating the 2 miles this side of Gloucester from the Cheltenham Branch would make it only 120½.——The Basing is 73 miles from Basing to Bristol, and it is 81 miles 3 furlongs from Tring to Gloucester by Oxford and Cheltenham, making together 154 miles 3 furlongs. The distance from London to Tring would be 34, thence to Oxford 29½, thence to Cheltenham 44¾, making 108¼, thence to Gloucester 7½, making 115¾, therefore the distance from London to Gloucester by way of Swindon is nearly 2 miles shorter than by Tring, and to Cheltenham it is 14 miles in favor of the Tring line.——I consider that the several lines of the Southampton, Basing, Tring, and London and Birmingham would divide the country in the most equal manner, the Great Western on the contrary would not make as equal a division of it.——The cost of the Great Western from Willesden to Bath is £2,100,000. including the Branches, (which are about £120,000.) and from Bath to Bristol is £400,000.; the Total cost, including the requisite portion of the London and Birmingham, will therefore be nearly £2,750,000., exclusive of the expense of sending the Goods down the river by the Regent's Canal.—The Total cost of the Basing line to Bath is £1,617,294., viz. £600,000. from London to Basing, or about £13,000. per mile, and £1,017,294, from Basing to Bath, which is full £16,000. per mile, (the Newcastle and Carlisle cost only £8,000. per mile); if Mr. Brunel's Estimate from Bath to Bristol was added to our's, the amount would not exceed £2,000,000. from London to Bristol,—therefore the Great Western will cost, in round numbers, ⅓rd more than the Basing, which the public must of course pay for;—I have included Stone Blocks, except where upon embankment, as Stone is met with upon the line.——Mr. Brunel may save £250,000. upon Earth Work, Ballasting and Fencing by taking my prices, as he has allowed 1*s*. for Excavation, which I average at 7½*d*. (it was 7*d*. only upon the Newcastle and Carlisle, including rails and sleepers, although the rails and waggons were of an inferior description compared with those upon the Southampton. I also know an instance of a man having taken a contract at 9*d*. and failing, upon which his

He offered to make a Prelimin. Survey to Brist. for £1500.

Formation of the Tunnel at Willesden

From Willesden to Bristol 115 miles.
Br. to Oxford, 12 m.
Br. to Glouc. 35 m. and 7 more to Chel.
Br. to Bradford and Trowbridge, 11 m.

Bas. to Bris, 73 m.

Tring to Gloucester 81 Miles 3 Furlon.

The South. Basing, Tring, L. & B. divide the Coun. equal. but the G. W. would not.

Total Estimate of G. W £2,750,000.

Details of same.

Southam. £13,000. per Mile.

Basing £16,000.

New. & C. £8,000.

Total Estimate of Basing £2,000,000.

Details of same.

Mr. Brunel might save £250,000.
Account of same.

Mr. B. all. 1*s* for Ex.
Mr G. do. 7½*d*.

Inst. of a Man failg. in a Contract at 9*d*. which was relet for 6*d*. & his debts paid.

Mr. B. Ballast 10*s.* Mr. G. do. 4*s.* Mr. B. Fenci. 5*s.* Mr. G. do. 2*s.* 6*d.* and 2½*d.*

Observ. upon same.

Do. Ballasting.

assignees let it at 6*d.* and paid the whole of his debts out of the profits,*) he has also priced his Ballasting at 10*s.* and his Fencing at 5*s.*—I proposed last year to Fence the Southampton line with oak posts and rails, and quick hedges, at 1*s.* 6*d.* the double yard, but I have since paid 2*s.* 6*d.* for the same; (I was led at the time to believe that I could get oak posts for 8*d.*) but a great portion will not cost above 2½*d.*; Fir is best for Sandy Soil, and Oak for Clay.—I have laid down some Ballasting at 5*s.*, but I have let most of it at 4*s.* per running yard of Railway, which makes it 10*d.* per cubic yard, including the leading; I stated in my Evidence originally that we should have to carry it 10 miles, but I am now of a different opinion; one mile of Railway takes about 10,000 cubic yards.—I priced my Masonry and Brickwork too high, viz. £11. per Rod, including centering, it is £10. upon the Great Western, and London and Birmingham, which is sufficient.——The Euston Square and the Vauxhall Depôts are about equal with respect to Passengers, but in reference to Goods Vauxhall is the best, the small Steam Boats and Wherries also offer some advantages with regard to passengers. I have always considered that a Railway between London and Bristol would be incomplete without a communication with the Thames; a person (Mr. Parkes) has offered to convey passengers from our Depôt at 6*d.* each to any part of London, and Booking Offices will be established for the facility of persons travelling by the Railway; our Depôt is an unoccupied spot, and comprehends about 4 acres, which is quite sufficient for a line of transit; I doubt whether many goods will remain long there; the Locomotives will not cross the road up to the wharf, (a space of about 1¼ acres of good hard ground,) but the goods will be taken by horses to the barges: it is about 290 yards from Vauxhall Bridge, the breadth of the river being about 130 or 140 feet.—I consider that such large communications as Holborn, Cheapside, Oxford Street, &c. should be crossed rather than passed down, in order not to further encumber them.—The Goods upon the Great Western will have to be taken from Camden Town to the River by the Regent's Canal, the Tolls upon which are very high, and it is also much obstructed by Locks and Bridges, which occasion a very considerable delay, and they have then to pass out of the Tide Basin into the Thames, and the time occupied will be quite as long as the whole journey from Bristol.—— ——It cannot be doubted that a line from London to Bristol would be a less eligible Investment than the Southampton and Basing.—In the event of the Great Western not passing, I am confident that the Subscription of the Basing and Bath line would be paid up immediately.—I have incurred great expence in making the several surveys, and for which I have not received any consideration; but I am willing to give them up unconditionally (to serve the Southampton) merely to be the Consulting Engineer, as the Basing line would be a public benefit.

Mr. B.'s Masonry and Brickwk. £10. Mr. G.'s £11.

Comparison of the 2 London Depôts. He considers the Southam. the best.

The latter compreh. about 4 Acres, and the Wharf 1¼ more.

Disadv of the Pasg. of Goods from the G.W. to the River.

Comparison of the Prospects of the competing Lines.

* No doubt the case of Mr. Wallace is here alluded to, the details of which are given in Mr. William Chadwell Mylne's Evidence, p. 228..—*Editor.*

Ex. MR. WILLIAM CHADWELL MYLNE, *Civil Engineer and Architect.*

There has been a succession of Engineers and Architects in our family from Father to Son for the last 400 years. (We were originally connected with the Royal Family of Scotland, and emigrated to England at the time the Court was removed.)——I have been Engineer of the New River Company for the last 25 years, having succeeded my Father at his decease in 1811, (who held the appointment 50 years,) and I managed his business for the 4 or 5 previous years. In consequence of an increase in the price of Elm in the year 1810, we adopted Scotch Fir for the purpose of Pipes, which were sent from Scotland and of about 30 years growth, and we used about 633 Loads annually until 1813, when we abandoned the use of them (we laid down at that time about 20 miles a year) for Iron, in consequence of their being unable to bear the weight caused by the introduction of Machinery necessary to supply the upper stories. I occasionally meet with old pieces of Fir pipe that have been laid down at least 20 years, very little of which is rotten, unless it happened to lay over a Sewer or a baker's Oven, although it has been exposed alternately to wet and dry.——My Father was the Architect of Blackfriars Bridge, the cost of which, including steps and abutments, was £152,000., (each Trade being let separately,) which is an infinitely less sum than such works usually cost. It is now being repaired, not on account of any fault in the construction, but in consequence of the stone decaying, the defects are principally between High and Low Water Mark. (Portland stone was the only stone used at that time for such works.) ——I have examined the country between London and Basing, having been over it twice last year, and three times this; I consider the Terminus at Vauxhall very commodious both for Passengers and Trade. I also think that Mr. Giles's Prices are quite sufficient, his rate of Wages being above the average paid to the labouring classes, even when clothing and food were dearer.—His reserve of 20 per cent. is a full Security, perhaps a greater portion should be retained at the commencement, to prevent the Contractor throwing up the work at any future time, as the more it advances the more expensive it becomes. The Sureties which men of large capital offer are more secure, (which operates as a complete check to honest men of smaller capital).——I have made an Estimate of the price at which the work was being executed at Shapley (on the 25th inst.) in which I brought the whole expence to 3¼*d.* viz. there were 41 men and 9 boys employed, which together with horses, ropes, waggons, grease, &c. made a daily expenditure of £8. 17*s.* 1*d.*, which divided by 590, the quantity of yards removed, gave 3¼*d.*; the Teamers were therefore earning 2*s.* 6*d.* and the Fillers 3*s.*—This work is let to the Contractor for 6*d.* who must therefore be making at least 2*d.* clear profit, the length of the lead being 830 yards; as it increases the work will of course cost him more, (some of the work at Frimley is underlet at 3*d.*)—11 waggons passed down the incline (24 inst.) in half an hour, which I consider were teamed within that time: I took them for an hour the next day when they amounted to 22, and some of them held 3½ cubic yards, but taking them at 3¼, and allowing 16¾ hours per day, (the men were working 16 hours 50 min. and were in double shifts,) gives 1,197¼ yards per day, a man could therefore earn 2*s.* 6*d.* per day easy, and they could work longer and consequently make

His Family ha. been in the Prof. during the last 400 Years.

He is Engin. to the New River Compy.

Scot. Fir Pipes were laid down in 1810 instead of Elm.

Iron Do. intro. 1813

Superiority of same.

Instances of the durability of Fir.

His Father was the Archit. of Blackfri. Brdg. which was let in separ. Trades, and cost only £150,000.

Portl. was the genl. Stne. in use at that ti.

His opinion upon the Basing Line.

Considers Mr. G.'s Prices sufficient.

And his Secur. suff.

Remar. upon same.

He makes the Work at Shapley Heath come to 3¼*d.* c. yd.

Details of same.

The same is let at 6*d.*

With a Lead of 830 Yards.

He allo. 3¼ cu. yds. per Waggon. Making 1197 c. yds. per Day.

more if they worked in single shifts. 4 teaming places only were laid down, they could therefore team more by increasingt he number. The men are paid 1¼*d.* per yard for Digging and Filling, and ½*d.* for Teaming; I am aware that the expences would increase in short days and very bad weather, but men will always fetch up the loss of a wet day upon the following, so as to make their wages average 3*s.* per day.—I allowed 27 waggons at £ 16. each, making £ 432. and I calculate they would last out the work,* and allowing 1000 cubic yards per day, gives the expence of them at ¼*d.* a cubic yard, I also allowed 2*s.* for Oil and 1*s.* for Machinery; the cost of the Rope (weighing 13½ cwt.) being £ 23, I have also allowed 3*s.* for the use of same and renewal every 80 days. —The men were not aware of my visit, neither were they working very hard, and I thought they evinced a disposition to strike, a person was watching them at the time who I understood belonged to the Birmingham Railway——I have always taken my Father's Standard, viz. that 4 men can fill and wheel 60 cubic yards of Clay a distance of 20 yards per day for 14*s.*, which allows each 3*s.* 6*d.* per day, and averages 2¾*d.* per yard, but we usually take it at 3*d.* and allow an additional 1*d.* for every 30 yards.——Engineers like high prices as well as Contractors, as it is of no use giving a very low Estimate to get a Bill through Parliament, and afterwards loose the opportunity of carrying it into effect. I do not suppose such men as Jolliffe and Banks, or McIntosh, would have took this work at 6*d.*, as they are getting 9*d.* and 10*d.* all over the country. I have heard Contractors say they are satisfied with 5 per cent. clear profit, which I think sufficient, indeed it is high time some change was made.—Mr. Rennie always made his Estimates high, he was also obliged to have recourse to the system of large Contracts, (which were introduced about the time he entered the profession,) in order to get through his business, as he engrossed nearly all the Engineering practice of the Kingdom; his first Survey was made for my Father, under whom he learnt his profession and was first introduced into life.—It is customary for Contractors of this description of work to make very great profits; none of the parties who were spoken to last year regarding the above work at Shapley Heath would undertake it upon a less calculation than £ 30,000. profit.—Mr. Banks formerly wheeled a barrow, but died immensely rich, his partner Mr. Joliffe (a Clergyman) found capital, and they had for some time the execution of all the Government works, under the direction of Mr. Rennie; Mr. McIntosh, who is now repairing Blackfriars Bridge, also began in the same manner, and has also amassed a large fortune; I have heard Mr. Banks say that he let his work in Gangs, for instance, he would let ¼ of a mile of a Canal to one man, paying him by agreement, in fact they all underlet their works to others.——A person of the name of Wallace had a Contract for deepening the River Cam in Cambridgeshire, and stopped about the middle of the work for the want of money, his price was 10*d.*, and the man that worked for me at Sandy's Cut took the work of him for 6*d.*, and with the remaining 4*d.* paid his debts, and Wallace had £ 100. left after all law expences were paid, although the Commission alone cost 2*s.* 6*d.* or 3*s.* in the pound.——I was Engineer at Sandy's Cut on the River Ouse, which was executed by common agricultural labourers, (Canal is worse than Railway work, as it is wet,) and I paid 2*d.* for getting, filling and

Margin notes:

The Digg. and Fill. was done at 1¼*d.* and the Team. at ½*d.*

The Wag. cost £16.

He all. ¼*d.* for same.

The Rope cost £23

He all. 3*s.* per Day.
Oil . . . . 2*s.*
Machinery . . 1*s.*

The Men were not working hard.

His Father's Earth. Stand. was 3*d.* c. yd.

Details of same.

Reas. why Enginrs. like high Prices.

Rem. upon the large Profits Contr. make.

He cons. 5 per Cent. sufficient for them.

Mr. Rennie made his Est. high and let his Work in large Cont.

His reas. for the same

Insta. of the Profits calculated upon by Contr. (at Shapley)

Mr. Bankes origina. wheeled a Barrow.

Mr. M'Intosh also.

They amassed large Fortunes by underletting the Works.

Inst. of a Man failg. in a Contra. at 10*d.* which was re-let for 6*d.* & his dbts. paid.

Descr. of the Works at Sandy's Cut.

Canal is worse than Railway Work.

* The time alluded to is not clearly stated in the Evidence, whether it is 1½ or 3 years.—*Editor.*

wheeling Peat, which is light in the barrow, but being wet is worse to get than Sand, and 3*d.* for heavy Silt, the men earned about 3*s.* 6*d.* a day.——I have had a good opportunity for forming a Scale of Labour, as I always employ from 150 to 200 Men; I never take a Navigator upon the works, but always fill up my vacancies from the Country, and as they are very awkward upon entering, they only receive 2*s.*; (a man of 6 feet is worth 6*d.* a day more than one of 5 feet, as he has a greater power of Leverage) but they all earn their wages from the commencement, and execute the work quite as cheaply as the others, although less rapid; the whole is measured every half year, by which I can judge of their work, and as they come up to the "Standard" before mentioned, I raise them from 2*s.* to 2*s.* 3*d.*—2*s.* 6*d.*—3*s.* and ultimately 3*s.* 6*d.*, when they become proficients and leave me for the railroads. Contractors prefer Navigators, as they require the work done quickly and on account of their having no time to teach men. ——I have let work to Mr. Fireday, the great Ironmaster in Staffordshire, also to Mr. Jones, which merely extended to a price per ton for the Iron, and a price per yard for the laying, although it ultimately amounted to £90,000. the work being measured up monthly, and 20 per cent. was reserved until the completion; we could have stopped the Contract at the end of the first month had we wished. This mode of letting Contracts is different to the generality of Engineers, and requires great attention to the accounts. ——I am about letting a Contract for a Bridge at Cambridge, for which the parties were very anxious to have one contract, and accordingly advertized, in spite of my remonstrance, as I am persuaded it is the most expensive, owing to the work being thrown into fewer hands: It amounted by small Tenders to about £970. but the estimate was £1,000. The Butterly Company have undertaken it, although there is the following clause in the Contract, "Such condition shall not prevent the due performance of the "several works according to the true intent and meaning of the drawings, but a fair and "liberal construction is to be put upon the Drawings and Specifications where they are "not fully described, or have been left imperfect. All such works as may have been "omitted by the Architect or Engineer are to be required of the Contractor by the said "Mayor and Burgesses of Cambridge and their Successors, or their Architect, as if "they had been within the Specification"; a man unacquainted with me would not tender with such a clause, although they are quite customary; therefore a Contractor runs a great risk, and is obliged to put on an extra price to cover himself, whereas if a correct specification is made, and the Contract properly subdivided, a great saving would be made.——In looking over one of the Contracts of the London and Birmingham Railway I observed a clause thus, "That the Contractor should be bound to maintain the Slopes and keep the work in repair for a certain time," now as the Contractor does not determine the proportion of the slopes it is not right that he should take the risk, it is turning it into "a Speculation," and in order to cover himself he charges accordingly. Slips may occur even if the work is well executed.——Great facilities may be afforded for the Engineer and Contractor to play into each other's hands by my method, (but an active superintendance would serve as a check,) by the former certifying the work as properly executed when it was the contrary, but men of this character would not want opportunities in the other case, the Engineer would state the Contract was of a proper amount when it was not. I have known Contractors to meet at a public house

He paid 2*d.* for Peat.
3*d.* for heavy Silt.
Scale of Lab. Wages
A Man 6 ft. high is worth 25 per Cent. more than one of 5 ft.
He pays them from 2*s.* to 3*s.* 6*d.*
Reas of Contractors preferring Navigat.
He lets his Works by admeasurement at a fixed Price.
Illustration of same.
Acco of a Bridge he is buil. at Cambridg.
He cons. lettg. Wrk. by smll. Con. the best
Instance of the Arbitrary Nature of the Clauses in large Contracts.
Influence of same upon the Cost of the Works.
His Strict. upon the Contr. of the L. & B.
He consi. the Contr. should not take the risk of Slips, &c. as it is turning it into a Speculation.
Comparisn. between Contracts by the lump and the same by his system.

and agree upon the price, as there are seldom many competing for large works, with whom the Engineer could coalesce if he pleased.—It appears the object of Contract is to bring the expences to a certainty, which possibly may be a great point, but they have to pay dearly for it.——St. George's Hill can scarcely be called Clay, if it had been in a wet situation, or in the neighbourhood of the Fens, it would have been called Silt, it makes good bricks, it is very unlike the London Clay which is remarkably tenacious. ——I have examined the Bridges which have been built at Battersea, Wandsworth, Esher, &c. and I consider them very good Brick-work, and quite sufficient for the purpose.——I made a Survey of the Thames from Cricklade to Reading for the Thames Commissioners, which was my last survey for them, (my Father was engaged by them for many years,) I have all the sections and levels of the River, which acts as a General Drainage, and it is deepened and improved from the Funds raised by the Navigation; the several estates which it passes through are well drained and consequently improved: and if the works were abandoned, the drainage might be better for a time, but the river would at length become silt up.

He consid. Contract a very dear system. Soil of St. Geo. Hill is scarcely Clay. Descriptn. of same. Memo. the Bridges upon the Line. He has been empl. by the Tha. Commis. The Thames drains the Country effectu. Remar. upon same.

Ex. Mr. HENRY SMITH, *of Bath, Land Surveyor.*

I have been connected with agricultural pursuits for nearly 30 years, and I have resided in the neighbourhood of Bath the last 50 years.——I have valued the Land required at Bath for the Basing and Bath Railway, which was pointed out to me by Mr. Fawcett, the gentleman who took the levels.——I commenced my valuation at the commencement of the Claverton Tunnel, (at No. 6 on the Plan,) and continued from thence to the Depôt, which I included, together with the Approaches, and after deducting the saleable value of the materials remaining upon the ground, it amounts to £6,698. 4*s*. 6*d*., including Compensation both for severance and injury done to Tenants, and I consider it a very fair price.—From thence to St. Mary's Buildings amounts to £4,443. making a total of £11,141. 4*s*. 6*d*. The whole quantity is about equal to the length required by the Great Western in passing through Bath, which is valued by Mr. Goodridge at £40,864., and exceeds mine by about £25,000.—— The number of Houses required to be taken down by the Basing Line, from No. 22 to St. Mary's Buildings, are 48, including the first two houses at Tiverton; and from Bathwick through Lidcombe and Widcombe, on the Great Western, 375 Houses are scheduled to be removed; Tiverton is a straggling village, perhaps 1 mile long, and about 1¼ miles from the bridge at Bath, I consider the Land required by the competing lines in passing through it of about equal value.——The Skew Bridge upon the Great Western passes near Claverton Street, where some houses have been pulled down to increase the width of the water way, the water flows out of this street at flood time, also from Ham Gardens through Dorchester Street, where there is generally a great quantity, it then enters Southgate Street, and crosses passing between the buildings adjoining the old bridge, and flows down the broad quay in a sheet into the River. I understand the Great Western again crosses the river at Ham Gardens by another

He has Valued the Land requir. by the Basing at Bath. From the Clav. Tun. to the Depôt, inclu. same and Approach. amounts to £6698. including Compens. Frm. thence to St. M. Buildings £4443. Total £11,141. Being ra. more than a ¼ of the val. of the Land req. by G. W. in passg. thro. Bath. 48 Houses in the above distance upon the Basing. And 375 by the G. W. The G. W. crosses the River by a Skew Brid. at Clavert. St. The Floods at Ham Gardens, &c. The G. W. cross. the Riv. again at Ham G.

bridge.—The ground for the Depôt must be raised above the flood mark of 1809 upon Widcombe Poor-house, which is 4 feet 9 inches above the surface of the pavement, and the latter is some height above the mud, owing to its having been raised of late years, by which the water will be intercepted, and will consequently rise above the bridges, (particularly as two bridges are intended).——The Great Western will injure Sydney Gardens very much, as it passes through them, (I have understood that 5000 people have been there at the same time.)——The Colliery Railway and Canal is nearly in a straight line from Dundas to Midford, (the Company possessed power by their Act to make either a canal or a railway) and when it arrives at the Aqueduct by Midford, it is joined by the Dunkerton and the Radstock Canals in the shape of the letter Y, and they were obliged to have a railway run down the hill, in order to get to the Aqueduct.—I would not allow them to pass through my land, by the side of the canal at Midford, on account of the Radstock Branch taking the water from my mill; but I would consent to their running a railway over it to the Basing, provided the public were allowed the use of it; by which the towns of Trowbridge and Bradford would be opened to us, (which we cannot at present get to, as the country is very hilly) also the country beyond them up to Berkshire, and it would likewise be very advantageous to the Collieries.—— ——There has been a great expenditure lately upon the four roads at Holloway, as the Wells and Claverton Roads, (where the Great Western passes) owing to the difficulty of uniting them.——A Line of Railway ascending Box Hill by a tunnel, and passing on near Chippenham, then turning back by the Branch to Trowbridge and Bradford, would be very little advantage as a Line of Communication between these towns and Bath, Bristol, &c.—The Great Western passes the spot where the Bradford and Chippenham roads join by a Bridge, 17 feet high, and the Embankment crossing the meadows from Bathampton to the same being very high, would consequently pen back the flood water in Winter. ——Some of the best houses in Bath are situated in the South and North Parades, which are the neighbourhoods most frequented by invalids, (the South Parade has been considered the Montpelier of Bath) their being flagged also removes all noise and traffic, but the Great Western would completely destroy their quietude.——The trees between this railway and the back of Pulteney Street will not in my opinion form a proper screen, as they do not exceed 20 feet in height.——The houses in Bathwick Terrace (under which the Great Western passes in a tunnel) are of a very superior class, and let for £60. or £70. per annum: the front prospect is very extensive, and extends over the whole of the hills which surround the city.——It also tunnels under Raby Place, which is a highly respectable neighbourhood, combining both town and country: the houses are letting at £80. to £100. per annum, and are in much request.——It also passes the top of New and Old Sydney Place, which contain superior houses.—— ——The Houses interfered with in Forefield Place by the Basing are of far less value than those of Raby Place.——Mr. Goodridge, whom I have known some years, is an Architect of great practice, but I doubt his knowledge as a Land Surveyor, (when the Church at Coombe Down was built, the Committee purchased the land, and sent for me to measure it, also to lay down the true meridian, in order to get it due East and West, although Mr. Goodridge was employed to super-

The raising of the Grnd. for the G.W. Depôt will increase the Floods.

Reason of same.

The G. W. will injure Sydney Gar.

Description of the Coliery Railw. and Canal at Midford.

A Branch from the above Railw. to the Basi. would be very adv. to the Pub. also to Bradf. Trowb. &c.

Objec. to the G.W. At Holloway.

Inconv. of the G.W. as a commun. betw. Bath & Trowbr. &c.

The G. W. would be inju. to the Meadows from Bathampton.

Do. do. Parades at Bath.

Do. do. Pulteney Street.

Do. do. Bathwick Terrace.

Do. do. Raby Place.

Do. Sydney Place.

He considers Mr. Goodridge more an Architect, than a Land Surveyor.

The Basing Line will not injure Prior Park.

intend it).——In order to ascertain whether his statement given in evidence was correct, "that the Basing Line would injure Prior Park," I took a plan of the line, and walked over the ground, and I ascertained that it is about 1 mile off, and considering the depth of the cutting is about 60 feet, and protected by a wall, 8 feet high, and that a shrubbery is at present upon the spot, I consider it will be impossible to see the railway from Prior Park, (it is a Catholic College) or from the temple at the head of the

His Accou. of same.

pond. The property belongs to Dr. Baines, (the Bishop of the College) who has built a new Lodge 200 yards above the old one, which is a considerable distance from the railway; it is also 80 feet from the fish ponds, which is quite sufficient to preserve them: the banks of the Kennet and Avon, also the Somerset Coal Canal, are not half that

Do. do. Mr. Tugwell's Hou.

thickness.——The line runs within ¼ of a mile of the back (where the Offices are situated) of Mr. Tugwell's House, and there will be 2 roads and a garden intervening.

Do. do. Col. Wrench's Prop.

——It also cuts off the corner of Colonel Wrench's Pleasure Grounds, passing in 40 feet cutting, but it will be hidden from the view of the house by a 12 feet wall,

He received a Letter from the Town Clk. of Bath, stating that the most resp. Inha. object to the G. W.

several high elm trees and a shrubbery will also be left.—— ——I have a Letter from Mr. Philip George, Town Clerk of Bath, stating that the most sensible and respectable people in the city object to the Great Western, and in reference to the Mayor and Corporation approving of it, I certainly do not consider they are competent judges of the merits of the competing lines.

Ex. Mr. CHARLES FOWLER, *Architect.*

He was the Archit. of Covent Garden and Hungerf. Mar.

I have practised in London for the last 16 years. I have had occasion to examine into the circumstances connected with all the Markets in London, having superintended the erection of Covent Garden and Hungerford Markets. I consider Vauxhall Depôt

Vauxhall Depôt.

very eligible for the diffusion of Provisions throughout the Metropolis, and it does not crowd the East or Western thoroughfares, as the traffic from thence will cross them. I

Remarks upon the choice of a Depôt.

think a Depôt desirable in proportion to its connection with the several markets, Country meat may thus be brought to town from very remotes distances, also Poultry, Butter, Eggs, &c.; the recent improvements in roads and spring carriages have already brought

Leadenhall formerl. but Newgate now the Principal Mark. for Butcher's Meat.

much to market which could not before reach London. Leadenhall was formerly, but Newgate is now the principal Market for Butcher's meat. Hungerford also possesses many advantages for the same; the proximity of the Depôt to the River is a great advantage, it is also a remarkably unincumbered spot, I am not aware of any direction which can equal it in this respect; it likewise offers a facility of communication with

Adv. of a communi. from the above Dep. with Greenwich.

Greenwich and Deptford, and a connection might easily be made with the Greenwich Railway, which would be advantageous as the large Steam Packets lie down the River on account of their not being able to get up.——I have also Surveyed the Great

Eust. Squa. Depôt.

Western Depôt (the London and Birmingham) at Euston Square, and I have made a Table of the comparative distances of the principal points of London from both Depôts, which I measured from Crutchley's large Map, as follows:

Table of Distances from the Southampton and the L. & B. Depôts.

DISTANCES of the respective RAILWAY TERMINI from Markets and other important Points.

	Southampton.		Birmingham. Camden Town.		Birmingham. Euston Square.	
	Miles.	*Furlongs.*	*Miles.*	*Furlongs.*	*Miles.*	*Furlongs.*
Hungerford Market—						
by Westminster Bridge	2	6	2	7	2	1
Vauxhall Bridge	2	0½	—		—	
The River	1	7	—		—	
Covent Garden Market—						
by Westminster Bridge	2	6	2	6	2	0
Vauxhall Bridge	2	0½	—		—	
Newport Market, by Westminster Bridge	2	3½	2	3½	1	5½
Farringdon Market	3	0½	3	0	1	7
Newgate Market	3	1	3	0½	1	7½
Billingsgate Market	2	7	4	2	3	1
Leadenhall Market	3	1	4	1	3	0
Spitalfields Market	3	6	4	5	3	4
Borough Market	2	4	4	1	3	0
Oxford Market	3	5	2	2	1	4
Smithfield Market	3	3	3	0½	1	7½
Bull Inn, Aldgate	3	4	4	5	3	4
Spread Eagle, Gracechurch Street	3	0	4	5	3	4
Swan with Two Necks, Lad Lane	3	0	4	2½	3	1½
Bolt in Tun, Fleet Street	3	0	3	0	1	7
Golden Cross, Charing Cross	2	6	2	6½	2	0½
White Horse Cellar, Piccadilly	3	3	2	6	2	0
Green Man and Still, Oxford Street	3	5	2	0	1	2
To Greenwich	5	5	—		—	
To Deptford	5	2	—		—	
The Houses of Parliament	1	2	3	3	2	5
Saint Paul's	2	7	3	3½	2	5½
Eaton Square	1	5	3	7	3	1
Bank	3	0	—		3	2
Hyde Park Corner	1	7	3	3	2	5
Saint James' Street	2	2	—		2	0
Kensington Barracks	2	3	3	7	3	1

N.B.—As Goods are not to be brought to the Birmingham new Station the Comparison as regards Markets does not apply to it.

Remarks upon same.

I measured the distance by a string in order to take all bends of the road.———I have since ascertained that the distance to St. Paul's is 2½ furlongs in favour of the Euston Square Depôt, instead of 1½, as stated:—Regarding the distance to the Bank, I went down the City Road from Euston Square, but if I had taken Gray's Inn Lane and that way, it would have made it ¼ of a mile in favour instead of against the latter Depôt.—— —— I am one of the Commissioners of Paving for the Bedford Estate in the Parish of St. Pancras, in which I reside, and can state that there are Gates at the North End of the Boundary towards the New Road, between the Bedford and Southampton Estates, which are closed against omnibuses, carts, waggons, &c. therefore some of the distances by these vehicles will be increased; it is our understanding with the Duke of Bedford that the gates shall be closed, and there is a strong local feeling in favour of the continuance of same, as they render the neighbourhood quiet and safe; the tenants of course have a key.

Ex. DANIEL MALLET.

Lighterman, of King's Arms Stairs, Westminster Bridge.

His Experience. I have been employed by the Waterloo, Vauxhall, and Southwark Bridge Companies, and by Government, in works upon the Thames.——The Wharf at Vauxhall Bridge, at the termination at the Southampton Railway was formerly a Dock, and Boats drawing 3 feet water (about 40 tons burden) can get along the wharf at Half-tide (or 2 hours tide).——The Flood tide runs 5, and the Ebb 7 hours.——There is a Hard (gravelly ground) in front of the wharf, which will bear carts, goods can thus be carted alongside the boats, and room can be made for a first-rate ship to come alongside if required, as the whole of the gravel might be excavated at dead low water: I could do it in 2 months, (30 or 40 feet might be removed if required) but 6 feet might be removed in 2 or 3 weeks, which I could undertake at 1*s.* 6*d.* per yard.——

Descr. of the Wharf at the Vaux. Depôt.

The Tide flows 5 ho. and ebbs 7 hours.

The Ground of the Wharf is a "Hard."

He would remove it at 1*s.* 6*d.* per c. yd.

List of Distances.

It is 163 feet from the wharf to low water mark, where the depth is 3 feet at the edge, and then 9 or 10 feet. ——

And about 300 yards from the wharf to the abutment of Vauxhall Bridge.——

152 yards from the Stone Yard to the end of the wharf adjoining the River Thames.——

21 yards 6 inches from the Stone Yard to the post that separates Lambeth and Battersea parishes.——

12 yards 2 feet 6 inches from one side to the other of Nine Elms' Road.——

And 3 yards 2 feet 11 inches from the post that separates Lambeth and Battersea parishes to the wall that stands in Lambeth parish.——

It is difficult (at certain times) to get into the Basin of the Regent's Canal, which enters the Thames below Limehouse; they will not let you in at half-ebb; the barges are sometimes 7 hours getting out.—The Water Carriage from the entrance of the Regent's Canal to St. Katharine's Dock is about 1*s.* a ton, and about the same from St. Katharine's Dock to Nine Elms, although the latter may be about 6 or 700 yards longer; but I have nothing to do with the upper part, the river below bridge is more crowded; and the navigation consequently more dangerous, particularly for loaded craft: we frequently ship a ton or two of water, as the steam boats will not give way for any thing.——We can work Barges against Tide, provided we have the Wind.

Reg. Canal entr. the Riv. below Limeho.

The Wat. Car. from R. Can. to the St. K. Docks abo. the same as from the latter to Vaux. viz. 1*s.* pr. Ton

Difficulty of the Naviga. below Brid.

Ex. Mr. ZACHARIAH ALLNUTT.

Gen. Receiver of the Thames Navigation

He is also employed as a Solicitor by the Oppon. to the Bill.

I have been General Receiver of the Thames Navigation for the last 35 years, and I was formerly their Surveyor; I also practise as a Solicitor generally, and was employed by the Opponents to the Great Western to draw up some Statements, which were circulated among their Lordships, from which the following is a quotation, "Then "comes the case of the Schemers, presumed to be made out by *Extracts* from the "*partial Evidence of Mr. Brunel, their own Engineer,* before a House of Commons

"Committee, *as well as the Hirelings and Helpers they have picked up,* who merely "*state their opinions,* and give *hearsay evidence* to questions previously arranged, *but* "*not a word on oath,*" by which I did not mean to cast any imputation, but merely to state that evidence given in a conversational style, and not upon oath *en masse*, is of less value than the latter. (The House of Commons did not then allow the oath to be administered.)——Very considerable improvements have been made of late years both in the Drainage and Navigation of the Thames, and it now forms a very considerable means of drainage to the neighbouring country.——The traffic in connection with the Canal has increased, arising from the Locks, Pounds, and Artificial Works, which have been carried to a very great extent.——The Thames Commission is composed o fthe Noblemen and Gentlemen residing on the banks of the Thames, but they are not Proprietors like a company, therefore they do not divide any profit. The Funds arising from the Navigation are employed exclusively upon the improvements, some of it is spent in forming extra embankments, which benefit the owners of the adjoining land although they do not study the neighbouring lands, or any individual's interest; they have borrowed about £ 94,000. at 5 per cent., for which the tolls are mortgaged. ——The General Clerk is obliged by the Act to present Parliament annually with an account of the Receipts and Expenditure, including the interest paid to the Bond-holders, Salaries to the Clerks, viz. The General Clerk, (Mr. Payne, who is Solicitor to the Opponents of this measure,) the Surveyor, and the Assistant Surveyor, (Messrs. Treacher and Son,) and the Receiver General, (myself,) also 28 Pound-keepers and 6 Ferrymen, and there are from 15 to 20 workmen regularly employed, and sometimes 40 additional; (I think it would be bad policy to reduce our Salaries); also Storehouses, yards, and other conveniencies, which have been purchased, sums paid for Rent, also the sums expended for repairing and ballasting the River and Cuts, also Towing Paths, &c.——The total Expenditure in 1832 was £ 12,695. 7*s.* 2*d.*, and the total Receipts £ 13,168. 18*s.* 5*d.*; (the sum paid for Interest amounted to £ 5420. 12*s.*) ——I conceive that a Railway between London and Reading would draw off a considerable portion of the Traffic of the Thames, and that the funds would not be sufficient to keep the River in order; more funds are required for the works at the present time, as there is no surplus left.——The Commissioners intend removing 2 or 3 dangerous turnings by fresh cuts, as soon as they get sufficient funds, (they have had to wait nearly 4 months for the loan of £ 4,000. although they advertized).—— The River would grow up and at length get dry, if it was not navigated; we could barely afford to lose a barge, as their passage assists in keeping it open.——The Light goods amount to very nearly ⅕th of our income, in the event of the Barge-masters losing which, they would be obliged to put a heavier toll upon the remaining Heavy goods, and the loss of the latter would be the consequence. The Light goods pay the Barge-masters for the Heavy, and they are very likely to go by the Railway, on account of the greater expedition. It is also very likely that goods liable to injury from the heat of the weather in the Summer, as Grocery, &c., would go by Railway, as we do not put ourselves upon the same footing as a Railway or a Canal. Low price will also influence parties who are ignorant of our speed, but no advantage that these people could

Evidence at first was not giv. upon Oath.

Improvements on the Thames.

The whole of the Funds are expended upon them.

Account of the Expenses of the Management.

Expenses in 1832 £12,695. 7*s.* 2*d.* and Receipts £13,168. 18*s.* 5*d.*

He cons. the G. W. wo. affect the Funds of the Navigation.

The Com. have not sufficient Funds.

There is difficulty in obtaining Loans. The River would cons. fall into decay.

The Lt. Goo. ⅕th of the traff. on the Tha.

Light Goods would probably go by Ry.

derive from the Railway would compensate for the destruction of the River. I believe our Commissioners opposed the Grand Junction Canal when in Parliament, and it certainly has injured our Navigation, as the traffic has not improved in proportion to the increased Population of the several Towns. A Railway to Bristol, whether it passes through Reading or not, would be a serious loss to us, but it would not be as injurious as the Great Western, if it passed further off. Reading is 40 miles by land and 78 miles by the river from London; from Teddington to Reading it is against stream, and from Teddington downwards it is exposed to the vicissitudes of the tides.——I do not remember any case of the detention of a Barge by Frost longer than 3 or 4 days for the last 14 years; when it reaches a certain degree we draw the Sluices, which breaks the ice all the way down, and the cuts being short it passes away. We had rather a severe Winter in 1832, but I do not think the frost continued more than 7 or 10 days, and the Navigation was not stopped above 2 or 3.—Stoppages of 2 or 3 days from Floods may have taken place—Stoppages by Drought are occasioned by the Bargemen, I have known innumerable cases of it, the Commissioners allow them to go to any depth they please, except at certain times, and they are aware of the depth of water, yet they overload their barges, (they take in too great a Load in London, notwithstanding the Bye-laws, hoping to get up, but at length the water fails them,) and stop one another; last Friday a Barge was loaded 3 inches higher than the Commissioners allow, and if I had not seen it 20 Barges would have been stopped; we intend employing a person to inspect them. We also avoid much of the obstruction which formerly arose from Drought, and we do not resort to Flashes (the passing of a certain quantity of Water at certain times of the day to assist the downward passage when the water is scarce) as frequently as formerly, in consequence of the improvements, and it is the intention of the Commissioners to do away with them.——The Drainage on the banks of the Thames is immense, the adjoining meadows being sometimes a mile in extent, the whole of which are affected. The Towing Paths must be kept in good order, as they protect the country from the floods; the flooding of such places as Oxford, Abingdon, Wallingford, Reading, Marlow, Maidenhead, and Windsor, would be a great injury, and would make them damp and unhealthy, the Navigation of the Thames is therefore indispensably necessary for the drainage.——The several Bridges and Embankments that will be required to carry the Great Western railway across the Thames and adjoining meadows will be very injurious: and must affect the river as a source of drainage, (although not perhaps in a very great degree) even if they use every precaution in the several works.—Provision is always made to let the Water pass off quickly by arches in the adjoining roads.——A very large quantity of water passes into the river at the first place of crossing at Maidenhead, spreading over the Flats on each side of the Bridge, which are very extensive.—The Second crossing is at Reading, at about 2½ miles from Sunning, near the mouth of the Kennet, where there is a large flat, also a contraction by the hill half a mile from the Railway Bridge, which will cause great inconvenience, notwithstanding any number of arches and culverts they may intend erecting.—And the Third place of crossing is about 5 or 6 miles above Reading, in the direction of the Valley of the Thames, where it crosses one of the branches of the river Kennet, which is a part

Margin notes:

The Commissioners opposed the Grand Junc. Canal, which has inj. the Thames.

Stop. by Frost very scar. upon the Tha.

The precautions against same.

1832 was a sev. Wint.

Stoppa. by Floods of 2 or 3 days have occ.

Stoppa. by Drought are occasioned by the Bargemen loadi. the Barges too full.

Instan. of the same.

They seldom resort to Flashes now.

The Thames indisp. as a General Drain. to the Country.

The Works upon the G.W. will cau. Floo.

Observ. upon same.

Inj. effect of the 1st Crossg. at Maidenh.

Do. the 2d Crossing at Reading.

Do. the 3d Crossing 6 Mil. below Readi.

of our Navigation, (we have the control up to the High Bridge at Reading.)——— Eton College is supplied with water by a mill on the Buckinghamshire side of the river, and Windsor Castle by a mill on the Berkshire side, (as the water is obliged to be raised to a considerable elevation,) they belong to each party respectively, and are nearly opposite each other, and any neglect arising from a failure of our funds would injure, and eventually destroy them; the River is dammed up by a Weir, or artificial head, across the river, (which we are at the expense of) which creates a fall and turns the mills, the water is then thrown into a sort of channel, and driven up pipes to the Castle, a diminution of our funds would disenable us from keeping the Weir in repair, and we should therefore be obliged to remove it, the power which supplies the Castle will thus be gone, (the water would certainly pass by Windsor better in the event of the removal of the mills,) and artificial means must consequently be resorted to, as the water would not retain sufficient power to turn the mills, and it would not pay the inhabitants of Windsor to keep the Weir in repair, as it is very expensive, (neither haveth ey the power to do it without going to Parliament); the cost originally was about £2,000. and I have known the Repairs amount to £50. or £60.: £60. per annum may be sufficient for some time, if it was put in a perfect state, but it is opened and closed every hour, therefore depends on the Wear and Tear. There is a Pound Lock at the bottom of the channel which supplies the King's Engine with water, which cost nearly £5,000.; when it was erected the power was increased, and more work has consequently been done, and it has acted better.—— ——Reading contains 11 or 12,000 inhabitants, but it is not a manufacturing Town; there are a great quantity of Provisions, Coals, and Building Materials constantly travelling to and from it, for which the Thames affords a suitable communication, and people can easily provide against delays from floods and frosts by a little foresight; I am at Reading 3 or 4 times a week, and can therefore state that there is no inconvenience arising from the want of another conveyance, and I am fully convinced it would suffer considerably by losing the advantages of the Thames. The Town of Henley, which is 7 miles from Reading, would also suffer considerably. The Traffic does not require unusual speed, with the exception of very light articles, which the inhabitants get up and down in one day with ease, besides having the principal part of it to transact business.———Large Barges are getting out of use, and smaller ones adopted, and the communication between London and Reading by them has increased very much lately, they go every day, and when loaded occupy 2½ days up and 2 days down; the light boats get down about 2 or 3 hours sooner, but if the Barges were better worked, they would occupy less time; the men are in the habit of going ashore and getting intoxicated, instead of attending to their duties; I remember the case of a Barge-master, who was a careful industrious man, making a fortune, although he did not receive more than a fair remuneration for his outlay of time, labor, &c. but I believe they are generally losers, and any diminution would be ruinous to them as well as to the river.———We have not had an instance of a horse being drawn into the river for some years, as our improvements have in a great degree obviated it.

Eton Coll. is supp. by a Mill & Windsor Cas. by anoth. both of which are put into action by the Works conn. with the Nav.

Explanati. of same.

If the Navi. is negl. they will lose their power and supply of water.

It would not pay the Inhabitan. for keep. the Works up.

Cost of the Weir about £2000. and £60. pr.an. for Rep.

The King's Engine cost nearly £5000.

Popula. of Reading 11,000 or 12,000.

Traffic of same.

The Thames is quite sufficient for it.

The loss of the Nav. of the Thames will be inju. to Reading, Henley, &c.

Traffic by Coach.

Small Barges are prefer. to larger do.

Barges occupy 2½ days in going to Read. and 2 in retu.

The Bargemen are very irregular.

Horses are seldom drawn into the Riv.

Ex. Mr. GEORGE TREACHER,

Assistant Surveyor to the Thames Navigation.

His Experience.

My father has been Surveyor to the Thames Navigation nearly 40 years, (as far as I know, at a salary of £300. per annum). I know every part of the river, having been on it all my life. I have been appointed Assistant Surveyor above 12 years, (at a salary of £150. per annum, which includes travelling expenses over a distance of 118 miles of the river).——

Benefit of the Tham. to the Towns upon it.

The Towns of Oxford, Abingdon, Wallingford, Pangbourne, Reading, Henley, Marlow, Maidenhead, and Windsor, have derived much of their prosperity from their communication with the river: the formation of the Great Western would cause a considerable falling off of trade upon the same, as there is much traffic to those places, although the canals carry some of it.——

Descrip. of Canals dependent upon it. The Severn. The Oxford. The Wilts & Berks. The Kennet & Avon

There are a considerable number of Canals dependent upon the River, viz: the Severn, which enters the Thames at Lichlade in Gloucestershire; the Oxford Canal, which enters at Oxford; the Wilts and Berks at Abingdon, and the Kennet and Avon at Reading.——

Amo. of Tong. upon the Tha. 240,000 To. The Com. charge 2*d.* per Ton at the Locks.

The Tonnage upon the Thames amounts to about 240,000 tons, and the Tolls (from which the Thames Commissioners principally derive their revenue) are about 2*d.* a ton at each Pound Lock each way; they derive very little of their income from rents.——There have been very few obstructions by Drought lately.

Speed of the Navi.

A Boat is 3 Days going to Abingdon: A Boat went to Mapledurham a short time back in 36 Hours. The obstructions are mostly above Reading.——

A Barge is 4 or 4½ days going to Brist. and 2 to Reading.

A Barge occupies about 4 or 4½ Days in passing between Bristol and London, and about 2 between Reading and London: I cannot therefore call 4 or 5 days a favorable passage, as stated by Mr. Davis in his evidence.

4 to 6 hours differ. betw. up and down.

There is from 4 to 6 hours difference in the passage between up and down: they perform the voyage as quickly in Winter as in Summer, unless there are floods, which occur oftener in the winter, arising from the rain.——

Memo. Pilfering.

I have not heard of any Pilfering to much extent, but the Bargemasters are responsible for it.——

Stoppa. from Floods are very rare.

I do not remember any stoppage of late years equal to a month owing to Floods, as the several extensive improvements have rendered it impossible, except in cases of extremely high floods.——

Precau. against same

Very high banks have been erected, in the nature of mounds, to prevent the flooding of the country, (independent of the towing path) particularly at Maidenhead and Windsor, which benefit the adjoining lands, (the proprietors of which do not contribute towards the same).——

The Receipts in 1832 was £13,168. The Exp. generally equal same.

The Receipts for the year 1832 were £13,168. 13*s.* 4*d.*, and the Expenditure generally equals the receipts.——

All the Funds exp. on the Works.

The appointment of the Commissioners is merely honorary, and they do not derive any benefit from their office, the whole of the Funds being appropriated to the support of the Navigation, with the exception of the salaries of the officers, &c.——

Windsor Castle is sup. from the Navi.

Windsor Castle is entirely supplied with Water by the Water-wheel at Romney Weir, below Windsor Bridge; the water being penned to a certain height for the Navigation, and also to give power to the wheel: the whole of the Commissioners works are absolutely essential for the working of the engine.

Also Eton and the Town of Windsor.

The Towns of Windsor and Eton are also supplied by another engine on the other side; the same water power being used for both.———The Cost of the Romney Weir, including the Pound Lock, was £4000. or £5000.; large sums have also been spent upon it since; perhaps a sum of £10,000. may have been expended on the spot from the commencement.——— If the Navigation of the river was deserted, it would alter its course, and run in different channels, heavy barges tend to keep it in order, and it is necessary to maintain a certain depth, that they may travel at all times, and the Drainage of the surrounding country could not be effected without them; the river would also in many cases run upon a higher level, instead of draining the country, and the floods would not run off as they do now.——— A constant expence is therefore incurred in cleansing the channels, also in preserving and maintaining the towing paths, and the works in the nature of artificial navigation, as pounds, locks, &c.—— ——There is an Embankment upon the Great Western Railway in the neighbourhood of Reading, which extends from a little below the mouth of the Kennet up the Vale of the Thames westward, towards Tyler's Lane, and is upwards of 3 miles in length. It crosses the mouth of the Kennet at about 7 chains from the Thames, and about 21 feet above the former, by 2 arches of 40 feet span each: there would therefore be a pier in the centre of the navigation belonging to the Commissioners, who will be put to great expence in ballasting a new channel, &c. (there are only 2 arches shewn upon the plan, but even if a great number should be erected, yet the water in flood time could not escape through them so fast as it does now.)———The extent of the Jurisdiction of the Thames Commissioners is from Lichlade, in Gloucestershire, where the Thames first becomes navigable, extending down to the City Stone, at the foot of Staines Bridge; the City of London then become the Conservators of the river, whose jurisdiction extends down below London: part of the Kennet is also under the jurisdiction of the Thames Commissioners.———It would be of very little use to keep the works below in repair, if those above were neglected. ———The Embankment is 15 or 16 feet high where it crosses the road from Reading to Caversham, which must either pass in the water below the railway, or over it, which would be inconvenient, and the water in flood time could not escape through those arches so fast as it does now.———It next crosses the Valley of the Loddon, in the neighbourhood of Twyford, at nearly right angles. The Loddon is a very sluggish river, and there is sometimes a great quantity of water in it; when any thing placed across the low lands will pen it back, and it is frequently flooded after heavy rains: wherever there is a fall in the river any piers or obstructions (boughs of trees, &c. would be intercepted) must pen the water back, even if the water course is left ten times wider, but if it is level it will not affect it provided the waterway is made wider, but there is not always the same fall in the river.———The Loddon passes under the Great Turnpike Road, which I have known flooded both in 1821 and 1822, and as the railway passes considerably higher it must therefore pen the water back.—— ——There are several Mills directly below the embankment: and as the water will be penned back, the power of the first mill will consequently be removed, and a single inch of tail water will in some instances take away much of the power: suppose it was 1 inch, in order to gain which they must pen back the water another inch, which would injure the one above it, and so on one after another. In order to gain the inch, they take it from the mill above; but the power of 1 inch above is not equal to 1 inch below, because the wheel

The Romney Weir cost £10,000.

If the Navigation was neglec. the Riv. would be stopped.

And the Drainage would be ruined.

The G. W. will increase the Floods near Tyler's Lane.

At the cross. of the mouth of the Kennet

He considers Piers must pen the water.

The jurisdic. of the Commissio. extends to Lichlade, when it falls into the hands of the City of Lond.

The G. W. wld. also increase the Floods at the Road from Caversham to Read.

Do. at the cros. of the Lodd. nr. Twyford.

Remarks upon the Lodden.

Piers obstruct a running River but not Level water.

At the Turnp. Road over the Lodden.

It will also pen the Water back from several Mills.

Observ. upon same.

has more water to work through, (I am not aware that there is any law which would prevent them doing it, unless they penned it over the land). The above mills are above the railway, the nearest being about 1 mile distant: there is only one below, (which is

It will also injure the Land.

less than 1 mile distant) which will not be affected.—If the water is kept long upon this land it will injure it very much; it now remains a long time upon it at certain

Prevalen. of Floods in this Neighbourh.

times. I have known the haycocks in the Summer carried off the land, and floated down the river.——It next crosses the Valley of the Thames at Maidenhead, ¼ of a mile

Descr. of the crossg. at Maidenhead.

below the bridge, with 3 arches, each of 80 feet span, and 4 arches, each of 40 feet span; there will therefore be 2 piers in the river, which are sure to pen the water back,

Objections to same.

particularly as there is a considerable fall at this point of the river. The floods rise above 9 feet directly below Maidenhead, but there have not been any lately, owing to our having increased the widths of the openings, &c., and as the trade increases we shall

Difficul. of increas. the Water-way at Maidenhead.

be able to carry off the smaller floods, arising from heavy rains.——I have erected Bridges over the Thames, and I consider it very difficult to widen the Bridge at Maidenhead without injuring the navigation, as it makes a bend where the water will wash.

Prevalen. of Floods in this Neighbourh.

The lower parts of many of the houses between Maidenhead and Maidenhead Bridge would be under water: I have seen the water run through the doors of the coaches from one side to the other, (in the year 1821), and unless a free course is kept it will rise

The Commiss. are always consul. when any Bridg. are built

yet higher, and injure the land in the immediate neighbourhood very much.—When new Bridges are built it is customary to consult the Commissioners, in order to guard against any thing of the kind.—I remember when the Crown wished to build a Bridge

Instance of the same

over the Kennet, a considerable time was occupied in consulting the Commissioners,

But the G. W. have not consul. the Com.

who made some alterations in the plan submitted to them, but they have not been consulted by the Great Western regarding any of the proposed works.—— ——Some

Account of some Houses injured by the G. W. at Readi.

Houses have lately been built on the Crown land at Reading, which have a beautiful view of the Kennet and Thames, which the embankment will interfere with, by injuring

Do. Lord Stowell's at Early Mead.

their views, as the situation is very low: it will also injure Lord Stowell's House, at Early Mead, in front of the Thames, and block out the view of the river; I went and ascertained it last Monday, which is our lowest water day; the mills not being at work on Sunday throws more water back, and it is not sufficiently deep to pass the barges at full

Account of same.

depth. I put up a rod 18 feet long (which is the height of the embankment of the Great Western) immediately opposite the house, (the spot as near as I could tell was about 17 chains from the Thames) and upon looking from the river, from which a full view of the house is obtained, I found it cut across the middle of the building, totally destroying the view from the lower windows: upon looking from the house towards the river, it cut beneath the pole, shutting out from the sight the river and the whole of the meadows: but it did not interrupt the view down upon Sunning. The house is about ½ a mile from the river, the Turnpike road being between them; there is a high plantation skirting part of the road, but there are 2 openings in it: I could see the banks of the river but not the water from the windows; the boats can always be

Light Goods, which are the principal Profit, have increas.

seen; the meadow is also a pleasant object from the house.—— ——The trade in Light Goods has increased, owing to the improvements, and there are more carried in the proportion than heavy goods. I am not aware whether the Commissioners could impose any additional tolls, in order to increase their income, without going to Parliament.

Ex. Mr. JOHN CUTLER.

I am Proprietor of the Eton Engine, which supplies the Towns of Eton and Windsor, also the Barracks, with water, and it is situated precisely opposite His Majesty's Engine at Windsor, which supplies the Royal Apartments only, the remaining portion of the Castle being supplied by the former.—If the water at Romney Weir was only a few inches below the water mark placed thereon, it would be impossible to supply the Castle and Town of Windsor, except by a steam engine, which would impose a tax upon the town greater than it could bear, the present Taxes being heavy.———It is entirely owing to the strict attention of the Thames Commissioners, and their Officers, Surveyors, &c. that His Majesty has a supply of water at the present time.———If it drops 2 inches the Engines only partly work. I went to the King's Engine last Monday, and observed that it was 4 inches below the mark, the fountain on the parterre at the Castle consequently could not work, although some of the Kitchens, &c. were supplied. ———I supply 380 houses with water, only 60 of which pay me 1*d.* per day, (they average 26*s.* a house,) it is therefore with the utmost difficulty that I get enough to pay my expenses, the wear and tear of the machinery being very great, ($\frac{1}{3}$rd of the expences of the London water companies is for wear and tear); I certainly should not get a sixpence if it depended on the town, but the Dean and Canons, also the Government, assist me.———I received it from my Uncle, who had it 40 years; he made the people a present of the water, but he did not intend that I should do so, and I am bound to keep it up, although I have only a life interest in it, but I am in hopes of getting something by it ultimately.———If the Thames Commissioners should neglect the River, I imagine that I should be obliged to have recourse to a Steam Engine, which would be a loss to me.———If the River was not kept in proper order, it would in some places run in different channels from the present, and it would flood the country in others, the Drainage of the country would also be effected. I am positive no local arrangements could answer the purpose, as it must be kept in good order throughout.———The Thames Commissioners are not a trading Company, but a mere roving Commission, sitting at 8 Towns on the Thames, and paying all their own travelling expences, and I have observed that they are very assiduous in their attention to the duties of the office. ———The Proposed line of Railway would be a two-fold detriment, as it would injure the Navigation, by taking away a portion of the traffic, (even supposing it should be a considerable distance off,) and the several Bridges and Embankments would also impede the flood waters.———The whole of the several works upon the river are necessary.——— I could mention many places where the river might be improved, if the Commissioners possessed more funds.———The Weirs and works above Romney Pound are also instrumental in enabling me to supply the water with regularity, otherwise it would sometimes come down in driblets, and at other times in floods.———I should have been without water last year, if the Commissioners had not allowed me to throw up a weir, and to have a goit carried to the mill in order to turn the wheel: in the event of the river being destroyed I could turn the wheel by this plan for a few weeks. I might

Propri. of the Eton Engine, which supp. Eton, Windsor, &c.

His Majesty's Eng. supplies the Castle.

Both are supplied by Romney Weir.

Great atten. to same is req. to keep up the Works, if the Water drops a few inches it affects the Supply.

Instance of same.

The Engine hardly pays him.

Wear and Tear is $\frac{1}{3}$rd of the Expense of Water Works.

He is bound to keep it up.

If the Nav. is neglec. he would be oblig. to get a Steam Eng.

No local arrangements could keep the River in order.

The Thames Com. pay their own Exp.

The G. W. would affect the Traffic.

And injure the Riv.

Many parts of the Riv. req. to be impr.

The whole of the sev. Wks. are neces.

His remedy when Water is scarce.

possibly procure sufficient water by local arrangements for about 10 years, but not beyond that period.——The Windsor water is at too great a depth for pumps, and it is so extremely hard that when obtained it is useless.—A few years back the Government determined to sink wells for the benefit of the Infantry Barracks, but the men could not use the water, the Board of Ordnance therefore determined to have my water for them, (and obliged me to lay down pipes and keep them in repair,) and for which I receive 3*d*. for every 224 gallons.——They cannot well do without my water at Windsor, yet notwithstanding my monopoly I cannot raise the price, having tried several times.——There are tremendous floods at a place called Upper Hope, sometimes my wheel cannot move from the excess of water, (but it has not been stopped by floods for the last 2 or 3 years,) at other times it is at a perfect stand still from drought, I am therefore obliged to watch it closely, and it is the same with His Majesty's Engine. ——The Town of Windsor was without water at the time of the Ascot Heath Races in 1834 entirely from this cause, my wheel was also stopped by it 3 times last summer, the navigation was also partially affected at the same time. I can sometimes throw the water to one part of Windsor but not to another, although I have expended a great deal of money in improvements, as the Town expect water every day.——Large vessels can always pass through the Town of Windsor on Sundays and Thursdays, but on Mondays and Fridays, (the days upon which my wheel was affected,) the water is low, and the tonnage of the boats is consequently reduced, and they draw only 3 feet 6 inches of water, instead of 3 ft. 10 inches, until the water recovers itself.

The Windsor Water is too hard to use.

Notwithstandg. his Monopoly he cannot raise the Price.

Uncertainty of the River.

Instan. of his being without Water.

He has spent much in improvements.

Best Days for the Navigation.

Ex. CHARLES TULL, *Keeper of Romney Pound Lock and Weir, near Windsor.*

The Romney Weir pens up the Water, and throws it into the streams right and left of the old channel of the river, in one of which the Pound Lock is situated, which is the only course for the traffic.——The Weir is moveable, and it is my business to draw it off in time of Floods.——The joint operation of the Weir and Pound Lock tend to keep up a proper body of Water for the Navigation, also for the supply of His Majesty's Engine for Windsor Castle, and the Eton Engine at Tangier, on the other side of the river, which supplies the towns of Eton and Windsor, also the environs of the Castle and the Foot Barracks, and it is being laid on for the Horse Barracks; every house in the town, (except one, where a pump is used,) is supplied from this source.—— The Castle and both Towns would therefore be without water if the works at Romney Weir should be suspended.——The Engine-keeper often calls at the Weir to inform me of parts of the Castle which are without water, and I am obliged to resort to other methods of producing force: (the water is sometimes very difficult to keep up,) for instance, yesterday morning the wheels were very nearly stopped.—— ——The Passage of deeply laden Barges are very useful in keeping up the scour of the river. ——The following is a List of the delays that have occurred by Floods and Frosts, during the last 4 years:—

Description of Romney Weir, &c.

The Weir is drawn in the Winter.

It suppl. His Majes. and the Eton Engin.

Description of same

They depend entire. upon the River.

He is obliged to resort to evry. means to keep up the supp.

The Barges scour the River out.

Table of Delays at Romney Weir by Floods and Frosts the last Four Years.

STOPPAGES at ROMNEY.

1831.	5th to 16th February	. .	10 Days.
1832.	None		None.
1833.	13th to 25th February	. .	10 Days.
1834.	16th to 27th January	. .	9 Days.
	Total Stoppages in Four Years	:	29 Days.

Ex. CHARLES HARDING, *Turncock of Windsor Castle.*

Winds Cas. is supp. from Romney Weir.

I am in the service of the Woods and Forests, and it is my duty to turn the water supplied by the Engine for the express use of the Castle on and off.——Our supply depends entirely upon the Romney Weir, and it is necessary to keep the engine to its full power, as any small diminution of the water affects the supply.——If the Thames Commissioners should neglect their works on the river, I am not aware how we should obtain our supply.

He is not aware of any other source of procuring Water.

Ex. JAMES BARNES, *Turncock of the Eton Engine.*

The Eton Engine supplies all Windsr.

A Dam is fix. across River to supp. same

The whole of Windsor is supplied from the Eton Engine, (a pump being used at one house only between the same and the Long walk).——The Dam is fixed across the main water of the river, which it drives into the pounds, the cuts being the navigable part.——I have been 15 years upon the farm, and I am well acquainted with the River for miles up; if the water sinks only half an inch, it makes a considerable difference in the supply, and the people make complaints of it, although we are seldom in want of water, but if it was much reduced in the river our works could not proceed.——I have driven water carts about Windsor when the supply has not been sufficient.——We shall continue to be supplied from the main source of the Thames as long as it is kept open by the Navigation, and properly ballasted, &c., otherwise it would become choked in the course of time, and the whole body of water that now passes down the bed of the river would be scattered and lost.

A difference in the level of the River affects the supply.

If the River is neglected they will lose their supply.

Ex. WILLIAM HATCH, *Keeper of Shepperton Pound, (Two Miles above Walton Bridge) on the Thames.*

The Navigation of the Thames is almo. equal to a Canal.

The Improvements which have been made of late years in the River have been so extensive that the Navigation is almost upon the same footing as a Canal.——The following is an account of all the Stoppages which have occurred during the last 10 years:—

An Account of the Number of Days in each Year, viz., from 1825 'till 1834, during which the Navigation of the River Thames was stopped by Floods and Frosts, distinguishing each, at Shepperton Lock, in the City District of the Thames Navigation.

Account of the Stoppages at Shepperton Lock, on the City Division of the Thames, during the last Ten Years.

Year	By Floods.	By Frost.
1825.	No Stoppage.	
1826.	. . .	. . 7 Days.
1827.	. . .	. . 5 Days.
1828	{6, 9} 15 Days. . .	. {4, 5} 9 Days.
1829.		
1830.	. . .	. {3, 4, 6} 13 Days.
1831.	8 Days . .	
1832.	No Stoppage.	
1833.	5 Days.	
1834.	9 Days.	
	37 Days.	34 Days.

Total Stoppages in Ten Years {By Floods 37, By Frost 34} 71 Days from all Causes.

He has never known a Barge stopped by Drought.

This statement refers to the City District, from London Bridge to a little above Staines Bridge, within which distance I have never known a barge stopped from Drought: I have occasionally seen a barge ashore at low water, for half an hour or an hour, but even this is very rare.

Ex. Mr. BENJ. W. SCOTT, *Chief Clerk in the Chamber of London.*

Loan of the Naviga.

Avera. of the Tolls abo. £13,953. pr an. for the City District

Insufficien. of same.

The Light Goods bear the grea. Profit.

The existing debts under the various Acts of Parliament for the improvement of the portion of the Navigation of the Thames under the jurisdiction of the City of London, amounts to £155,000. (which is independant of £90,000. borrowed by the Commissioners of the Thames for their portion of the river).——The average produce of the Tolls for the last 5 years has been £13,953. per annum. The average sum paid annually for interest, at 4 per cent., is £6,200. and £960. for life annuities, and the surplus is expended upon the river, for which it is insufficient, we are therefore quite unable to bear any further reduction of income.——I have always understood that the principal profit arises from Light Goods, and judging from the various applications which are made to us, the Heavy Goods will not bear an increase of toll.

Ex. Mr. STEPHEN LEACH, *Clerk of the Works and Principal Receiver of the Tolls upon the Lower District of the Thames.*

Jurisd. of the City of Lond. upon the Tha.

His Experience.

The jurisdiction of the City of London commences at a little above Staines Bridge, and extends as far as Lee, near Sheerness.——I have held the appointment about 25 years, and I was the Assistant the 9 previous years.——The several extensive

Improvements, also Locks, Weirs, &c. are executed under my direction, and although of a very expensive description, they could be executed in a more substantial manner if the funds admitted of it.———The more complete the Navigation of the river the better the Drainage is secured.———A sum of £185,000. has been borrowed to execute the several works, all of which has been expended, independent of the balance arising from the tolls, after paying the interest of the Loan; no benefit being derived by any person but the necessary offices of the establishment.———The Towing Path occupies one bank only, which must be kept up, and the other bank is also required to be kept in repair, as it prevents the flooding of the river.———The River is deepened in order to make it sufficiently deep for barges drawing 3 feet 10 inches, it is also dammed up at different parts by weirs and locks to assist the navigation, which are removed during the winter, being merely temporary.———Temporary or moveable Piers and Embankments laid across rivers do not affect the water provided the sluices are properly attended to.———I do not consider the present penning back of the water, and raising of it by locks, &c. obstructs the drainage sufficiently to become injurious to the land; (the penning of it back is quite as important for the navigation as the deepening of the river.) ———I apprehend the Romney Weir is merely temporary, and removed in winter, with the exception of the Piles to which the tackle is fixed, which afford a very small obstruction, when placed judiciously.———There are two streams on each side Romney Weir which are used as mill streams, one for the King's Engine, and the other for the Towns of Windsor and Eton, but the spot where the weir is fixed is the proper course for the flood current to pass off; there is a channel on the right of the weir for barges to pass, by means of a Pound lock and cut, and it is absolutely necessary to let off the surplus water for the navigation.———If only ⅕th of our present trade was lost, there would not be sufficient funds to support the navigation, as it is now barely sufficient.———I have always understood that a large proportion of the profit is dependant upon the Light goods, the Heavy goods are charged to the utmost extent allowed by the Act, and if there should be a new Act for raising the tolls, I do not think it would operate as a sufficient compensation for the loss of the light goods.

The nature of the Wks. upon the Tha.

The Drainage is dependt. upon same

The Funds of the Navig. are employ. in the Works.

Towing Paths, &c.

The Works are removed in Winter.

The Works do not affect the Drainage.

Acct. of the removal of Romney Weir in Winter.

A loss of ⅕th of the Trade will affect the Funds.

Light Goods are the principal Profit.

Ex. MR. CHARLES BAKER, *Surveyor.*

I have Offices at both Cheltenham and Stroud, and am intimately acquainted with both places; I have directed my attention to the Line of the Great Western, also to the Branch to Gloucester from Swindon, and have sufficient knowledge of the country to know the latter would be extremely difficult.———In order to get to Stroud it must either pass the valley of Chalford, or that of Avening, both of which would be very difficult; Chalford is the most direct, but it is very steep and narrow for about 4 or 5 miles, which I have surveyed (up to the tunnel); there are also a great number of cross Roads and Mills in it, the latter are fed by Mill-ponds and Reservoirs; there is also much ornamental and resident property in the neighbourhood: the Soil is blue and yellow Clay, and is inclined to slip.———A Canal runs between them containing 28 Locks, (each 8 or 9 feet fall,) from Stroud to the Summit Level, the total rise being 242 ft.

He is well acqu. with Chelten. & Stroud.

A Line from Swindon to Stroud and Gloucest. must pass either Avening or Chalford Vallies.

Difficulty of same.

The Canal between them has 28-9-ft. Locks in 7½ Miles.

Some of which are in very difficult Soil.

9 inches, and the distance 7 miles, 4 furlongs, 170 yards; some of the Locks were found very difficult in the construction, on account of the nature of the soil, (one is or was obliged to be propped up at the side).——Part of the Country from Stroud to Gloucester would also be difficult, as it would be a continuation of the same valley.——

The Port of Gloucester is rising.

The Port of Gloucester has been rising the last few years, and the Duties have increased tenfold.——I have gone over part of the Line between Tring and Oxford,

The Line from Tring to Gloucestr. would be an easier Line.

And it does not interfere with Prop.

and I certainly think it would form an easier line of communication to Stroud than the line by Swindon, it passes principally through back country, by which it does not interfere much with private property; the Levels are also good, with the exception of Hales Hill, where there would be some difficulty, and it would require to be passed by a tunnel.——A

Comparison betwn. a Line from Chelten. and Glos. and a Line from Stroud to Glos.

The present Traffic is considerably in favor of the former.

line from Gloucester to London through Cheltenham would be preferable to a line through Stroud, it is in fact the present route, there being no traffic from Gloucester to Stroud and Swindon. There are 8 Coaches daily from Cheltenham to London, (some of which do not travel on Sundays,) and calculating that they take 9 Passengers each way, and 5 each way by the Mail, the Total number would amount to 44,024 per annum; but the coach traffic between Stroud and London amounts to only 8,366 per annum, the Mail being the only coach that runs all the year; there is a coach that runs every day except Sunday, for 8 months, and every other day for the remaining 4 months in the year.—I have taken my average of Passengers from the Liverpool and Manchester, although I am sure there are not 9 Passengers daily upon the latter coach.—There is therefore a difference of 35,658 Passengers to and from London in favour of Cheltenham.——

All the Coaches from the W. and N. W. upon arriv. at Glouc. take the Chel. Road.

The Coaches from Wales, and from the West and North West, upon arriving at Gloucester take the Cheltenham road in preference to the Stroud, and the Passengers between Cheltenham and Gloucester are considerably greater than those between Stroud and Gloucester, there being 14 or 15 Coaches running between the former places, independent of Flys, and only one Coach running every day between the latter; but it has the advantage in point of heavy traffic.—The Traffic in Carriages, Gigs, Posting, &c. is in

Stroud has a Canal to Lond. and Glou.

Cheltenh. has only a Tram Rd. to Glou.

Popu. of Chl. 22,942

Do. Paris. Strd. 8607

But there is a large add. Popu. in the Val.

the same proportion.——Stroud has a Water communication with Gloucester, also with London by the Thames and Severn Canal, but there is none from Cheltenham. (There is a horse Tramroad at present between Cheltenham and Gloucester.)——The Population of Cheltenham is 22,942, and the Parish of Stroud contains 8,607, but there is a considerable population in the Valley, perhaps 30,000, which are employed in the Clothing Manufactories, and are of a stationary description.——The best route to London is decidedly through Cheltenham, as it is the natural course.——I do not think the line by Swindon would increase the number of Passengers in that direction.

Ex. MR. BENJAMIN WINGROVE.

Surveyor of Roads at Trowbridge.

Where large sums have been expended on the Roads.

I have been a General Surveyor of Roads for the last 16 years, and have resided in the neighbourhood of Bath and Trowbridge all my life, I am therefore well acquainted with the neighbouring country, also with the country about Frome, and large sums have been expended in improving the roads in this part of the country.——

I consider the best line of Railway from the Metropolis to North Devon, would be by Trowbridge, Bradford, and Frome, and there is a practicable line in this direction, into which Exeter naturally falls, to which it would also form the best line of communication; a little Tunneling might be necessary at Frome, but you may leave Bruton Hill either to the North or South. I have not taken any levels, but I can partly judge by my eye. It would be a superior line to the Great Western for the North Western traffic, and would afford equal facilities of progress Eastward to London, and Westward to Bristol, forming an infinitely greater accommodation to this large tract of country; it would pass through the Bee-hives of the country at nearly a dead level, but I apprehend that much of the traffic in this direction would be lost by the Great Western.——The distance from Bath to Bradford by the turnpike road is 8 miles, and it is less by the Basing line, and by the Great Western it is 20.——The Population of the Basing for the first 11 miles out of Bath is double that of the Great Western, and the manufactures, trade, capital, &c. employed in it amounts to £800,000. but it is not above 100 or £150,000. on the latter, I am therefore astonished that any other line than the Basing should be proposed, as it is a manufacturing district, and its course Eastward passes through a richer and more populous country than the Great Western Railway, which goes through mere agricultural towns, (some malt is made at Wootton Basset and Wantage,) inhabited by small shopkeepers, who, as I am informed, are favourable to the measure, but I cannot conceive their motive, as it will take their trade from them.——I have no wish to impugn the Returns made by the Officer of the Crown of the number of Factories, but his allotting only 67 to the County of Wilts, (including the carpet manufactory at Wilton,) is a perfect fallacy. He also states the number of persons employed in them at 3,721; I suppose that his forms were filled up by the manufacturers, according to the number of persons employed in their factories, no notice being taken of the number actually employed out of them; there are 20 in Trowbridge alone, but every house is a Factory, it is a perfect bee-hive, there being a population of about 11,000 in a space of only 1,200 acres, which is the extent of the parish; perhaps there are 300 persons employed in agriculture; the aged paupers probably amount to 3,000, and I should think that 8,000 are employed in the factories, or at their own houses.—Every Village within 5 miles is engaged in the Cloth Trade, which they take to and from Trowbridge, but they do not require a railway for the same, perhaps it may amount to 1 or 2 cart loads per day.—The trade at Bradford has very much increased lately, I should think there are about 7 or 8 Factories at work, and together with Trowbridge makes 27, which leaves only 40 for all the remaining towns in Wiltshire; he must therefore have made his return from imperfect information.——The present Trade of Trowbridge and Bradford is principally with London, although they have much Irish trade.——The Gloucestershire trade is principally in broad cloths, and the Wiltshire in kerseymeres, which are less bulky, (a cloth of 1 yard wide of the latter is of less bulk than 2 yards of the former,) and although Wiltshire is a smaller manufacturing district than Gloucestershire, yet there are more people and capital employed in it than within a similar space of the latter.——The roads between Bradford and Bath have not been much improved lately, a road is being made round Claverton Hill, but it cannot be of any use to Bradford and Trowbridge, unless a Branch is made to them.——I have seen the Ground marked

The Basing is the best Line from Lon. to North Devon.

A little Tun. might be necess. at Frome.

And it affords equal Facil. as the G. W. East and Westward.

It pas. thro. the Bee-hives of the country.

From Bath to Brad. 8 Miles by the Ba. and 20 by the G. W.

The Population near Bath in fav. of Bas.

He is astonished that any other Line shou. be preferred.

The G. W. goes thro. Agricul. Towns.

He considers there are more Factories and Workmen in Wilts. than shewn by the Returns.

There are 20 in Trowbridge.

Pop. of same 11,000. 8,000 of which are empl. in the Facts.

There are 7 or 8 Facts. at Bradford.

Tra of Brad. & Tro. princ. with London.

Trade of Gloucester is prin. Broad Clths. Wilts. Kerseymere.

The communication betw. Bradford and Bath is very bad.

The Ba. Line passes within ½ to ⅝ of a mile of Trowbridge
And about the same from Bradford.
Remarks on same.

out for the Basing line at Trowbridge, which is from ½ to ⅝ths of a mile from the town, and the line passes about the same distance from the town of Bradford, where it is upon an undulating piece of ground, as at Bath, and about 30 feet elevation, but it is impossible to bring the line level with all parts of the town. I should prefer taking it to the lower part, where the trade is principally situated, but it is very subject to Floods, and all the houses at the bottom of the bank of the river are usually covered by them.——

Some of the Manu. of Brad. and Trow. are sent to Langley, Christian Malford, Chip. & Melksham, &c. to be milled.

Some of the manufactures are sent from Bradford and Trowbridge to Langley to be milled, some are also sent to a mill at Christian Malford, and some to Chippenham, but a great portion goes to Melksham, (which the Branch from the Great Western to Trowbridge will be 4 or 5 miles from) and Bromham, to the north east of the former, and a little out of the line from Bath to Devizes.—There are not above 6 cart loads sent to Langley per week; the Trade of Trowbridge is quite sufficient to support the mills all round the country.——

There are only 2 Facto. at Chippen.

There is one Silk manufactory at Chippenham, possessing very good trade, and one Cloth factory, which had been closed some years, but was restored by Mr. Gye for electioneering purposes.——

1 at Chris. Malford.
2 or 3 at Melksham.
Several at Heytesb.
1 or 2 at Bratton,
Some at Westbury.

There is 1 factory at Christian Malford, also 2 or 3 at Melksham, and there are several at Heytesbury, but I cannot say what number of factories may be in actual employment. There are 1 or 2 at work at Bratton, near Westbury, where a vast deal of the Trowbridge trade is sent to, it may be considered a branch of the trade of the latter; and there are manufactories at work at Westbury.——

Much of the Bristol Trade has gone to Exeter & Bridgewa.

I believe that much of the Bristol trade is gone westward, (the ports of Exeter and Bridgewater possess ships which formerly went there,) but it would be much frequented by the people of the Wiltshire district if there was an easy access to it,

The G. W. does not afford a good Com. betw. Wilts & Brist.

but I apprehend the Great Western would be very little used for that purpose, and the Box tunnel is considered a great objection by most of the manufacturers.——The Line taking the circuit of the manufacturing districts would be better, particularly as it would not be any longer.——Branch lines are very inconvenient.——

He would prefer a Line to Gloucester more North.
Thro. Cheltenham.

If a Railway is necessary for Gloucestershire and its vicinity, I consider that a better line than the Great Western might be obtained, by going through the county more north, which could be taken without crossing the Cotswold Hills; I should go from Gloucester to Cheltenham, which is a populous Town, and branch away to the right through the Vale of Evesham, in the line which opens there; the land at Winchcombe is very good, but Stow on the Wold is not a fertile place: the soil on the Wantage line is certainly superior to that upon the Winchcombe.——

The G. W. interfer. Floods near Balfo.

If the Great Western passes the country near Balford, above the turnpike roads, they must pass by a very high viaduct, (to allow the water to pass through,) in order to guard against the floods. (The Dundas Aqueduct crosses the road higher up.)——

The G. W. inteferes with the Property at the Angel at Bath very much.

There must be a great sacrifice of property connected with the Angel Inn at Bath, as considerable sums have been expended both by Public bodies and Private individuals in taking down houses in order to widen the road, and the crossing of same must also be attended with great expense, as extraordinary works must be constructed, the hill also rises very precipitately.——

Comparison of the Depôts at Bath.

A Depôt 300 yards from the old Bridge, and up a hill, cannot be equally convenient as another passing through the lower part of the town.

He is sure the Floods will disturb the town if not the Railway.

But if the floods should not disturb the railway, I am sure the lower part of the town would be affected, as it is already inundated, and many thousands of pounds have been spent in devising means to remedy the same, and a mound in the direction of the Great Western would increase the danger arising from them; the building of Pulteney Street materially altered their level.—The floods of 1809 were above the tops of the houses in Doll Mead, many people were ferried out of their windows in boats, and some of them were drowned; and two or three houses washed down at Walcot, and some people drowned.——— I understand that Lord Manvers, the proprietor of the lower part of Bath, has given his consent to the Bill, his land is let at present merely as common garden ground, and produces only 6 or £7. per acre, but for the purposes of a Depôt it would be worth £100. per acre, therefore it would not affect him as it would others, it would only be an exchange from £10. to 2 or £300. an acre.———He is also Ground Landlord of the South Parade, the cellars and kitchens of which are under water in flood time, although the habitable part is raised above them. I am not aware whether they are let upon building leases.———I once consulted with Mr. Telford about a line of road from London to Bath, and he suggested the line taken by the Basing Railway; I went over the country with him———I have no party feelings, all I wish is that the public may have the most advantageous line for Trade and manufactures, and that which will best promote social communications.

Account of the Floods at Bath.

Lord Manvers is the prin. Propr of the lower part of Bath.

As his Land at pres. is only £6. or £7. per A. and will then be worth £100. He assents to G. W.

He is also Propriet. of the South Parade.

Mr. Telford prefer. the Basing Line of country as a Communication to Bath.

Ex. Mr. GEORGE HENRY ELLIOTT.

I am Surveyor of the Roads in the vicinity of Hurst Lodge, in the neighbourhood of the Loddon, where I reside, which the line of the Great Western crosses at right angles. The adjacent Country is already inconvenienced, and subjected to Floods, on account of the difficulty of the water finding its way to the Thames; the Embankment will therefore further delay the passage of the flood water, and it is impossible to remedy the objection, as there must be piers occupying some space throughout the whole range of the valley near Twyford: the injury will be accordingly as the embankment is more or less open.———The Parish of Hurst is on the right, and the Parish of Sunning on the left bank of the Loddon.———The railway is not in sight of my house, but it passes through land of which I am Surveyor.———The Floods at present render the Roads impassable in Winter, and any further increase would incommode the Farmers, and further injure their land, as the flood water would remain on longer.——— Mr. Palmer's Land will be injured by it, and Lord Stowell's Property, which descends in one uninterrupted glacis down to the river, and I think that an 18 feet embankment, at right angles with the river, will also interrupt his view; at all events, it would considerably injure the property, especially if the embankment should be in chalk (as it does not admit of verdure).——— ———I have repeatedly attended Meetings at Reading on the subject of this line at which there were few Proprietors (in consequence of their passing through so much property belonging to one party; for instance, they continue on Mr. Palmer's estates for a length of 3, and on Mr. Leveson Gower's for 2 miles). ———I also attended a Meeting called to discuss the measure, and to consider what steps should be taken to obviate the nuisânce (as we considered it): I cannot say whether it was convened by advertisement, or from an understanding among the gentlemen of the neighbourhood; but about 15 or 20 persons attended, all of whom determined to oppose it by every fair way, and none spoke in fávour of it.

Surveyor of Roads at Hurst.

The G. W. will increase the Floods of the Riv. Lodden.

Explanati. of same.

The Floods are now very consi. in Wint.

They will injure Mr. Palmer's Property.

Also Lord Stowell's

Description of same

He has attended Meetings concern. the compet. Lines.

Account of same

Ex. Mr. WILLIAM ASHMAN.

Engineer of the Clan Down Colliery

Somers. Coal 20 per Cent. better than Coal Pit Heath Col.

Price of former 8*d.* per cwt. latter 10*d.* per cwt.

Descriptn. of same.

I have been 26 years Principal Engineer and Superintendent of the Clan Down Coal Works, near Bath, (I have likewise the charge of 4 other Collieries in the neighbourhood) which is the great coal field in that part of the country. Somersetshire Coal is 20 per cent. better than that of the Coal Pit Heath Colliery: the latter sells at Bath for 8*d.* and the former at 9*d.* or 10*d.* per cwt., and is used by the nobility, who prefer it, even if 3*d.* or 4*d.* a cwt. more than the other, which is sold to factories, poor people, &c. If the Great Western is made it will give the latter increased advantages over us, and no doubt more of it will be sold (there was not much competition between us until they made the Coal Pit Heath Railway).——

Description of the Clan Down Colliery Canal and Railway

Rate of Carriage to Bath.

Shifting of Coals is very injurious.

Our Coals are first passed along the Colliery Railway, (which is about 8 miles long,) when they are put upon the Somersetshire Coal Canal, which joins the Kennet and Avon, at Dundas, about 2 miles from Midford, from which junction there is a communication to Bradford and Trowbridge and another to Bath, and by those canals the coals are carried up the country. The large coal is sold at the pit's mouth for 10*s.* per ton, (our charge is 2*s.* 6*d.* for carrying it upon our railway, and loading it in the boats; when at Bath it is sold for 16*s.* 8*d.*, which gives 4*s.* 2*d.* for the carriage of the latter 6 miles): The small coal is sold for 3*s.*, and the several shiftings tend to break and injure it, which prevents our supplying the Bath market advantageously.

Advantages which would accrue to the Somerset Collieries by the Basing Line.

The Barges that navigate the Kennet and Avon Canal will not pass up the Somersetshire Coal Canal, in consequence of the narrowness of the latter, which obliges us frequently to shift them: if we could reach it by railway the carriage would be 2*s.* a ton less, and there would be a proportionate saving eastward towards Bradford and Trowbridge: we supply them at present by land carriage (occupying 1 day).——

The Coll. Ra. is a PlateRa. 8 Mil. long and work. by Hors.

If a Railway from London to Bristol passed down the Claverton Valley, between our line and the Dundas Aqueduct, we should carry ours on to meet it at the latter place. Our railway has been laid down about 17 years, and is not an edge but a plate railway, and is worked by horses: there was an intention about 5 years back of laying it down according to the improved method, and I inspected some lines in the North, and made a Report upon the subject.——

They rai. 6 to 10,000 Ton of Coal per Dy.

The Pit 200 fath. dp.

They emp. 170 Pers.

We raise about 600 to 1000 Tons per Day, and about 400 tons are sent to Bath daily (half of which goes by the railway). The Pit is 200 Fathoms deep by a perpendicular shaft, and the coal lies considerably lower; we could therefore increase the supply if required.——We employ about 170 persons, but we have had 200 at work; there is also a proportionate number employed at Radstock Colliery, which is about 1 mile off.——

Their Goods are sent from Bristol, 15*s.* per Ton.

We purchase our iron, fire bricks, foreign timber, and grocery at Bristol, which is sent by waggons, and the lowest price charged is 15*s.* per ton; they would not be above half that amount by a railway.——I believe we supply the greater quantity of the coals consumed at Reading, and we send a small portion as far as Maidenhead.

The G. W. would be of no advantage to Bradf. and Trow. regarding Coals.

Do Frome.

They partly supply the Salisbury Mar.

The Great Western would be of no advantage to Bradford or Trowbridge as a communication for coals; and they will never pay for Branches: I consider that a line passing through them would be much better. Frome and that district would also get their coals cheaper by the Basing line, which are also at present sent by waggon.——We supply part of Salisbury Market by Frome and Warminster; the remainder is sent from Redbridge, and is sea borne from Southampton.

Ex. Mr. GEO. BAILLIE, *Magistrate of the County of Middlesex.*

I am one of the Trustees of Hobbin's Charity lands at Hanwell, (where I have resided 16 years).——Having had my attention called to the Proposed line, I found that it crossed the valley of the Brent from the Lunatic Asylum, and passed through some accommodation or pasture land belonging to the above Charity, which is in the occupation of Mr. Turner of Lawn House, to whom it was let by auction, at a very high price (£15. per acre,) for a term of 7 years.—The land belonging to the Charity is a long strip, only 80 or 90 yards wide, and the value of it as occupation land will be very much destroyed.—But neither I nor any other Trustee (to my knowledge) have been consulted by the Great Western.——Lawn House is on the high road to Uxbridge, and is the property of Mr. Commerell, who dissents; the above field which is crossed by the railway lies between Lawn House and Mr. Bridges' property, which is ornamental, and some of it is very near the Church, part of the Glebe belonging to the Rector also joins it, (the Rectory overlooks the latter,) there is also a cottage near the Church occupied by Lady Hope, which is well fitted for the residence of a person of rank.——The railway would pass this spot upon arches 60 feet high, which would therefore impede the view and be objectionable to the Rectory.—It would be the ruin of Mr. Bridges' property, upon which he has laid out 4 or £5,000. (he gave £700. for 2 acres of it, on account of its eligibility for building, which will be destroyed by the proposed line.)—There are several ornamental villas, cottages and houses near the line. ——The Brent is a winding river, and the houses upon it would be very much injured by the Proposed line, and their value lessened: very small cottages at present let for £50. a year, in consequence of the retired situation; the tradesmen must also sustain a corresponding loss: the ornamental part lies to the north, and they have a prospect towards Harrow in one direction, and towards Osterly Park and Richmond in the other. ——The Floods in the neighbourhood of Hanwell seldom continue above 24 hours, but I have known them last 48.——The works would be a great injury to the people of Greenford, (where there is a ford and a bridge,) as it would pen the waters back upon the meadows. It is now impassable in flood time.——I have signed the Petition against the Proposed line.

He is a Trustee of Hobbin's Char. Lds.

Part of which the Propo. Line crosses.

Account of same.

Account of Property interfered with by the G.W. in the Neighbourhood of Hanwell.

The Floods at Han. last from 24 to 48 ho.

Greenford is someti. impass. by Floods.

Ex. The Rev. THOMAS CARTER.

I am a Fellow and the Bursar of Eton, and Master of the lower school. I am also Vicar of Burnham, in Buckinghamshire.——The Provost and Fellows of Eton entertain a strong feeling of opposition to the Proposed line, as they expect it will injure the College by affording facilities to the boys of reaching town, and as the establishment of a Depôt, which would be a very great objection on account of the Cabs and Omnibuses which it would bring.——The attendance of the Provost and 4 Fellows is required at all our Meetings, if a sufficient number of Fellows should not attend, I summon some of

He is a Mast. of Eton

Account of the Objections of the Provost and College to the G.W.

The Collg. consists of 7 Members.

the juniors; (our whole Body consists of 7,) and in all our meetings regarding the proposed line we were always unanimous.——The Company made an offer of Police to keep the boys off the line, but I do not think that any precautions would be effective. —The Railway will bring us within such a distance of London that a boy will be able to get there and back in his absence, but it does not affect the Harrow, Rugby, or Winchester Schools.—(I think that a Harrow boy could get to town in less time by the road than by the railway.)——The Line of last year passed under the Long Walk at Windsor, and the Depôt was at Clewer fields, but it is now situated near Slough, which is rather more than a mile from our fields, the former would have been the least injurious, as the spot is not much frequented by the boys, and the town of Windsor lies between it and Eton; they are both about the same distance from the quadrangle.——I do not think the railway would be objectionable if it did not pass within a distance of 4 miles, (instead of 2) of Eton.

A Railway is not so injurious to Harrow.

The Depôt of last year was preferable to the present.

If the Railway was 4 Miles off Eton it would not be objec.

Ex. The Rev. WILLIAM GIFFORD COOKESLEY.

A Master of Eton.

He considers the G. W. will be inju. to the School.

Compa. of Eton with the Winchestr. Coll.

The prosp. of Eton owing to the freedom allowed to the Boys

Which could not be continued if the G. W. was made.

As it would afford facili. to the Boys.

Account of the Absences.

If the G. W. Depôt was 4 Miles off it wd. remed. the evil.

I have been one of the Masters of Eton for about 6 years, I was also an Etonian for about 10 years.——I think that the Proposed line of Railway will be injurious to Eton School, on account of the tone, habit, and more especially the wealth and rank of the boys, and it is difficult to avoid the evil, although it might not be so injurious to other schools,——The system at Winchester is one of great restraint and coercion compared with ours, the boys are enclosed within certain walls, and taken out to a particular hill for cricket under superintendence.——I consider that the prosperity of Eton is mainly owing to the practice of giving the boys as much freedom as possible consistent with discipline, and in the event of a railway passing within a short distance of us, we should be compelled to alter the system, in order to prevent the boys travelling upon it, the principle upon which our prosperity is founded would thus be subverted, which I am sure would be objected to.——A railway would afford facilities which the boys would be unable to resist, as they could get out of the reach of School discipline within 30 minutes.—I do not think any system of Police could prevent them getting upon it, as the boys do not carry badges upon their backs. A proposal was made to us of not having a Depôt within a certain distance, but we objected to it, also the building of a wall to keep them off: I am satisfied they did not contemplate making it of sufficient length.——2½ hours is the longest space of time that can elapse between the absences, and 1½ or 1¾ hours is the shortest.—A boy may be absent 3 times during the day upon a whole holiday, and the head boys in the upper cricket club are allowed to miss a call.——If a Clause was inserted stating that no passengers should be taken up or put down within 4 miles, it would not be objectionable. ——The line passes within about ⅛th of a mile of the front of Baylis's Mansion, at about midway between the house and the road.

Ev. Mr. WALTER LONG, M.P. *for the Northern Division of the County of Wilts.*

I am a Subscriber to the Basing and Bath line to the amount of 50 shares, but I have no personal interest in the undertaking, I took them with a view of benefiting the Manufacturers and Agriculturists, as they conceive this line to be the most favorable, (and any one may take them of me); I shall be equally benefited by either line, as I have nearly the same property contiguous to each, but I have most on the Basing, which runs through above 4 or 5 miles of it; I do not think it would benefit my land, as I should be obliged to make a different arrangement of my farms, but it would afford me the means of procuring manure. (I have one farm only crossed by the Great Western line.)———I believe the great majority of land owners assent to the Basing line.—I am averse to Railways generally, but I have assented to the latter, being of opinion that it will be much more advantageous to the county of Wilts than the Great Western, as it passes through a very busy manufacturing district; and it would likewise be very useful in bringing down manure, (as bones and peat ashes,) from London and Berkshire to places where it is required, as the Wiltshire Downs, (or Salisbury Plain); peat ashes are not to be had in an inland county like Wiltshire, the waggons could come once or twice a week with corn from Pusey and take back manure, which they now obtain by flocks of sheep, (I have 5 or 6000 sheep at my farms on the Downs, as they are necessary for the production of the crops,) a less number of sheep would consequently be kept, and a greater portion of land appropriated for corn, &c.; one acre of land produces 8 or 10 sacks of wheat. I should not have any objection to my Tenants ploughing the ground, provided they could get bones or other light manure to answer the same purpose as sheep.———The Kennet and Avon Canal might be used in bringing peat ashes from Newbury as far as Semington, as they require it to be conveyed to Pusey, and to the Wiltshire Downs, the land carriage would therefore be very great.———I cannot say whether a railway to Salisbury would pay, but the conveyance of Coals, &c. from the coal pit would be a great advantage.———The country upon the Great Western does not require manure, as it is principally dairy and pasture land, but the line would be useful for the transmission of butter, eggs, and other perishable articles.———The Manufacturing towns on the Basing line are Bradford, Trowbridge, Melksham, and Westbury.———My principal constituents lie in the manufacturing towns, but they have not influenced my opinion.———Chippenham is a decayed manufacturing town, I do not know more than one cloth factory in it: (manufactures were formerly carried on in it to a great extent).—Melksham is about 5 miles from the Great Western, and 2 or 3 from the Basing.———I have had Requisitions sent me from Chippenham, Wotton Basset, and Swindon.—I presided at a meeting, on which occasion it was made a *sine quâ non* that the main line should pass through Bradford and Trowbridge, upon which understanding the meeting sanctioned the Great Western, and the Directors privately stated before they left the town that they had no doubt it would, but that they were not authorized to say so.—The Advocates of the Southern line stated that they had taken professional advice upon the subject, and Mr. Brunel remarked that although he was satisfied the Northern line was the best, yet there was

He has 50 Shares in the Basing.

And has Property on both Lines.

Descriptn. of same.

He Ass. to the Ba.

Comparis. betw. the advan. of both Lines.

The Basing passes through the Manufacturing Districts.

It would also be useful in conveying Manure to Salis. Pla.

Advantages of same.

The G. W. passes thro' Dairy Land.

List of the Manufac. Towns on the Basi.

Chippenham on the G. W. contai. only 1 Cloth Factory.

He presided at a Meet. which appro. of the G. W. on cond. that in run through Bradfd. & Trowbr.

Account of same.

nothing to prevent their taking the other if it should be considered the most advantageous.—They also offered to submit the case to competent and impartial judges, and promised to take whichever was considered the best line.———An Advertisement appeared in the Salisbury and Winchester Journal, about a week after the second reading of the Bill in the House of Commons, stating that it was determined the Great Western Railway should adhere to the Northern line, and from that moment we withheld our support.———I was the Chairman of meetings held at Trowbridge and Frome in the autumn of last year, to consider the merits of the Basing and Bath line, and the resolutions were carried unanimously in its favor: I have no doubt that the feeling of those towns is in favor of it, and that they are averse to the proposed Branches.———I attended the above meetings nearly one year before I represented the Borough in Parliament, the meeting in favor of the Great Western was one of the first, and I was willing to support the latter provided it passed through Trowbridge and the Vale of Pusey.———I have heard that a Branch is contemplated from the Great Western at Swindon, passing through the Valley of Stroud, which is a manufacturing district of Gloucestershire, and would certainly be shut out of a communication with the South by the Basing line.—The Basing line passes along the Vale of Pusey, between the Marlborough and the Wiltshire Downs, by which any lateral communication or branches are rendered difficult throughout that distance, on account of their height; but there is very little traffic in the direction of Swindon, Wotton Basset, &c., perhaps one coach can accommodate the whole of it. I do not put my opinion in competition with Engineers, but I think it would be difficult to make, as the country is very hilly. I am aware that there is a Canal to Stroud, but it has a tunnel of 3 or 4 miles upon it.———I do not see why the Basing line should not be equally advantageous as the Great Western in forming a junction with Devonshire and Exeter, although they have sent up Petitions in favor of the latter, which is the longest by 10 miles; which difference is not great as regards steam power, yet the nearest line must be the best.———I believe the Gradients of the Great Western are the best, but those of the Basing line are quite practicable, and would therefore answer the purpose sufficiently well.———A Branch from the London and Birmingham at Tring through Cheltenham (which is a place of sufficient importance to require a railway) to Gloucester, appears to be a more direct line.———The greater part of the manufactures of Trowbridge and Bradford goes to London, and they have continually loaded the Coaches with cloth, when they have been unable to procure any other conveyance.———The Branch from the Great Western would be as good for the London trade as the Basing, but the people have expressed their fears that it will not be made, as the traffic may be considered insufficient, (the traffic with Bath would continue to go by the turnpike road, in consequence of the railway being so circuitous). The Great Western Branch from Chippenham to Bradford forms a very acute angle, and the circuit from these towns to the South and West of England towards Bristol would be 19 or 20 miles, the road would consequently be a preferable communication, as it is only 6 or 8 miles. The Basing line would take the traffic both ways; and a communication with Bristol would be desirable, as some of their raw material enters the western ports, although I believe the greater part is sent from London.

The G. W. did not adhere to their Agr. and we withdrew from them.

He presi. at Meetgs. which approved of the Basing.

Account of same.

If the G. W. had passed thro. Bradf. and Trow. he would have supported it.

Account of the Branch from the G. W. at Swindon.

The Bas. goes down the Vale of Pusey.

There is very little Trade in the directn. of Swindon.

The Basing would be the shortest Line to Devon by 10 Mi.

He consi. the Gradi. of the G.W. the best.

Advantage of a Line from Tring through Chelten. to Glouces.

The greater part of the Trd. from Bradf. and Trowbridge is with London.

Comparison of the Lines in reference to the above Towns.

Objections to the G. W. Branches to Trowbr. & Bradfd.

Ex. MR. JOHN BLEECK, *Wool Broker, of Warminster, Wiltshire.*

I am acquainted with Gloucestershire generally, and I consider that the relative importance of the Gloucestershire trade (which is situated at Stroud) has been declining, compared with that of Wiltshire and Somersetshire, I include Melksham and Tiverton in the latter, but not Calne, Chippenham, Malmesbury, and Christian Malford; there is a large factory at the latter, but very little work is done in it; and most of the factories at the latter places are open upon commission; there is but one factory at Chippenham, which belongs to Mr. Sanders of Bradford, and is rented by Mr. Rawlins; Mr. Gye gave a stimulus to the trade at the time he was candidate for the borough, but it has since been almost deserted.—The Gloucestershire trade may possibly be greater than the Wiltshire and Somersetshire, and it may also possess more Steam Power, although it is on the decline.——I have been acquainted with the Trade and Manufactures carried on at Bradford and Trowbridge for the last 20 or 30 years, it has improved very considerably of late years, and is yet increasing, I therefore consider that a direct communication between those towns and Bristol would be very desirable.—Trowbridge and Frome do much in kerseymeres, which is a very improving trade, and quite distinct from the Gloucestershire and Yorkshire trade.——There is a considerable trade carried on at Frome, and some of the manufactories are at a short distance from the town; but they are all within the town at Trowbridge: they are situated on the neighbouring streams at Bradford, the largest factory in the division being at Staverton.——Dyeing was formerly (about 35 years back) one of the excellencies of the Gloucestershire trade, the cloth was first manufactured and then dyed, but the manufacturers of Wiltshire have introduced, (within the last 20 years) a dyed wool, which has obtained the preference, (I allude to black cloth generally, scarlet is dyed in the piece at the present time) and is in great demand. It does not look so well at first, although it is more durable and never looks white at the seams, but retains the colour to the last. (The coat I now have on is made of dyed black wool, and although I have worn it for 10 years, very little effect has been produced upon it.)——The Great Western Railway Company applied to Parliament the early part of last year for a bill to make the two ends of their line only, (from London to Reading, and from Bath to Bristol) and while it was in the house they sent to Trowbridge and Bradford soliciting Petitions, and I was instrumental in supporting it; the Manufacturers also passed favourable resolutions upon it, but with an understanding that the main line should pass through the above towns, and if they had adhered to the same the inhabitants would have continued their support, but they do not consider the Proposed Branches sufficiently available for them; for instance, they anticipated that Dealers from Ireland, &c. in visiting this country for the purpose of purchasing woollen cloths, might have been induced to make their purchases in passing through, by which they would save the expense of transmitting their goods to London, and the commission upon same; instead of which, in travelling from Bath upon the Proposed line, they will first have to pass through the Box Tunnel, which is 1¾ miles long and continue up the plane nearly as far as Chippenham, and then return 9 miles to Bradford by the branch line, the proposed communication in this direction would there-

Trade in Gloucester is declining, compa. with Wilts. & Somer.

Manufactur. Towns in the above Distri.

Trade of Bradford and Trowbridge is improving.

Manufact. of same.

Dist. of the Factori. from the Towns.

Wilts. is superior to Glou. in Dyeing.

Account of same.

The G W. proposed origi. to go through Bradf. and Trowbr.

Advant. of the same to those Towns.

Disadvantages of the Prop. Branches to them.

fore be useless for commercial purposes, and a railway carried direct along the road would be far more advantageous, as the distance is only 8 miles.——A line taken from Basing, through Newbury or Reading, to Bradford would be nearly level all the way, except the hill by Bruton and Somerton.—The country between Bath and Wells is particularly hilly, but it is unnecessary to pass to Wells.——It is about 10 or 11 miles from Trowbridge to Bath, following the Valley of the Avon.—The present means of communication between Bradford, Trowbridge and Bath consists of the Kennet and Avon Canal, upon which I believe there are no locks until it arrives at Bath, but it requires a great number to let the Canal down into the river Avon, which consequently creates considerable delay; there are two Swift Boats on the Canal, but they are mere pleasure boats drawn by Post horses, and two people placed opposite each other occupy the whole width, they convey very little goods. The turnpike road between the same is not bad, although there is considerable hill upon it; (a new road is being made from Warminster to Bath, but it will not be any advantage to Bradford or Trowbridge.) The goods are sent from Trowbridge to London by waggon. They are generally sent by waggon from Bristol to Bath, as the communication by the river is not very eligible——I imagine that $\frac{4}{5}$ths of the working Population of Bradford and Trowbridge are dependant upon the clothing trade, and it is the practice of the masters when slack to divide the work amongst their hands instead of turning them off. ——The weaving of Wiltshire is done at the houses of the workmen, and not in the factories, and the greater portion of the dressing and some of the spinning is also done out. And some of the weaving and dressing of Gloucestershire may be done at the houses of the workmen, but there is more of it done in the factories compared with Wiltshire.——Trowbridge and Staverton have much trade with Ireland, which is forwarded to Bristol by waggon, (I do not think the Proposed line would be adapted for the conveyance of same, unless it was cheaper and more expeditious,) but they have a greater trade with London; there was formerly a great import of Spanish wool at Bristol, but very little of it is now used, German wool having superseded it, which is imported at London and Hull.

A Line from Basing would be superior to the G. W.

Description of the Commu. betw. Brad. and Trowb. to Bath.

By the Canal.

By the Road.

Do. Bristol & Bath.

$\frac{4}{5}$ths Popul. of Brad. and Trowb. engag. in the Cloth Trade.

Much of the Work is done at the Workm. houses in Wiltshire.

And a port. of same in Gloucestershire.

Most of the Trade of Bradford, and Trowbridge is with London.

Account of same.

Ex. Mr. WILLIAM GRAY, *of Newbury.*

I was born at Newbury, and I have lived there all my life.——The Proposed line of the Great Western goes through the neighbourhood of Wantage and Reading, and will therefore be very prejudicial to Newbury, which possesses the largest market within a circle of at least 40 miles round it, consisting of Agricultural produce entirely, people from Marlborough and from Abingdon attend it, and some even from below Andover, Amesbury, and Winchester; all the farmers within 30 miles attend it, and I consider that they will be affected if the market should be injured——There is an idea of establishing a Market at Streatley on the Proposed line of railway, which would be very injurious to Newbury.—There is a large Sheep Market at Market Ilsley, which is frequented by the same farmers who attend Newbury Market, perhaps there are at par-

The Proposed Line will inj. Newbury,

Which is the largest Mark. within 40 M.

Account of same.

Do. Market Ilsley Sheep Market.

ticular seasons 100,000 Sheep at market, which are disposed of to the Agriculturalists, and some of them are sent to London and some to Hampshire.———We consume 11,000 Tons of Coals annually, which are chiefly from Somersetshire and Gloucestershire.———The Population of Newbury in 1831 amounted to 5,967, and in Speen to about 3,000, and in Greenham to 1,061. Speenhamland is an integral part of the town of Newbury, there being only a brook between them. The Population of Reading is 15 or 16,000, but the market is not as large as Newbury.———All the Bristol, Bath and Frome Coaches pass through Newbury, also a cross Coach from Oxford to Southampton, and another from Reading to Southampton; we have also 2 or 3 Carriers.———We have a great trade with Bristol and London in Malt, but Flour is the staple commodity of the town.—There are about 26 Corn Mills in the neighbourhood of Newbury, and 5 Paper Mills, making 31 Mills in a distance of 7 miles, and there are several beyond which are principally supplied from Newbury.—There are 70 Malt Houses within a circuit of 7 miles round Newbury, all of which are dependant upon the market: and reckoning the Malt at 12 quarters a week, makes 32,760 quarters, which will give a duty of £33,852.—The following is—

11,000 Tons of Coals consu. ann at Newb.

Popul. of Newbury.

Do. . . Reading.

Traf. thro' Newbur.

The principal Trade of same is in Flour and Malt.

Account of same.

"An Average Statement of the different Sorts of Grain sold in Newbury Market, 1834."

Sellers Names.		Wheat.	Barley.	Oats.	Beans.	Peas.
Abraham Clayton	.	15,925	14,560	9,100	4,095	1,290
George Bayley	.	12,779	5,915	2,535	910	659
Thomas Langton	.	10,010	3,640	1,820	910	472
Richard Gough	.	21,151	9,100	3,640	1,365	2,840
		59,865	33,215	17,095	7,280	5,261

Table of Grain, &c. at Newbury Market.

This includes the Sales effected by Farmers as well as Corn Porters; these latter acting as Agents for the former.

Seeds sold, 500 Tons and upwards. Pigs sold, 26,000, upon an average of 500 a Week.

Ex. Mr. WILLIAM STONE,

Of Streatley House, Berkshire, (on the Banks of the Thames).

The Present line of the Great Western passes through three or four large fields of mine at Streatley, but the line last year crossed both my farms, and went through a greater quantity of my land. I was not aware that the alterations had been made to meet my wishes until my assent was solicited; at which time enquiries were made whether a large sum of money would satisfy me, I replied that I would not take any sum (Mr. Brunel was present when this proposition was made to me).———The Proposed line passes near Basildon Park, at Caversham, the property of Sir Francis Sykes.—It then passes a Farm of mine, (Basildon Farm,) consisting of 450 acres: the Farm-house was occupied by my father many years, but my servants at present live in it.—It then passes the Grotto belonging to Sir Francis Sykes, which is inhabited by Mr. Arthur Thelluson.—

The Proposed line passes through his Proper. at Streatley.

Do. Basildon Park at Caversham.

Do. the Grotto.

And on the opposite side of the river, leaving Reading, it first passes Mr. Blunt's, then Mr. Powis's, and then Mr. Gardiner's.—It crosses the River twice in my immediate neighboorhood, at Pangbourne, passing over my land to one of the bridges: (The inhabitants of Pangbourne consist principally of retired persons: and I have known the Floods from the Thames come up into the village.)—It then passes to the Village of Goring.—The whole of this part of the Thames is subject to Floods, and I think the embankment crossing the meadows at this spot will have a tendency to increase them, to the injury of the adjacent mills and land.—I own 2 adjoining Mills: and their power is affected by a very small pen.—They again cross below Mongewell, where the embankment is sure to impede the flood waters, and prove injurious. Mongewell House is very likely to be affected by the bridges, as it stands low: I have seen the water very near it.—Purley Park, the property of Mr. Storer, is very ornamental and extensive: the proposed line passes under some of the lands adjoining the house in a tunnel, and Mr. Fitzgerald (the party to whom it is let) objects to the measure.—The railway passes on the opposite side of the river to Purley Hall, (to which an old park is attached) the property of Mr. Wilder, between the turnpike road and the river.—I have been told that the railway crosses Early Meadow, in front of Early Court and farm, which are occupied by Lord Stowel, Lord Sidmouth, and Mr. Shackell, where I think it must interfere with the floods.—And the Embankment on the road to Caversham must interfere, and pen the water back, particularly if carried along the mead: I have known these meadows covered with water for weeks together, and as late in the season as May.——I am one of the Commissioners of the River Thames, and I have been very active in the Opposition to the proposed line: I subscribed towards the expenses last year, but I have not this time, (which has not arisen from any understanding that the Southampton Railway Company was to find the money).——I never suggested that Sir Francis Sykes might be prevailed upon to dissent, although his Steward declared that Sir Francis would dissent, but he has not.——I am intimate with Mr. Gardiner, who was anxious to oppose the proposed line, without any persuasion on my part, but he has since altered his mind.——The railway embankments would not protect any part of the country from Floods, as the water does not merely overflow, but in the low grounds it springs up from the earth, in consequence of the long duration of the floods, although we make drains and the like to carry them off.—Heavy drains on each side of the embankment may possibly carry the water off.——I attend Reading Market, which is one of the best in England, and is well supplied with all articles of consumption for the general purposes of the neighbourhood.——The River Kennet runs through Reading, and there is a wharf in the middle of the town; and I believe it joins the Thames at Caversham, which completes the water communication to London; the Kennet and Avon, and the Wilts and Berks Canals, form the water communications to Bristol and other directions; and the roads are equal to any in England. Goods may be sent from Reading with the greatest facility to any town in Berkshire. I therefore do not think that a Railroad would be of much benefit to it: the Tradesmen say "*If a Railroad must be made let us have it;*" but the most influential people object to it.

It crosses the River twice at Pangbourne

Whi. is sub. to Flds.

Do. at Goring.

The G. W. will incrs. the Floods, and inj. the Land and Mills.

Do. at Mongewell.

Do. Purley Park.

Do. Purley Hall.

Do. Early Court and Meadows.

At Caversham, whi. is subject to Floods.

He strongly objects to the G. W.

The Floods rise up out of the Ground in the Low Lands.

Remedies for same.

Excell. of Reading Market.

Description of the Rivers and Canals to Reading.

Communica. to same quite sufficient.

A Railroad would injure it.

Ex. JAMES HAMMOND, *Miller, Berkshire.*

I am the occupier of Streatley Mill, upon the Thames, which is the property of Mr. Stone, and I understand that the Proposed line passes in embankment at about a quarter of a mile below the mill tail; which, (together with the bridge,) will be very likely to stop the water, and render my mill useless, as a very small impediment in the river or mill tail has the effect of penning the water.—And I do not think they can carry the arches across the river without penning back the water.———The Corn is brought to my mill by the farmers' waggons, and it is taken away in barges. The present water conveyance is quite sufficient; for instance, last Monday fortnight, at eight o'clock in the evening, I put 100 sacks of flour on board a barge, and I had advices on Saturday morning informing me that they were sold, which is sufficiently quick for me, and I believe it is equally convenient for all the intermediate towns up to London, and the other way as far as the Thames goes.———The Railway would be very likely to cut up the trade on the river, by which we should lose the advantages of it. I can send flour now a tany hour, but if I am to convey it across the water to the railway, or drive it a mile or two, I must keep a team of horses.

If the Proposed line pens the Water, it will injure his Mill at Streatley.

The value of the Thames Navigation in this direction.

The Railway would injure the River.

Ex. Mr. JOHN CHILD, *Miller, Streatley.*

I once occupied Streatley and Goring Mills under Mr. Stone.———I consider that the bridge and the embankment of the Proposed line will be very likely to injure these mills, and to flood the meadows on each side of the river.———I believe the communication by the road and canal was quite sufficient for the neighbourhood at the time that I rented the mills.

He cons. the G. W. will injure the Mills

He cons. the present Conveya. sufficient.

Ex. Mr. WILLIAM PINNEGAR, *of South Marston, Farmer.*

I hold land of Lord Caernarvon, which is crossed by the Proposed line, (the greater part in embankment, 25 feet high in the highest part) and will be very much injured, but it does not affect any other farm in South Marston.———It is a Dairy farm, about 1 mile long, but very narrow in proportion to its length, and I keep 60 or 70 cows upon it: the land is wet and rather poachy, being sometimes flooded.———The Proposed line cuts across the run of the water, and crosses the canal, which will prevent the floods passing off, but it may not increase them, provided sufficient drains are made; the arches under the embankment will also be very inconvenient.———I consider the communication by the road and canal quite sufficient: (which run down the side of my farm, and we have only one crossing over the road.) The River Cole passes under the canal through one of my fields.———I do not know how we are to carry on the farm if the railway should be made.—I do not pretend to be a judge of land, but I think it

The Proposed line will injure his Farm

It will also affect the Floods.

He considers the present Communica. quite sufficient.

Amount of the Injury to his Farm.

would lessen the value of it £50. per annum, (perhaps it would make a difference of £2000. in its saleable value): it would not be so injurious if it was arable land.——— We make Cheese, (North Wiltshire cheese) and Fresh Butter, most of which goes to Abingdon and Henley: there is not always a demand for cheese, but the butter is sent to market every week.———I do not think that either of the Proposed lines of Railway would be beneficial to the farming interest; perhaps the bringing of a quantity of provisions from Ireland, would make things cheaper, which would be any thing but advantageous to us, (there is already too much sent, which spoils the breed of pigs) and our produce will not fetch its value in London, as every thing there is sold very low.

Produ. of his Farm.

He consi. that either Line would inju. the Farmers generally.

Ex. MR. JOSEPH HULL, *of Christian Malford, Dairy Farmer.*

The Proposed line passes through 2½ miles of the Manor of Christian Malford, and cuts across it diagonally, and some part between the canal and the railway will be left very narrow.———I rent a Farm of Lord Caernarvon, and about half a mile of the richest part of the same will be crossed by the embankment, (the land is not so good at the top of Christian Malford Wood as at the bottom) consisting of ploughed, wet, and dairy land, it will separate some of it from the homestead, and divide it in a very inconvenient manner: it will also have the same effect upon Mr. Cullamore's Farm.———My farm consists of nearly 300 acres, 36 or 37 of which are severed by the railway: 2 archways might be sufficient, but it could not then be worked so well as it is at present: I should require more persons to drive the cows, as they would have to go in two different directions.———I have 40 Cows on my farm, and the Archways under the embankment would be injurious to them, and it is necessary to drive them to the homestead and back again twice a day. (When large numbers of cows are driven through a small space they injure each other; perhaps 2 or 3 of them would be killed in the course of a month. I should consider 15 feet very narrow for such archways, although accidents do not frequently happen in passing through gates of about 10 feet wide, as the cows go right and left after getting through.) I consider that the extra Haulage to different parts of the farm which the proposed line would cause, and the injury it would be to the cattle, fully equal to £12. or £14. per annum.———The water from these lands drains towards the Avon, and the embankment will have the effect of keeping the water on the lands, as it cuts across the same; but if there are plenty of culverts, and a dyke on each side of the embankment, I think it would improve the drainage.———The Proposed line will also be very injurious to the land generally, particularly as it is upon an embankment: (some of the fields will be divided into small triangular pieces.) I should say that the 2 other farms upon the estate would not be so good by £20. per annum each, and that the value of the manor (which contains 3,000 acres) would be lessened £6,000. ———There is a very good Turnpike Road through the manor, communicating with Chippenham, which market I attend.—The Wilts and Berks Canal passes through my farm, and I can get a hundred tons of manure, &c. by it in a week, either way: we have a drawbridge over it.

The Proposed line is very injurious to Christian Malford.

Do. to his Farm.

Account of same.

The Cows will inju. each other in passg. thro. the Archways.

Account of same.

Amount of injury to his Farm.

The Proposed line will also affect the Drainage.

It will also injure all the adjacent Lan.

Extent of injury to the Borough.

The communicatio. by Road and Water are sufficient.

Ex. Mr. ROBERT HULL.

I am Lord Caernarvon's Bailiff, and occupy one of his Lordship's farms at Christian Malford. I have known the Manor 40 years, and I am well acquainted with the land throughout it.——I perfectly agree with the several Statements of the witness Joseph Hull, of the damage which the Proposed railway would inflict upon it.

He holds a Farm at Christian Malford.

He considers the G. W. will injure the above Manor.

Ex. Mr. EDWARD SHERWOOD, *Farmer, Purley.*

The Proposed Line passes within ½ a mile of Mr. Storer's residence, and would therefore be very prejudicial to his property.——I rent the Farm which is situated near it, and in the event of the railway being made, I should not expect to pay the same for it.—The land required by the Railway in this neighbourhood has been valued at £60. per acre, which I do not think sufficient, but if another £60. is added for Compensation, it will be quite enough; yet I should not like to dispose of it at that price, taking the inconvenience of severance, &c. into consideration.——I attend Reading Market, which is a ready money market, and one of the best in England; I find the present communications with London quite sufficient; if the more distant towns get equal facilities of communication as Reading, I think that the prosperity of the latter must proportionally diminish.——Reading supplies the greater part of the neighbourhood on the Oxfordshire and Berkshire sides of the country, but in the event of the Railway being made, some of these places must draw their supplies through it instead of through Reading.——I also frequent Newbury market occasionally, but it is not equal to Reading; if there was a Railway to Newbury, although Reading would lose, I do not think that Newbury would gain much, as there is a Canal to it.——I took some land about 3 months since belonging to Mr. Wilder, and some belonging to Mr. Powis, which I hold on condition that I shall be remunerated in the event of the railway passing through it, as I consider the land will be injured thereby; (it would pass through a part of it in a tunnel, by which the surface would be preserved.)——I object to Railways generally, as I think they will lessen the Agricultural interests, and the importation of Irish Cattle, Pigs, &c. would also be a great detriment to our trade.—I do not think I should be benefited even if the proposed line opened an additional market for our produce, as I am satisfied with the present.

The Proposed line injures Mr. Storer's property in Purley.

Also his Farm.

He consid. the valu. of same at £120. per Acre sufficient.

The Excellency of Reading Market.

The Communica. of same quite sufficient

The Railway will injure Reading.

Newbury Market notequ. to Reading.

He holds land subj. to Compensa. if the Railway is made.

He con. Railw. will inj. English Farm.

Account of same.

Ex. Mr. WILLIAM SHACKELL, *Farmer.*

I am a tenant of Lord Stowel's, and hold a Farm of 300 acres near Early Court, it goes almost up to the house, and about 70 acres of it, consisting of the best arable land, will be intersected by the Proposed line. The embankment runs exactly in front of my house, and although I am several feet higher than it, yet the view will be inter-

The Prop. line will injure his Farm at Early Mead.

Account of same.

cepted, it will have the effect of reducing the rent considerably, indeed I should not care about having the farm in the event of it being carried.——The Arch under the embankment must be 16 or 17 feet high, as a load of hay is above 15 feet high, I have no doubt that many of the loads are 17 feet, (they load high to save a journey, as much of it has to be carried 4 miles). It is customary when a wet night is anticipated to draw the hay into the barn, and I have frequently known it lowered 2 or 3 feet in order to get it in, although the doors are 14 feet high.——I attend Reading market, which is abundantly supplied with all necessaries, and many of the respectable tradesmen are of opinion that the railway will injure the Town. Its communications with London are also very good, a man by starting at 5 o'clock in the morning may have the bulk of the day to transact his business, and return by 8 in the evening. The trade is small except in corn and coals.—The Grocery is supplied from London.—The Navigation by the river Thames is very good, but I have known it impeded by Frost, (a river takes a longer time to freeze than a canal,) but never from Drought; they regulate the tonnage of the boats to suit the depth of water; I have had 14 tons of coal sent me from London within 3 or 4 days from the time of shipment.—It is a pull against stream from Teddington up to Reading.——I do not think that facilities for sending the produce of my farm up to London in 1½ or 2 hours will be any advantage, and if all the produce of the country is to be sent to London, it will be the worst market for us; I get my oil cake from London, and my manure from Reading, a distance of two miles, and the employment of a railway for the same would be of no use, as the road is very good a horse can draw almost any load over it.——The portions of my meadows crossed by the railway, (but not my arable land,) are covered with water at times of general Floods, (but the Thames does not flood so soon as the Loddon,) and the railway embankment will further impede them.——My land is situated next to Mr. Palmer's, which is subject to the floods of the Loddon, (and the soil is very bad for an embankment, as it is a running quicksand,) and will also be injured by the Proposed line, but not to the same extent as mine.——The River forms a very fine object from Lord Stowel's residence, and you cannot look in a direct line from the front windows without seeing it, although there is a turnpike road intervening, but if an 18 feet embankment passes between them it will be partially hidden, which his Lordship has a great objection to.—There was at one time an intention of enclosing the parish, which is Lordship would also have objected to, if Early Mead had been included in it, as it would have obstructed his view of the river.——I have a great objection to railways generally, and the facilities which a line in this direction would afford for the conveyance of Irish produce would be very injurious to English farmers, we have plenty of it already.—My principal objection to the Proposed line is on account of its interference with my farm and Lord Stowel's residence. The estate is small but very beautiful, comfortable, and near the road. I do not think that any sum, even though it exceeded the value of the land, would be a sufficient compensation; if it was my property I would rather give it up.——The Proposed line will also intercept the view of some houses which have been recently built in the King's Road, Reading, upon some ground formerly in the possession of the Crown, which was sold by auction, and the catalogues described the beauty of the prospects, which extend over both the Kennet and the Thames.

A Load of Hay is 16 to 17 ft. high, the Archways sho. therefore be more.

The Excellency of Reading Market.

The Railway will injure the same.

The Communicatio. quite sufficient.

By the Road.

By the Thames.

Descriptionofsame.

The Railw. will be of no use to Farmers

The Thames does not flood so soon as the Loddon.

The Prop. line will inj. Mr. Palmer's la.

The Prop. line is inj. to Early Court.

As Lord Stowel obj. to the blocking out of the view of the Thames.

He cons. Railw. inj. to English Farmers.

Early Ct. is a small but beautiful estate.

The Prop. line will also inj. some houses at King's Rd. Read.

Ex. Mr. JOHN BULLEY, *of Reading, Surgeon.*

I have practised at Reading the last 40 years, and I am in daily attendance upon Lord Stowel at Early Court.—The room in which his Lordship usually sits looks down upon the River, which forms a fine prospect from the windows; the reflection of the sunset on the surface of the water is exceedingly beautiful, and is seen from the windows daily, even at extremely low water, it therefore forms an important feature of the landscape. Mr. Scott removed some trees about 2 years back to make this view of the Thames more distinct.——The embankment of the Proposed line passes between the house and the river, and thereby intercepts the above prospect.——There is a piece of artificial water in front of the house, which is small but very ornamental.

He attends Lord Stowel at Early Ct.

The River forms a fine object from the windows of same.

The Prop. line will intercept the view of same.

Ex. Mr. ROBERT PALMER, M.P. *for the County of Berks.*

The Proposed line enters my Property at a farm in my own occupation, (adjoining Lord Stowel's at Early Court,) and it will be in both cutting and embankment for the first 2 miles, it then meets the crest of rather a sharp hill, and tunnels through it for ½ a mile, it passes through the whole of the parish of Sunning, every acre of which, commencing from Lord Stowels, belongs to me.——I objected to the line last year; but the present line is more disadvantageous to me, and the tunnel is also shorter.—The nearest point of the Railway to my house is within ¼ of a mile.——The embankment crosses the Loddon at right angles, the Valley of which is about ½ a mile wide, and the whole of it is occasionally subjected to very rapid floods, I have often seen the whole valley 1 foot deep in water after a 3 days rain, and I have known floods to come down and wash the hay out of the meadows in the spring of the year, some of the land also is so nearly upon the same level that the water oozes up through the soil, it is therefore constantly subjected to floods.——I have farming land on the banks of the Loddon and the Thames which is subject to floods, and I think that an embankment would increase them, and prove injurious to the same and to the land above me, which is also liable to floods.——The Embankment in front of Lord Stowel's house will obstruct the view of the river and the meadows, and I understand that it is to be made of the excavation from the Tunnel, which is composed first of Gravel, then a mixture of Sand and Quicksand, and the like, the substratum of the hill being Chalk, and it would be a long time before there could be any thing green upon the banks with such materials.——I suggested a small Deviation at the early part of the year: commencing at the top of Early Meadow, it would have brought the commencement of the Tunnel upon the turnpikeroad (provided it could be done) instead of upon my farm, which would have partly removed my objections. It was to be taken into consideration, and I do not know whether they have yet determined upon it, (of course the parties whose land would be affected

The Prop. line will injure his property at Sunning.

Account of same.

Description of the Flooding of the Loddon.

He cons. an Emban. wou. incr. the latter.

It will also obstruct the view from Lord Stowel's House.

Nature of the Soil of the Tunnel.

His Proposed Deviation.

He objects to a Pecuniary consider.

by it must be consulted.)———I would never listen to any thing in the shape of a pecuniary consideration, although I have had hints given me.—I know several who have objected to the line, and afterwards assented without any ostensible reason.—One of the greatest inconveniences arising from it is the carrying out of the works, and the nuisance of the men; none of the lands would be worth occupying at the time, as they would be open to the public in all directions.———The Proposed line last year only went as far as Reading, and 13 miles of it passed through Berkshire, upon which distance there were only 6 or 7 Proprietors: it occupies 3 miles of my estate, and 2 of Mr. Gowers ———Reading has both road and water communications equal to any town, and I do not know the feeling of the Inhabitants towards the measure further than from the Petitions upon the subject, and there are 800 names attached to a Petition in favor of it, out of 14 or 15,000 Inhabitants.———Reading is not a Manufacturing town in any point of view, the chief passage through it consists of stage coaches, gentlemen's carriages, and stage waggons, which carry such goods that are not sent by the river.—If you assume that the Railway went through Basingstoke, and that the traffic was diverted from Reading to the same, it follows as a matter of course that it would injure the latter town.—A Railway going through Reading would also affect the River, (I am a Commissioner of the Navigation of the Thames): the Thames Navigation is equal to and perhaps superior to any other navigable river of like extent, but I conceive that a Canal is preferable to a River, provided that it is well constructed and has a good supply of water.—I purchase some of my coals in London, which are sent by the river, and I buy some inland.—If a Railway is made to Bristol by Reading, the Light goods will be lost to the river. The heavy traffic upon the river consists of Coals, which are chiefly sent upwards, and Corn, Timber and Stone, (all the Bath Stone is sent down by the Kennet and Avon Canal,) downwards from Bath. If any material reduction was to take place in the traffic upon the Thames, it would be very difficult for the Commissioners to pay the interest of their loan, and keep the river in repair.—If the Tolls are increased the parties sending will endeavour to find some other conveyance.—A Railway passing at a greater distance from the river would not be so injurious to it.

Description of the Line of last year. — Reading has excel. communications. — Trade of same. — The Prop. line wou. affect the Naviga. of the Thames. — It is almost equal to a Canal. — An Account of the Traffic upon same.

Ex. Mr. ALBERT DANGERFIELD, *of Cheltenham, Coach Proprietor.*

The Proposed line would be very inju. to the Traffic thro' Cheltenham.

The Proposed line of Railway from Swindon to Cheltenham would be very injurious to the Coaches passing through Cheltenham; it would take away all traffic that at present passes through it from Wales and Herefordshire, and ⅞ths of the traffic between London and South Wales, Herefordshire, &c. passes through it; it is also the commanding line of road to North Wales, but not to Swansea, and a considerable number of persons also pass through it from the North West and the Aberystwith line of road; there is also a direct line of road through it from Liverpool to Southampton, and the total number of Coaches which pass through from all directions do not amount to less than 90 to 100 daily. The Passengers meet at Cheltenham, and I think that the

Account of same. — 90 to 100 Coaches pass thro' it daily.

travelling from it is equal to the whole of Gloucester, ⅔rds of Herefordshire, and ¾ths of South Wales.——It is 9 miles on the London side of Gloucester by the road, but it would be 14 miles further off by the railway, independent of being upon a branch line, although Gloucester would be brought 9 miles nearer London.——Much of the South West traffic passes through Gloucestershire; the Old road from thence to London omitted Cheltenham, but the present road, which was made to avoid difficult parts, includes it.——Cheltenham is a very large place, and much frequented by people for the good of their health, and for the sake of quietude, and they are not of a fixed class as those of South Wales, (*i. e.* they travel more): I think that a railroad would introduce a very different description of people.——An offer was made me of the exclusive privilege of running Conveyances to communicate with the Steam Carriages at Swindon, (which is the nearest point of the Proposed line to Cheltenham,) provided I took a certain number of Shares, but I declined, as I did not think it a good investment; I told Mr. Pitt, (an eminent Banker at Cheltenham, through whom the proposition was made,) that I would take 30 shares if he would guarantee me 4 per cent. for them.—I was satisfied that there would be a branch from Swindon. (Many people subscribed with a view of selling their shares.)——In the event of there being a Branch line from the Birmingham railway to Oxford, we should have an additional inducement to go there, and I consider that a line by Tring would be much the most direct, and it would be ten times more advantageous than the Branch from Swindon, (even if it did not extend to Gloucester, as the distance is trifling and the conveyances very numerous,) and it would preserve the communication between Cheltenham and Gloucester. There is no direct traffic to London by Swindon; a line to Tring *via* Oxford would be preferable, but I have no doubt the former branch would be used: the distance would be done in the same time by both lines. The traffic from Cheltenham to Oxford is in the proportion of 10 to 1 compared with that of Swindon.——There are only two extensive Coach Proprietors in Cheltenham, of which I am one; 5 Coaches start from our office for London daily, independent of those from other offices, and 1 to Oxford, and there are a number of short coaches, (there are 4 coaches daily from Cheltenham to Bath).——I believe that the Population of Cheltenham amounts to 30 or 35,000.——Stroud is a place of some importance, but I should not think the Population amounted to more than 10 or 12,000. I do not know any thing of the traffic from thence to London, but there is only one Coach to London: the chief support of Coaches going from or passing through it is derived from Cheltenham.——The Public opinion of Cheltenham was originally in favour of the Proposed line, but it has since changed.——I did not sign the Petition, as I am not an advocate for either line of Railway.

Cheltenham would be 14 Miles further off, and Gloucester 9 Mi. nearer Lond. by the Propo. line.

Cheltenham is much resor. to by Invalids.

He refused a great advanta. offered him provided he subscri. to the Propo. line.

Account of same.

He cons. the Tring line by Oxfo. much better than the Swindon branch.

The Traffic much in favor of the former.

Acct. of Coaches from Cheltenham.

Population of same 30 or 35,000.

Coaches fr. Stroud.

Population of same 10 or 12,000.

The Propo. line is disliked at Chelten.

Ex. Mr. WILLIAM SEAL EVANS.

I attended a meeting at Cheltenham convened by the promoters of the Great Western Railway, and I seconded one of the resolutions. (Their Agent stated that the only way of obtaining a railway communication between Cheltenham and London was

He sanctioned the G.W. orig. from an understan. it was to pass nr. Cheltenham.

by a junction with the Great Western, which statement influenced the meeting.)—I was not aware at that period of Mr. Giles's line from Gloucester to Tring, through Cheltenham, or I should have opposed the Great Western. I afterwards saw a Plan of the Tring line, (which takes the Northern part of the town by Winchcombe,) and I was Chairman of a Meeting to discuss the same, and after Mr. Giles had concluded his address I made a few observations on the injury which the Great Western line would do to Cheltenham, but no resolutions were passed, as most of the gentlemen attending were Subscribers to the Great Western. (I remember Mr. Gardiner said his object in subscribing to the Proposed line was to serve the town of Cheltenham.)——I consider that a direct communication between Gloucester, Cheltenham, and London would be advantageous to us, but a Branch from Swindon to Gloucester, and another from thence to Cheltenham would be highly injurious, as it would remove nearly all the traffic from it, and would run us about 14 miles directly away from London without making a turn. ——I believe that about 14 miles is gained between Cheltenham and London by the Tring line over the other, and 4 between Gloucester and London.——It would also carry us 14 miles in the best direction for a junction with Birmingham, which is of vast importance to Gloucester, as it is the very centre of their market, to which they have at present a very imperfect communication.——I am a Director of the Cheltenham and Gloucester Bank, and I attended a meeting at which I met several influential Merchants and other gentlemen of Gloucester, (who were not Subscribers to the Great Western,) who expressed their astonishment that any one in Gloucester should encourage the Proposed line by subscribing to it.——Mr. Baker has much underrated the Traffic of Cheltenham in his evidence, as the Coach Proprietors put on an extra coach 2 or 3 times a week, and he has also greatly underrated the average number of Passengers by each Coach, (as I am quite sure that they exceed 9). ——Cheltenham is the first and Oxford the next point where the roads branch off to Worcester and other places.—I believe the best line of communication from South Wales is through Gloucester, inasmuch as all the present traffic comes through it, (except the Bristol traffic) as it is 4 miles nearer.——Stroud is a manufacturing place of considerable importance, and about 9 or 10 miles from Gloucester; it would make rather a long circuit from Stroud to London by the Tring line.——I have a knowledge of the country upon the Tring line between Gloucester and Oxford, having travelled it both on horseback and on foot; I have also hunted and fished over it, including the Valley of the Winrush, which I think is exceedingly well calculated for a line of railway, as the stream is slow, excepting its source at the hill, and I judge of the inclination by the rapidity of the stream.—There would be a tunnel 2 miles long through Hailes Hill, ——There are some manufacturing towns upon the line, as Witney.—I have estates at Winchcombe, which is a farming district, Cheltenham is the principal market near it; I do not think that a railway would be of any advantage to my property in the neighbourhood, as I am only 3½ miles from Cheltenham, and the suggested line, (Mr. Giles's) is 4 miles from my estate.

Margin notes:

He now approves of Mr. Giles's branch from Tring.

Comparis. between the Tring and the Swindon Branches.

The latter wo. make Chel. 14 Mi. farther from London, and Glouc. 4 Mi. nearer.

And it would remo. the Traf. from Chel.

The Basing would facilitate com. with Birmingham, which is very desirable.

Mr. Baker has much under rate. the Traf. thro' Cheltenham.

Remarks on same.

Remarks on Stroud and its importance.

Description of the Tring line.

The Rapidity of a Stream is a good index of the levels.

His Est. at Winchc. wou. not be benefit. by the Tring line.

Ex. Mr. PEARSON THOMPSON, *of Hatherly Court, Cheltenham.*

I am a Barrister and Magistrate of the County,—and I am the Owner of the Montpelier Spa, and some other public buildings at Cheltenham.———I was introduced to Messrs. Saunders and Hunt, (the Agents of the Great Western railway,) at the Magistrate's Office, by Mr. Griffiths, the Company's Clerk, who was engaged in promoting the undertaking at Cheltenham,—and I had a long conversation the next morning with Mr. Hunt upon the subject, at my house, when he informed me that the former opposition had cost nearly £60,000. and was all paid. I stated that any further opposition would be fatal, and I inquired whether it was likely, as I should object joining under such circumstances; he replied that the Lords Jersey, Cadogan, and Caernarvon, also Eton College, were the principal Opponents, and that he was satisfied they would withdraw, and I understood him to say that the Parliamentary expenses would not exceed £10,000.—He also informed me that the line would be on the Cheltenham side of Chosen Hill, (which is also known by the name of Church Down Hill, there being a Church on the top of it,) which is midway between Cheltenham and Gloucester, and I think Stroud is about 5 or 6 miles on the north east side of the hill, I therefore concluded that the junction with the Gloucester Line could not be far from Cheltenham. Mr. Hunt gave me a written guarantee, a week after I subscribed, stating that the line was to pass from Swindon to or near to Cheltenham.———I was present at a Meeting in November last, and at the request of the promoters of the Proposed line I moved the first resolution; Mr. Saunders spoke at considerable length respecting the nature of the undertaking, and Mr. Hunt also. Mr. Brunel did not say much, but he stated that a line of Railway from Cheltenham over the Cotswold Hills was impossible, on account of the great height of the hills, and the depth of the valleys.——— Having read Mr. Brunel's Evidence since my arrival in London, I found (to my great astonishment) that his description of the line was totally different to Mr. Hunt's statement.—It now appears that the Cheltenham Branch is to leave the line upon arriving within 2 miles of Gloucester, by which we shall be sent 14 miles round from London, and Gloucester, which is at present 10 miles further from London than Cheltenham, will be made 7 miles nearer than the latter.———(I understood at the time I subscribed that the branch was to pass through Stroud, but they have not taken the most direct line from it. A Branch might have been made to it on the North side of Chosen Hill, the Gloucester branch being kept on the other side of it.)———I have 50 Shares in the Proposed line, but I should have opposed instead of subscribing to it, if I had known Cheltenham was to be thus cut off, as it is impossible to calculate the injury it would be to it; I am satisfied the people would not have supported it if they had been aware of it.—The Proposed line would remove the whole of the Traffic from the West of England and Exeter, and South Wales from Gloucester and Cheltenham, also the Traffic from North Wales and the North of England, from Leeds, Manchester and Liverpool which passes through Tewkesbury.———Tewkesbury, Cheltenham, and Gloucester represent a triangle, and no person would travel along the two sides of a triangle when they could pass along its base.———In the year

Magistr. of the Co.

He has Property at Cheltenham.

Acc. of his connex. with the G. W.

Mr. Hunt's acc. of the procdgs. last yr.

He stated that there wo. not be any Opp.

and that the Pro. li. wo. pass near Chel.

Des. of Chosen Hill

Was pres. at a Meet. in fav. of the G. W.

Account of same.

Mr. Brunel's des. of the line at tot. varia. with the represen. made to him.

The Prop. line wo. make Glouc. 7 mi. nr. Lond. than Chel. (it is at pres. 10 mi. further).

He wo. have Oppo. the mea. had he been aware of it before.

Acc. of the Traffic it wo. remove from Cheltenham.

Ac. of the Road whi. he made at Chelten.

And the Opposition he met with.

1828 I was desirous of making a road to the new town of Cheltenham, to avoid going round by High Street, which is a very circuitous route from Gloucester, (upon which occasion Lord Wharncliffe kindly assisted me,) but the inhabitants of the old part of the town, being afraid that it would remove the traffic from it, opposed me *en masse*, and presented a Petition of immense size against it, and I was obliged to give a Bond, in a Penalty of £20,000. not to afford any facility for the extension of the line further into the High Street.——Mr. Giles's line keeps more in the present course, and I consider that there would be a great deal of traffic upon it.——The Swindon line would be the least injurious to me, as Mr. Giles's could not possibly serve me.——Cheltenham is only 28 miles from Swindon by the road, but it would be 42 or 43 miles by the Proposed branch.—And it is 12 miles from Cheltenham to Stroud by the road, but it would be 19 by the Proposed branch.——I received my Summons to attend and give Evidence this day week, but I was not made acquainted with the purpose of it. (I have not seen Mr. Giles since he called at my house, and solicited my attendance at a Meeting, which I did not go to.)——Having informed Mr. Gardiner, the Banker, (at a Meeting of the Trustees of the Gloucester Road) of my summons, he remarked that he had done a very foolish thing with regard to the interests of the Town in subscribing to the Great Western, (he had 50 Shares).——Cheltenham is 9 miles directly east of Gloucester, although they charge 10 miles of posting; the distance is only 8 miles by the old road, (by Birdlip to Andover ford,) and which is almost in the immediate direction to London.—I believe that every Gloucester Coach passes through Cheltenham.

Mr. Giles's Plan does not interf. with the Traf. of Chelten.

Distance by the Pro. line & by the Road from Cheltenham to Swindon.

Do. Stroud.

Do. Gloucester.

Ex. LIEUT. THOMAS FRANCILLON, R.N.

Dock Master to the Port of Gloucester, which is rising.

Account of Duties.

Do. the Dues upon the Gloucester and Berkeley Canal.

Trade prin. Medite. and some W. Indian

The Imports are Corn and Timber.

Destination of same.

And Exports Salt, Iron, Bark, Oak, &c.

Much Welsh Iron passes thro' Glouc.

I have been Dock Master of the Port of Gloucester nearly 3¾ years, and I have been acquainted with the town about 6 years.——The Custom House Duties are increasing every year, they amounted to £19,000. in the year 1826, and last year to £131,000., and they are expected to reach £150,000. this year.—The Income of the Berkeley and Gloucester Canal is also increasing: the dues amounted to £3,500. in 1829, which was £1,000. more than the previous year, and their income last year was upwards of £12,000.——We have much Mediterranean trade, and a little West Indian trade, within the last 2 years, which is increasing.——The principal Import trade consists of corn and timber, and by far the greater portion goes up the Severn to the North, a very large proportion of it to Birmingham, and some towards Warwick, Leamington, and their vicinities, a large quantity of it also goes to Stroud and Cirencester. We have within the last year had 2 or 3 cargoes from Hambro' imported for Stroud (but I have never observed any of the Stroud manufactures sent to Gloucester for exportation).—The principal Exports are salt, iron, and iron manufactures, bark, and oak timber; some Birmingham manufactures are also exported coastways.—A large quantity of Iron also passes through Gloucestershire (from South Wales) upwards to Shropshire and Staffordshire, the principal part of which returns in a manufactured state.——

There is a Canal (for barges 14 feet wide) from Gloucester to Stroud, which communicates with the Thames and Severn, thereby forming a connection with London, but the latter Canal (between Stroud and the Thames) and the upper part of the Thames, are sometimes short of water, (they generally make use of land carriage). ——I consider that about $\frac{9}{10}$ths of the Gloucester trade goes northward and north east.—The Coach traffic between Gloucester and London (including all which passes through, together with that which originates in Gloucester) is very great.—There are 3 day and 3 night Coaches daily through Gloucester and Cheltenham to London, and there is much traffic in two-horse Coaches, Flys, &c. between them.—The communication between Stroud and Gloucester is trifling, there being only a two-horse coach, (which is in connection with the Mail,) and there are sometimes detentions of two hours by it.——I consider that about half the waggons from Gloucester to London pass through Stroud, and the other half through Cheltenham, but a much greater portion of the goods conveyed by them stops at Gloucester and Cheltenham, than at Gloucester and Stroud.——Stroud has a communication with Bristol, but its chief traffic is with London, (in expensive woollen goods) from whence it also derives its foreign supplies.——A great quantity of heavy goods are sent round from London to Cheltenham through Gloucester, (which is the natural port to it,) and there is much traffic in heavy goods on the Tramroad between Cheltenham and Gloucester.——It is about 1 or 2 miles nearer to London by the Stroud road, (there is a much nearer road than by Cheltenham, over Birdlip Hill).——The Navigation of the Severn upwards is very much impeded by drought.——I believe the Merchants are more desirous for a Railway to Birmingham than to any other place. ——I was present at a Meeting held at Gloucester about 12 months back, for the purpose of discussing the merits of the Proposed line, and I remember Mr. Saunders stating that a railway between Oxford and Cheltenham was out of question, as the ground had been examined, which had the effect of producing subscriptions to the Great Western, and the resolutions were carried unanimously in favour of it.——I am well acquainted with the Eastward towards Winchcombe, and thence by the Valley of the Winrush, (There has been a considerable fall of Timber at Guiting Wood, in the upper part of the val eof Winrush, which was sent to Gloucester for the Royal Navy and Dock yards.) and having observed Mr. Giles's Assistant at Gloucester, when engaged upon the survey of the Tring line, I suggested to him what I considered the proper direction, which he followed, and I am satisfied it is the only practicable line to Oxford, as the country is very difficult. This line would present great facilities to a large portion of the present trade of Gloucester, instead of being merely available for the lesser portion, and it would certainly be the most advantageous for Cheltenham. After leaving Cheltenham, it goes nearly north, and takes a *détour* round Nottingham and Cleveland hills, and then arrives at Hailes Hill, where there is some difficulty, but the country beyond is favourable, (but it does not include the towns of Tewkesbury and Worcester).—I have understood that Mr. Brunel has made a survey of a Line from Gloucester to Birmingham; I have also seen a Plan and Section of a line between the same by Capt. Moorsom, which terminated near Mr. Giles's Depôt, and upon the River Twiver at Gloucester, which is a very convenient spot for a terminus, (and almost the only one practicable for a line from Cheltenham or Birmingham,) it would take the line

There is a Canal from Glo. to Stroud to the Thames.

$\frac{9}{10}$ths of the Glouc. trad. goes N.& N.E.

Coach trade thro' Glouc. is immense.

Do. by Swindon is trifling.

Traf. by Waggon to Chelten. and Stroud about equal.

Communi. between Stroud and London.

Heavy goods are all sent from London. by

Glouc. the natural Port to Cheltenham.

And the nearest road to London is by Stroud.

The Naviga. of the Severn wants repa.

Glouces. requires a Railw. to Birming.

He attend. a Meet. of the G. W.

Account of same.

(Memo. Guiting Wood.)

He suggested the Tring line.

Advan. of same to Glouc. and Chelten.

Route of same.

Descrip. of Capt, Moorsom's line fro. Glouc. to Birming.

It starts at Glouc. from the same spot as Mr. Giles's.

of the tramroad to the river, and would not interfere with property to much extent. (The fact of there being a tramroad shews that it could not be an ineligible line).—Captain Moorsom's terminus was about ⅓rd of a mile from the Gloucester and Berkeley Canal basin, and rather more from the Quay: It passes near Tewkesbury, and within 2 miles of Worcester, by which he gets into a very difficult country when near Birmingham.—(There is some trade at Tewkesbury, but Stroud is the principal Manufacturing district of any importance in Gloucestershire.)—It crosses a ridge of hills connecting the Lickey, for which 6 different lines were surveyed, No. 6 being chosen; and it has a Tunnel at a short distance from Birmingham, where the line is 150 feet above the terminus of the London and Birmingham railway.—If he had taken a line more to the right, and availed himself of the levels up the Valley of the Avon, following Mr. Giles's line, and then kept as much as possible to the left to Birmingham, (as it would shorten the distance,) it would have enabled him to join the London and Birmingham between Birmingham and Coventry with great facility, (If taken direct to Birmingham the line would not be so direct, but I cannot say whether it is practicable, without a survey being first made,) which I consider preferable to terminating at Birmingham, on account of the difficulties.———The Clothing district is situated in the Chalford Valley and a passage through it would be difficult, as it is very narrow and the hills very steep. (The whole rise of the Canal along it, from the Berkeley Canal, is upwards of 200 feet,) which is the case with all the numerous branches of the Stroudwater, upon which there are a great number of Mills. The hills are formed of Stones, having large vacancies within, and the supply of water is very plentiful.—I think that the line passing through Hailes Hill would be the easiest of execution.———The Vale of the Thames is of very great importance, but not equal to the Vale of Gloucester, which is the principal agricultural district in the county.

Descr. of the Depôt. It passes Tewkesb. and Worcester. Stroud is the princ. Manu. Dis. in Glou. It ends 150 ft. above the L. & B. at Birm. Capt. M. shou. have kept more to the Tring in this direct. and join. the L. & B. Remarks on same. Diffic. of a li. down the Chalford Valley. The Branches of the Stroudwat. are very narrow and steep. He thinks the Tring line the easiest. Description of the Vale of Gloucester.

Ex. MR. MAURICE SWABEY, *of Langley Mearish, Buckinghamshire.*

I am a Barrister, attending the Oxford Circuit.———The Proposed line crosses some property, (at about ¼ of a mile from my house,) called "The Grove," belonging to my stepmother, who resides in Ireland, and is very aged; the property will become mine at her decease.—There is a Paddock of 40 Acres adjoining the house, with ornamental timber thereon, about 8 acres of which is enclosed by a Ha Ha, or sunk fence, (but it appears all one paddock,) and the railway crosses the same upon a level, and I have no hesitation in saying that it will destroy the property as a Gentlemen's residence.—It is thickly studded with trees, from which it takes its name, (there are 10 or 12 oaks and elms).—The house is in very good repair, but has been untenanted the last 3 years, and I believe the circumstance of a railway crossing the grounds has prevented the letting of it; it was said to be worth £300. per annum at the period of my marriage, but its value is much reduced.—It was occupied some years back by Lady Frances Coningsby, after which my father reduced the size of the house, and let it to Mr. Gosling, the Banker, and afterwards to his widow, who occupied it together

The Pro. li. crosses "the Grove," belon. to his Step-mother. Descrip. of same. The Pad. is thickly studded with Trees. It was once worth £300. per Annum.

21 years to my knowledge.——The promoters of the undertaking have offered to treat with me, and one of the gentlemen proposed purchasing it, which I declined, as I do not wish it turned into a brick ground.——The Proposed line also runs through my brother-in-law's Estate (Mr. Clewer) at Delafont, which is bounded by the Colne, (a beautiful trout stream,) and I think it crosses the river 3 times.——The feeling of the Land Owners in my neighbourhood is decidedly against the measure.—I have been very active in the opposition to this bill; and I have taxed my purse very willingly for the same both last year and this.

The Prop. li. crosses the Colne 3 times.

He has been very active in the Oppo.

Ex. MR. SAMUEL FOSS DESSIAU, *Shipping Agent of Portsmouth.*

There is a considerable trade between Bristol and Portsmouth in butter and salt provisions, which are very often detained in winter, as they are obliged to be sent round by the Land's End, where the navigation is very difficult and dangerous, on account of the South West winds.—I have been in the Portland roads with a fleet of 700 merchantmen wind bound to different ports, the West Indies, and other places, and they would all want provisions more or less; (about ⅔ds of the ships require fresh meat every day,) which a Railway from Bristol to Southampton would obviate.—There is no occasion for Irish provisions to travel quickly, but the detention of merchandize in time of war is of some importance.——There are many head of Cattle sent from Wales to Smithfield, where they are slaughtered, and then forwarded to Portsmouth for the supply of the Navy Contracts, and the above line would be very advantageous for the conveyance of Cattle from Bristol and Salisbury. (We have some from Salisbury and Devizes, and even some from London.)——Three Irish families within the last 5 weeks, travelling from Ireland to Portsmouth by Bristol, have preferred going round by land to avoid the probable detention at sea by the London Packets, (which call off Portsmouth,) notwithstanding the great extra expense.—And I have known a person in the Iron trade order goods in London, at an advance of 10 per cent. compared with Bristol, as he could not get them sufficiently quick from the latter. They would doubtless cheerfully pay the increased freight of the railway for the advantages of the increased speed.——A great deal of Coasting trade passes by Portsmouth, round the Foreland, perhaps that portion of it which goes to London would be taken by the railway.—The Basing line would be equally beneficial to Bristol as to Portsmouth and Southampton.—There is no Canal between Bristol and Portsmouth, and Waggons are very rare. There are equal to 5½ daily Coaches, besides one which travels alternate days.—There is a Canal from London to Portsmouth.——There is a direct communication from Southampton to Havre, and other parts of France.——A Railway has been agitated from Gosport to Southampton, a distance of 14 miles, (it is 4 more to Portsmouth).

Trade betw. Bristol and Portsmouth.

It is conveyed by Sea, which is very uncertain.

The Basi. li. would remedy the same.

Cattle are sent from Wales to Lond. and the Meat is then sent to Portsmouth.

The Basing could be advanta. used for convey. the same.

Proof of the want of a line from Bristol to Southampton.

Effect of the Basing on the Coast. Trade.

The Basing would be equally beneficial to Bris. as to Ports. and Southampton.

Traffic to Portsmo.

A line is proj from South. to Gosport.

Ex. CAPTAIN JAMES WEEKS.

Of the Souamp. & HavreSteam Boats.

40 the Average of Passeng. each way.

(Passen. to Guerns. & Jers. doub. that)

Most of his Passen. are from Bristol.

There are 4½ Coach. betw. Bris. and Sou.

The Passage occu. 12 Hours.

They carry 90 or 100 Letters ea. way.

⅔ds. of which are for Bristol & Liver.

There is much Trade betw. Havre and the United States.

Havre is a very flourishing Port.

I am Captain of the Apollo Steam Packet, which runs between Southampton and Havre; taking alternate days with the Camilla, it makes 3 Voyages each way, (on Mondays, Wednesdays, and Fridays.)——There are at present 4 Passages weekly to Jersey each way, and 4 daily to the Isle of Wight and Portsmouth.—We average 40 Passengers each voyage to Havre, but we had more last season, (the fares were lower in consequence of an opposition); our seasons are from April to the end of October, which gives about 7000 annually, and I do not think we have been under that amount for some years.—The Passengers between Guernsey and Jersey double that amount, and they are principally from London.—But most of the Passengers from Southampton to Havre come through Bristol, (we have taken a great number of Irish Passengers since the Peace.)——There are 3 Coaches besides the Mail running from Bristol and Bath to Southampton, one goes every other day, and the remainder daily, but there is not much Posting; a Railway between the above places would be advantageous to both.—There is also a considerable communication between Liverpool and Havre.——We occupy 12 hours on the Passage, and are permitted to carry Letters to and from Havre, which average about 90 to 100 each packet, and they sometimes amount to 250 or 300; perhaps ⅔rds of them are for Bristol and Liverpool. ——There is a considerable trade with the United States and Havre, which is surpassing Bordeaux as a commercial port, and I am not aware of any town in Europe which is progressing equal to it.—I have been employed on the line since the present Peace, and I have never got damaged to the extent of £5. in entering or coming out of Port, as it is very safe.—It is becoming a very favourite passage to Paris.——There are several Steam Boats on the River to Rouen, two new ones were put on lately, (which brought over 400 each way last time, and more are building.—A new Steam Packet Company, is also in contemplation, with a view of facilitating the communication. —It is the general opinion that there will be a Railroad from Rouen to Paris.

Ex. JAMES WILLIAM DEALE, *of Southampton.*

He is connect. with the South. Coaches and Packets.

He agrees wi. Capt. Weeks' Statement.

The Port of Havre is increasing.

A Railw. between South. and Bristol would increase the Passengers.

I am connected with the Coaching trade, and I am Bookkeeper of the Isle of Wight Steam Boat Company, I am also connected with the Portsmouth Steam boats. ——The Evidence given by the last witness (Captain Weeks) is tolerably correct. ——The greater number of Passengers from Bristol pass on to Havre, which port is increasing in traffic, and I have no doubt that a direct communication between Southampton and Bristol would increase them even more.—Complaints are continually being made of the want of conveyances.—The Packets from Bristol to Dublin only sail on Tuesdays and Saturdays, and passengers for Waterford or Cork cannot arrive there until the Saturday following, if they do not get in Bristol by Tuesday.—If the above communication was opened, I have no doubt the Packets would be accommodated to it; we are at present absolutely crowded with Passengers, but our Packets do not run from November to April.——A Railway would be the making of the town of Southampton, as it would cause a greater number of Irish Families to pass our way.——I think that the London and Southampton Railway is a very good undertaking.——The Fares to the Islands of Guernsey and Jersey are at present very low, being only 5*s.* and 2*s.* 6*d.*, in consequence of which they sometimes take out 230, and return with 70 to 80 Passengers.

The Packets do not run fro. Nov. to Ap.

He considers the Lond. & Southamp. a good undertaking.

Packets to Guerns. and Jersey.

Ex. JOHN ROUSE, *Lighterman.*

I have been in the practice of letting horses for the purpose of Towing barges up the Regent's Canal for the last 10 years, I am therefore well acquainted with the Canal and the traffic upon it.——The present Traffic upon the Canal frequently occasions considerable delays, and if the traffic of the London and Birmingham Railway should pass upon it the delays might be increased.——There are about 30 Bridges between Camden Town Bridge and Limehouse Basin, also 12 Locks, and a Tunnel under High Street, Islington, about ¼ of a mile long, through which only one barge can pass at a time, and the same with the bridges.——A Stationary Engine is employed to take the barges through the Tunnel, horses being used upon the other parts. It is very seldom that 2 barges meet under the bridges, but great delays frequently occur at the tunnel, when any thing happens to the Engine, or in the event of the Chain breaking, as the boats are then pushed through by Leggers, (men lying on the backs and pushing through with their feet, 2 men can take a light barge through in ½ an hour, and a loaded barge in ¾ of an hour) and frequently occupy 1 hour in passing through it, but they sometimes accomplish it in 17 minutes; much time is also lost in waiting, sometimes 6 go through together——The average time consumed in going down the Canal from Camden Town to the basin at Limehouse is about 6 or 7 hours, (I have done it in 5 hours) and they are about the same time in returning, the traffic each way being nearly equal.——We are obliged to wait for the Tide at the Basin before we pass into the Thames, when the Tide is whole we have to wait 6 hours, at other times we get out directly; perhaps 3½ hours is about the average time of waiting.——Some time is lost in getting from thence to the London and St. Katharine Docks, perhaps 1½ hours, according to the tide, as barges cannot make against it;—and I believe it takes about 2 or 3 hours to pass from thence to Nine Elms, but I do not belong to the river.——We sometimes get stopped by Frost.——The Expence of Towing up to the Hampstead Road is 14*s.*, exclusive of Lighterage.—The following are the Rates of Tonnage:

He Towes Barg. up the Regent's Canal.

Rem. on the delays.

Do. in the Tunnel.

Bridges upon it 30.

Locks . . . 12.

and 1 Tun. ¼ of a mi. long, upon which an Engine is used.

It occu. 6 or 7 hours in passing the Canal.

And 3½ hours the ave. time of passing out of the Tide Bas.

Thence to Lond. & St. K. Docks, 1½ h.

And to Nine Elms, 2 or 3 hours.

The Exp. of Towin. on the Canal, 14*s.*

1835.

REGENTS CANAL.

RATES of TONNAGE chargeable 'till further Notice.

1. Cement-stone, Chalk, Clay, Coke-breeze, Fire-bricks, Gravel, Refuse Ashes and Cullet, in whole Cargoes, Road Materials, Sand, and Tiles.
2. Coals and Coke.

3. Cement, Earthen Pipes, empty Sacks, Flour, Grain, Lead, Lime, Mangel Wurzel, Marble, Plaster of Paris, Potatoes, Pottery Flints, Oilcake, Pulse, Salt, Slates, Stone, and Timber.

4. Iron (Pig and Scrap).

5. Iron (Bar, Hoop, Plate, Rod, Sheet, and Strip), and Cast Iron Pipes.

6. Iron (except as above), Brass, Copper, Nails, Spelter, Tin Plates, Hay, Straw, and Live Stock.

7. All other Goods.

Rates of Tonnage upon the Regent's Canal.

From Paddington to the following Points; viz.				From the Thames to the following Points; viz.			
No.	Per Ton. s.	d.		No.	Per Ton. s.	d.	
1	0	4		1	0	5	
2	0	7		2	0	5	
3	0	4		3	0	5	
4	0	6	Head of Hampstead Lock, also Regent's Park Basin.	4	0	7	Tail of Johnson's Lock.
5	0	9		5	0	8	
6	0	9		6	0	8	
7	1	0		7	0	11	
1	0	4		1	0	5	
2	0	7		2	0	7	
3	0	6		3	0	7	
4	0	6	Head of Pancras Lock.	4	0	8	Tail of Old Ford Lock.
5	0	10		5	0	9	
5	0	10		6	0	9	
7	1	4		7	1	0	
1	0	6		1	0	7	
2	0	7		2	0	8	
3	0	7		3	0	9	
4	0	8	Head of City Lock, also King's Cross Basin.	4	0	9	Tail of Acton's Lock, also Cambridge Heath.
5	1	0		5	0	11	
6	1	0		6	0	11	
7	2	0		7	1	10	
1	0	6		1	0	7	
2	0	7		2	0	8	
3	0	9		3	0	10	
4	0	8	Head of Sturt's Lock, also City and Wenlock Basins.	4	0	9	Tail of Sturt's Lock, also Shoreditch Basin, King's-land Road.
5	1	0		5	1	1	
6	1	6		6	1	3	
7	2	6		7	2	4	
1	0	6		1	0	7	
2	0	7		2	0	8	
3	0	9		3	0	10	
4	0	8	Head of Acton's Lock, also Shoreditch Basin, Kings-land Road.	4	0	9	Tail of City Lock, also City and Wenlock Basins, and Islington Tunnel (East).
5	1	0		5	1	1	
6	1	6		6	1	3	
7	2	7		7	2	4	
1	0	6		1	0	8	
2	0	7		2	0	8	
3	0	9		3	0	11	
4	0	8	Head of Old Ford Lock, also Cambridge Heath.	4	0	9	Tail of Pancras Lock, also King's Cross Basin.
5	1	0		5	1	1	
6	1	6		6	1	7	
7	2	8		7	2	8	

RATES of TONNAGE chargeable—continued.

Rates of Tonnage upon the Regent's Canal.

From Paddington to the following Points; viz.				From the Thames to the following Points; viz.			
No.	Per Ton. s.	d.		No.	Per Ton. s.	d.	
1	0	6	Head of Johnson's Lock.	1	0	8	Tail of Hampstead Lock.
2	0	7		2	0	8	
3	0	9		3	0	11	
4	0	8		4	0	9	
5	1	0		5	1	1	
6	1	6		6	1	7	
7	2	9		7	2	9	
1	0	6	Head of Commercial Road Lock.	1	0	8	Tail of Stop Lock, Paddington, also Regent's Park Basin.
2	0	7		2	0	8	
3	0	9		3	0	11	
4	0	8		4	0	9	
5	1	0		5	1	1	
6	1	6		6	1	7	
7	2	10		7	2	11	
1	0	6	Thames.	1	0	8	Paddington.
2	0	7		2	0	8	
3	0	9		3	0	11	
4	0	6		4	0	7	
5	1	0		5	1	1	
6	1	6		6	1	7	
7	3	0		7	3	1	

N.B.—The Numbers in the First Column refer to the Classes of Articles enumerated at the Head of the Rates. The Locks are Eighty-three Feet in Length, clear of the Gates, and Fourteen Feet wide. Barges to be charged for not less than Ten Tons (except as Back-carriage).

The under-mentioned Articles to pass at the following reduced Rates:

	Per Ton. s.	d.
Iron of all Sorts, and Nails, if conveyed the whole Length of the Canal . .	0	6
Ashes for Manufacture, Alum, Barilla, Saltpetre, and Tallow, ditto . . .	2	0
Cheese from Paddington to the City Basin	2	0
Hay and Straw from Paddington to the Regent's Park and King's Cross Basin . .	0	6
Ditto Ditto City Basin	1	0
Manure and Ashes and Breeze for Brick-making	0	4
Bricks from Paddington to the Regent's Park Basin and to the Hampstead Road .	0	3
Ditto from Paddington or the Thames to any other Point on the Canal . ,	0	4
Articles conveyed as Back-carriage, upon which the full Tonnage has been charged .	0	1
Articles loaded or unloaded in the Passage or Water-way of Canal to be subject to an extra Toll of	0	6

There is a little alteration in the charge for Tin Plates, they are now 1*s.* 9*d.* to Hampstead Road.

Ex. MR. HENRY SCRIVENER, *Secretary to the British Iron Works.*

The Co. contracted for the Rails of the Southamp. Railw.

50 lbs. the yard.

They were delayed owing to a Strike.

They are allo. 2 yrs. to complete the Ord.

The Company entered into a Contract with the Southampton Railway Company, at the latter end of last year, for supplying them with 6,000 Tons of Rails at 50 lbs. to the yard, the first delivery was to have been made on the 1st of March, and succeeding deliveries at intervals of two months; and applications were repeatedly made for them by Mr. Giles, but great delay occurred, owing to a strike in the Colleries in February, and the works were not again in operation until March.——We have delivered about 700 Tons, and about 4 or 500 Tons are now on their passage.—We are allowed 2 years for the completion of the order.

ADDENDA TO MR. GILES'S EVIDENCE.

ACCOUNT of WAGGONS teamed over one end of the Embankment at Shapley* from 4th August to 15th August.

No. of Waggons teamed at Shapley Heath.

4th August	270 Waggons.
5th —	309 —
6th —	300 —
7th —	324 —
8th —	343 —
10th —	314 —
11th —	248 —
12th —	316 —
13th —	307 —
14th —	294 —
15th —	291 —
	3,316

19th August, 1835. (Signed) FRANCIS GILES.

The following is a true EXTRACT from the BOOKS of the LONDON and SOUTHAMPTON RAILWAY COMPANY.

Intention of joining the Lond. & Southamp. with the Basing.

" At the General Half-yearly Meeting of the London and Southampton Railway Company, held pursuant to Notice at the City of London Tavern on Friday the 27th February, 1835, John Wright, Esquire, in the Chair, it was resolved unanimously, That this Meeting having taken into Consideration the Subject of the Incorporation of the London and Southampton Railway Company with the Basing and Bath Railway Company, is of opinion that such Incorporation will be advantageous to the Shareholders of both Companies, it is therefore the Opinion of this Meeting that the Directors of the London and Southampton Railway Company in due Time do take such Steps as may be advised and deemed to be needful by their legal Assistants for effecting the Incorporation of the Two Companies."

20th August, 1835. (Signed) FRANCIS GILES.

* This Table has reference to Mr. Giles's assertion that 30,000 Cubic Yards had been teamed at Shapley Heath, (vide page 215) —*Editor.*

A GLOSSARY

OF

TECHNICAL TERMS

USED IN

CIVIL ENGINEERING;

WITH

REFERENCES TO SUCH PARTS OF THE EVIDENCE

WHERE THE SEVERAL WORDS OCCUR,

THEREBY FORMING AN INDEX TO THE SAME.

GLOSSARY.

A.

PAGE.

ACCIDENTS. On inclined plane, Canterbury and Whitstable Railway, described 84, 85
„ On Liverpool and Manchester Railway 23
„ In tunnel on ditto 90
„ On inclined planes, &c. 131

ADDENDA. To Mr. Giles's evidence, Great Western Railway 276

ADHESION.

The force acting on the surface of two bodies in contact with each other, which prevents one sliding over the other:—as the force which prevents the wheels of a Locomotive engine from slipping on a railway, the adhesion of which is greatest when the rails are either quite dry, or quite wet, as the surface is then most free from obstruction: when partially wet it is much reduced, as they are more apt to catch up the dust.

The adhesion of the best modern Locomotive engines, exclusive of the power to drive the engine itself, is supposed to be capable of overcoming a resistance equal to $\frac{1}{15}$th part of the insistent weight upon a level plane (or $\frac{1}{10}$th in fine weather, and $\frac{1}{20}$th in very bad weather); and that of common Locomotives, working with vertical cylinders, to $\frac{1}{20}$th part of the weight pressing on the rails by the driving wheels; or taking the friction as equal to $8\frac{1}{2}$ lbs. per ton, or the 263rd part of the weight, a load equal to $\frac{263}{15}$ or $\frac{263}{20}$ of its weight respectively, or the weight acting upon the driving wheels.

„ Effect of dew or snow upon the rails 92
„ Of an engine about $\frac{1}{11}$th to $\frac{1}{13}$th of its weight 127

AGENTS, CLERKS, AND WORKMEN. On Liverpool and Manchester Railway . . 24, 28
„ Particulars of same, with salaries 31

AGRICULTURAL INTERESTS. Lessened by railways 261

AIR. Consumption of same in tunnels 194

ALLNUTT, ZACHARIAH, receiver of tolls of the Thames navigation. His evidence descriptive of the improvements thereon,—Expences of management,—and the injuries it would sustain by the proposed Line (Great Western Railway) . . 234, 237

ANGLE OF REPOSE. PAGE.

The utmost inclination at which a carriage will stand at rest upon a railway or road, and when upon the least increase of slope it is put in motion by the gravity of its weight. It consequently occurs when the gravity of the load and friction upon the road are equal.

The Angle of Repose therefore varies according to the amount of friction: taking the friction at 9 lbs. per ton, makes it 1 in 250, or about 21 feet per mile, which is generally considered the Angle of Repose upon a railway. Taking the friction at 8½ lbs. per ton, gives it at 1 in 263¼, or 20 feet per mile.

The Angle of Repose upon a turnpike road is about 1 in 40, with a good description of carriage, and supposing the road to be perfectly hard.

The slope at which the soil of a cutting or embankment will stand without slipping is also called the Angle of Repose.

AQUEDUCT.

A term applied either to a bridge over a valley or road, or to a tunnel through the earth; either expedient being used for the conveyance of a canal or other body of water.

The ancient Roman Aqueducts, some of which remain at the present day, consisted of several tiers of arches, supporting the water-way. (See BRIDGE).

ARCH.

A circular arrangement of overlapping stones or bricks, with radiating beds, commencing from two fixed points or abutments, and meeting in the centre; thereby forming an equilibrium upon the removal of the wooden frame or centre upon which the arch is turned. Arches are of various shapes; semi-circular, segmental, elliptical, or pointed. The joints of all Arches should be perpendicular to the surface of the soffits. The top of an Arch is called the extrados, and the underside the intrados: the line from which they commence is called the springing line, and first arch stone on each side the springers, which rest on the imposts or abutments: the extreme width from springer is called the span of the Arch, and the rise of the curve the versed sine. Arches are either cylindrical, or groined. The former is an elongation of the same curve throughout its length,—and when intersected by other arches, cutting across it transversely, the point of junction is termed a groin; such Arches being described as groined.

All Arches should be well sustained by backing. Spandril walls are generally built over arches upon railway works, to prevent any irregular pressure of earth upon same.

The general size of the Arches of the occupation bridges over the London and Birmingham Railway is 30 feet in width and 17 feet in heighth to the crown, elliptic arches being adopted, the rise of which is 9 feet. And the arches under the railway are made 15 feet wide, and of various heights, according to the embankment. The bridges erected over the metropolitan roads by railway com-

BREAK OR CONVOY TO RAILWAY CARRIAGES.

A contrivance to check the velocity of a train of carriages upon a railway, consisting of a piece of wood, which is pressed upon the rim of the wheels of the carriages by a hand lever, worked by the breaksman. The break of the tender alone affords sufficient resistance to stop a train under ordinary circumstances.

BREAKWATER.

A kind of artificial embankment, at the entrance to a sea port, erected for the purpose of counteracting the effects of violent winds and waves, being formed of large stones. The Plymouth breakwater is one of the most celebrated in this country.

BREAST-WALL.

A retaining wall at the foot of a slope only.

BRICK.

A preparation of clay, sand, and ashes, burnt in a kiln or clamp. Good brick earth is sometimes found in a natural state. A brick is 9 inches long, 4½ inches wide, and 2½ inches thick.

Brickwork is measured in London by the rod, and was taken from the original standard of 16½ feet cube, which gives 272¼ square feet as the superficial contents of 1 rod of reduced brickwork; but as the standard thickness of a brick wall is taken at 1½ bricks, or 13½ inches thick, instead of 12 inches, there are consequently 306 cubic feet in a rod of brickwork, and a standard rod will require about 4500 bricks, allowing for waste, but it depends on the closeness of the joints, and the exact size of the bricks; and 1 rod of brickwork will take

PAGE.

Brick *(continued)*.

1½ yards of chalk lime, or 1 yard of stone lime, and 2½ yards of sand with stone lime, or 2 yards with chalk lime, for the mortar. 1 foot of reduced brick-work will require 17 bricks.

Brickwork is generally measured by the yard in the country. It is therefore the general custom of Engineers to adopt the latter measure. There are about 11⅓ cubic yards in a rod.

London stocks, also those of Manchester, are the most durable. Suffolk bricks have been celebrated for their light color and close texture, which renders them nearly twice the weight of common bricks, and also for their even form, and the softest and most porous bricks are those made in the Midland Counties of England.

Bridge.

A very common engineering expedient for passing over rivers, canals, and roads.

Bridges are of various descriptions, but arches are mostly used. In some cases the road is suspended by rods from cast iron girders occupying the place of the parapet walls, as the bridge over the Regent's Canal, near Chalk Farm, on the London and Birmingham Railway. In other cases the road is carried over at once by iron or wood beams thrown across, and trussed according to the span. Where the span is very great, and there is not sufficient height for an arch, the road is some times suspended from an inverted cast iron bow by iron rods, supported upon piers at each end, and from thence carried down and secured in the ground, which are called iron suspension bridges, as Hammersmith suspension bridge.

The floods form the principal considerations to guard against in bridges connected with rivers and canals, and their effect upon the nearest adjacent bridges or arches should be carefully ascertained previous to deciding upon the width of the arches or openings of intended works. The traffic should be considered next, and sufficient space left for it between the parapets.

The number of bridges required for a railway varies in almost every instance. There are about two in a mile on the Liverpool and Manchester Railway, exclusive of the viaducts. The proportion of bridges on the Leeds and Selby Railway is about 2⅛th.

The projected railways before Parliament in the year 1837 averaged 2½ bridges per mile.

C.

CAISSON.

A large water-tight floating box, used for the purpose of putting in the foundations of the piers of bridges, &c. used generally in rapid rivers. A suitable pit is first dug to receive the caisson, and after one or two experiments are made to ascertain whether they perfectly suit each other, it is permanently sunk, and the masonry commenced from within it, and carried up level with the water, when, by a contrivance, the sides are removed, leaving the pier resting firmly upon the bottom grating.

CANAL.

A system of internal communication, principally confined to the conveyance of heavy articles.

Canals were not unknown to the ancients, although they were not introduced into this country until the year 1755, since which period they have spread throughout the whole kingdom; and the competition presented at the present day by the several railways has given an impetus to improvements upon them; the boats have been improved, and new machinery employed at the locks, in order to accelerate the traffic upon them.

The power of draught of a horse upon a canal has been stated to be from 20 to 30 tons, at about 2 miles an hour, and a horse can draw a greater weight on a wide canal than on a narrow one, viz. about $\frac{1}{5}$th more.

The following Table will shew "The cost of conveying goods and passengers on canals, at different rates of speed."—according to Mr. Macneil's Tables.

Description of boats.	Rate of speed, in miles per hour.	Resistance, per ton, in lbs.	Cost of haulage, per ton per mile.	Cost of boat-hire, &c. per ton per mile.	General expenses per ton per mile.	Aggregate charges.	
						Useful load, per ton per mile.	Gross load per ton per mile
			d.	d.	d.	d.	d.
Slow boats -	$2\frac{1}{2}$	2·73	0·18	0·32	0·86	1·36	1·02
Fly boats -	4	7·07	0·5	0·66	2·34	3·5	2·275
Swift boats -	10	56·8	0.275 per passenger, 3·5 per ton.	- -	9·7	1·08 per passenger, 13·25 per ton.	10· per ton.

PAGE.

CANAL *(continued)*

The following Table gives "The comparative cost of goods and passengers on Canals and upon Railroads, both with horse and with locomotive power on the latter."—*

CANALS.—HORSE POWER.				RAILROADS.—HORSE POWER.				RAILWAYS.—LOCOMOTIVE POWER.				
Rate of speed in miles per hour.	Resistance per ton in lbs.	Cost of haulage and boat hire per ton per mile.	Cost of conveyance per ton per mile.	Rate of speed in miles per hour.	Resistance per ton per mile.	Cost of haulage and carriages per ton per mile.	Cost of conveyance per ton per mile.	Rate of speed in miles per hour.	Resistance per ton in lbs.	Cost of haulage and carriages per ton per mile.	Cost of conveyance per ton per mile.	Charges of conveyance per ton per mile.
2½	2·73	0·5*d.*	1·36*d.*	2½	8·5	0·75*d.*	1·65*d.*	8	8·5	0·565*d.*	1·065*d.*	1·065*d.* 1·565*d.*
4	7·07	1.16*d.*	3·5*d.*	4	8·5	1·127*d.*	3.627*d.*	12	8·5	0.727*d.*	2·138*d.*	3·5*d.*
10	56.8	0·275*d.* per passenger, 3.5*d.* per ton.	1·08*d.* per passenger, 13·25*d.* per ton.	10	8·5	0·25*d.* per passenger, 2·24*d.* per ton.	1 to 1·5*d.* per passenger, 15*d.* per ton.	20	8·5	Haulage. 0·25*d.* per passenger, 0·73*d.* per ton.	0·675*d.* per passenger, 2.855*d.* per ton.	1*d.* to 1½d. per passenger, 12·37*d.* per ton.

* Arranged from Wood's Practical Treatise on Railroads, to which the Editor is indebted for much valuable data.

CEMENT

A composition of certain mineral substances, naturally or artificially prepared, which, become hard, upon mixture with lime and a small portion of water.

CENTRE (OF AN ARCH).

The wooden frame or mould used in the construction of arches, for the support of the arch stones during the course of execution.

The construction of the centres of bridges over rivers is of great importance, and it is usually necessary to form them in such a manner that the navigation shall not be impeded. The centres used in the construction of bridges were formerly removed piece by piece, upon the completion of the arches, the practice of "*striking them*," (as technically termed) by driving wedging pieces between two striking plates, fixed on each side, is employed at the present day, which have the effect of lowering the centre, thereby leaving the arch standing without support,—thus it may be gradually eased in every direction simultaneously, which prevents any unequal pressure or strain. The best way of supporting the striking plates upon which the whole rests, is by strutting or raking pieces resting upon sills laid upon the top of the footings.

The system of supporting a centre by piles driven in the bed of the river at two or more places, should not be resorted to, unless the span of the arches is so great as to prevent any other mode of execution, and the foundation is very safe; but even then the work is very likely to suffer unless the framing is exceedingly well balanced and secured together.

The centres of the arches of some bridges may be constructed with a level tie beam, provided a communication is not required through them during the execution of the works, which lessens the difficulties attending the same exceedingly.

PAGE.

CONTINUOUS BEARINGS.

The method originally employed of laying rails in this country, consisting of longitudinal sleepers, secured to transoms (as detailed under the article TRAM OR PLATE RAILWAY).

The system of continuous timber bearings has been considerably improved and much used in America, where it has been found very suitable, on account of the abundance of timber in that country, and the scarcity of iron.

The plan of forming the line of the Great Western Railway may also be described as a return to this system. The longitudinal or continuous bearings are from 5 to 7 inches in depth, and 12 to 14 inches in breadth, and laid down in about 30 feet lengths, securely bolted to cross transoms, 6 inches broad by 9 inches deep. There is a double transom at the joinings of the former, and a single one between them: thus they are alternate double and single. Within each track, at nearly midway between the rails, piles of beech are driven, 10 inches diameter and 12 feet long, to which the transoms are secured by a horizontal bolt: there are therefore two piles to the double transoms, (which are situated between them) and the same number to the single ones.

When the piles and timbers are properly fixed and secured together, sand or fine screened gravel is beat or packed underneath the longitudinal bearers, until the spaces between the piles are forced upwards, and a firm bed is obtained. Mr. Wood, in his excellent work upon railways, states, that "the whole stability or superiority of this railway over other wooden railways depends entirely upon the retaining power of these piles."

A pier or buttress generally applied at the back of retaining walls, in modern Civil Engineering, for the support of same, and likewise for the purpose of forming a tie to the material at the back of the wall. It is sometimes carried up upon the face of a wall.

A wire cap, covering the top of a Locomotive engine chimney. They are intended to prevent the escape of lighted flakes of fuel, &c., and are made of various shapes, but are not employed upon all railways.

A tie used for securing the stones of a wall together. Copper is the best material for them, particularly if used externally, but iron is generally employed. A vertical cramp is termed a dowel or plug. Each description of cramp should be well run and covered with lead.

PAGE.

CROSSING.

On a double line of railway.—The necessary arrangement of rails to form a communication from one trackway to the other. They are similar in construction to sidings, having switches and crossing points. (See SIDINGS.)

" Of a road by a railway on a level objectionable, (see LEVEL CROSSINGS) . 209

" Of the Thames, injurious effects of, at Maidenhead and Reading (Great Western Railway) 236

CULVERT.

A drain carried under a railway, or otherwise, and constructed of stone or brickwork. Iron Culverts are sometimes used, for the carrying of brooks from one side of a line to the other. After determining the best situation for a Culvert, it is necessary to ascertain the quantity of water that is likely to run in the direction of it, previous to determining the size of the bore.

(See BALLASTING).

" On Southampton Railway 88

" A sufficient provision against floods 70

CURVE.

A term applied to a sudden bend in a line of road, canal, or railway.

Curves upon railways of less than $\frac{3}{4}$ths of a mile radius should be avoided, as the centrifugal force arising upon them has a tendency to throw the train off the rails. Many expedients are resorted to of obviating the difficulty attending them.

The periphery of the wheels of railway carriages are always enlarged in diameter next the flanches, being made slightly conical, which compensates to a certain extent for the increased length of the curve of the outer rail, (the tire of the wheel is usually made about 1 inch more in diameter on the outside than on the inside, the breadth of same being $3\frac{1}{2}$ inches), and 1 inch is allowed upon each side of the rails for play in fixing the wheels to the axle, by which they are not strained in passing along a curve. An engine with wheels 3 feet diameter, and of the above description, will turn a curve t h of a mile radius, provided the outer rail is elevated sufficient to counteract the centrifugal force, by causing a gravitating power towards the centre of the curve. The degree of elevation necessary to balance the load depends upon the velocity with which the train is moving. Upon a curve of $\frac{3}{4}$ths of a mile radius, and traversed at a rate of 10 miles an hour, it should be .07 of an inch; and at 15 miles an hour .20 of an inch; at 20 miles .36 of an inch.

The least radius of curvature on the Liverpool and Manchester Railway is 13 chains; and on the Leeds and Selby $\frac{1}{2}$ a mile. There are several curves on the Newcastle and Carlisle Railway about $\frac{1}{4}$ of a mile radius.

" On railways 2

" " a dangerous one on Liverpool and Manchester Railway . 193

" " dangerous ones not always apparent 194

E.

PAGE.

EDGE RAILWAY.

A certain description of roadway, consisting of a succession of iron bars or girders, upon which the periphery of the wheels revolve, a flange being formed upon the inner edge of the same, projecting 1 inch, to prevent their getting off the rails.

The form of edge rails is considered much superior to that of plate rails, as it combines the least expenditure of material with the greatest possible strength, and the friction upon them is less than upon the latter, which they succeeded, having been first used in about the year 1785.

They were originally made of cast iron (in 3 or 4 feet lengths) with a flat base at each end, in which holes were left for the insertion of pins, by which they were secured to the sleepers. Cast iron chairs were ultimately adopted for this purpose. The rails were fish-bellied on the underside, which form they have retained until very recently; the head being made about 2½ inches wide, and rounded, and a cross section taken through the centre of a rail shewed a greater thickness of metal at the upper part than the lower.

Wrought iron edge rails were afterwards employed, consisting merely of flat bars at first, from 1 to 2 inches square, or bars 1 or 2 inches by 3 inches, which owing to their not having rounded heads similar to the cast iron rails, (on account of the difficulty of forging the same) and their narrow shape, damaged the periphery of the wheels of the carriages considerably, (neither case hardening the wheels, nor wrought iron ties, were invented at that time,) and they continued to labor under this disadvantage until 1820, when Mr. Birkinshaw, of Bedlington Iron Works, invented a method of rolling and manufacturing iron rails of a fish-bellied form, and with heads complete, similar to cast iron rails.

The increased velocity of the trains on public railways have rendered wrought iron rails absolutely necessary, they are therefore generally employed. Cast iron rails are also becoming less used every day upon private railways, as they are brittle and apt to snap upon a sudden shock, and wrought iron rails do not require as many joinings to be made, as they can be manufactured in longer lengths.

The wear and tear of the surface of the rails upon the Liverpool and Manchester Railway is stated by Mr. Dixon, the Resident Engineer, at $\frac{1}{10}$th of a lb. per yard per annum. And it is remarkable that good malleable rails do not oxydize when in use upon a line of railway, although similar rails thrown down at random by the side of the line will lose weight continually.

The surface of cast iron rails is harder than the interior, arising from the more

The superstructure upon which all erections rest, depending entirely upon the nature of the sub-soil. In the case of good firm ground, as rock, hard clay, or gravel, being met with, very little attention is required, except to rest the structure upon it square and regular throughout. When the soil is of a loose or yielding nature, recourse must be had to artificial means of consolidating it. York landings and wood sleepers, together with strong chain bond laid in the footings of the walls were formerly very generally employed for the foundations of large buildings; but in the present day *concrete* is the favorite expedient resorted to, which is composed of good lime, gravel, and sand, in the proportion of $\frac{1}{7}$th to $\frac{1}{9}$th of lime, and it should be laid in about 12 inch layers or courses, and pitched from a height of 10 or 12 feet, neither should it be disturbed until properly concreted and set—the footings may then be laid upon it, and the walls carried up.

It is generally necessary to drive a row of sheet piles, next the foundations of walls adjoining the sea, or rivers, canals, &c., to keep the water off, the space between them being well puddled in; and in very marshy or watery ground, the whole superstructure is obliged to rest on a timber platform, supported by piles and sleepers.

FRICTION.

The obstruction or resistance offered by the several parts of a carriage or vehicle to each other, upon the application of any force to move the same, arising from attrition.

The friction or resistance of the wheels of carriages arises first from the *friction of attrition*, or the pressure of the axles against the bearings resting upon them, which support the carriage,—and secondly, from the *rolling friction*, or the resistance offered to the revolution of the wheels by the roadway; the amount of which principally depends upon the degree of smoothness and hardness of the surface over which the wheels are run.

The friction of the axle forms by far the greater resistance, and it is very important to keep up a constant supply of oil or other lucubrating matter, in order to reduce it as much as possible.

At an early stage of railroad communication, the chairs or bearings resting upon the axles were made very narrow, under an erroneous idea of reducing the friction. They were not above 1¼ inches in length, and less than the diameter of the axles in breadth. The bearings are now made 3 inches long, and upwards. Brass bearings present the least friction; but as they are made narrower, nothing is gained in this respect by them. The bearings upon the axles were formerly situated upon the inner side of the wheels, but the bearings are now placed on the outside, and the stage or frame work of the waggon is elevated above the wheels, projecting beyond them on each side; the wheels are thus protected by the bearings, which are made very strong, and as the ends of the axles are not required to be as large in diameter as the other portions, the friction is reduced, compared with bearings on the inner side of the wheels, (an axle, 3½ inches diameter, need not be above 2 inches on the outside of the wheels). The invention of case hardening the wheels also reduced the friction very much.

The amount of friction generally averages about $\frac{1}{230}$th part of the weight of

PAGE.

H.

PAGE.

LOCK.

A small dock in the line of a canal, &c. provided with gates at each end, and sufficiently large to receive the boats navigating the canal, being a contrivance for passing the same from one level to another.

Where it is required to pass a boat from a higher to a lower level, it is

PAGE.

Lock *(continued.)*

first floated into the lock and the gates closed; the water is then allowed to escape from the lock chamber to the lower level, which is effected either by paddles formed in the gates, or by side culverts; the boat being thus sunk to the lower level, the lower gates are opened, and it is taken through. The boats are passed up by the same process, reversed.

The upper portion of the canal is generally called the *Upper Pond*, and the other the *Lower Pond*, the difference between the levels being termed the *Lift of the Lock*, (which is generally from 3 or 4 to 10 or 12 feet; the greater the lift the more water is consumed.) That portion of the lock enclosed by the gates is termed the *Lock Chamber*, the size of which is regulated by the boats employed upon the canal. The portions of a lock at each extremity of the lock chamber are termed *Bays*, (and are either fore or tail bays according to the situation,) and they are usually finished with circular wing walls extending to the full width of the canal, and carried down below the bottom of the same.

„ On the Warwick Canal 17

„ On the canals between London and Birmingham 39

Locomotive Engine.

A moving steam engine used upon railways for conveying the passenger carriages and waggons.

Locomotive engines differ considerably from other steam engines in their mode of construction; as numerous modifications became necessary to render the machine suitable for a rapid transit, the foremost of which was the combination of the engine and boiler in one.

It also became necessary to form the boiler of much smaller dimensions, in proportion to its power, than customary, and to reduce the size of the cylinders; a greater degree of strength was also required in securing the several parts of the framing together, in order to render the whole proof against the sudden shocks and strains to which it is subjected.

Locomotives were in a very imperfect state previous to the opening of the Liverpool and Manchester Railway, having merely one flue passing through the boiler, and returned again to the fire box, at which end the chimney was situated. A greater velocity than 8 miles an hour could never be attained by them, owing to the small extent of evaporating surface. They did not possess above $\frac{1}{14}$th the power of the present locomotives.

The Directors of that railway having, in the year 1829, offered a premium of £500. for the best locomotive engine, gave the first stimulus to the subject, and the Rocket engine, by Mr. G. Stephenson proved the successful one, in the boiler of which tubes were introduced for the first time, which greatly increased the evaporating powers of the engine, and formed a considerable improvement, and although locomotives have since been considerably modified, yet it has formed the basis of all the many great improvements which have taken place in them.

LOCOMOTIVE ENGINE (*continued.*)

The boiler of the Rocket engine was 3 feet 4 inches in diameter and 6 feet long, having flat ends; the lower half of the boiler was kept constantly filled with water, and 25 copper tubes, 3 inches in diameter, were passed along its whole length, the further ends of which were open to the chimney, and at the other ends to the furnace, and fixed water tight; (the tubes employed in locomotives at the present time are of much smaller diameter, but three or four times the number,) the upper half of the boiler was appropriated as a reservoir for steam. A square furnace box, 3 feet long by 2 feet broad and 3 feet deep, was fixed at the extremity of the boiler, the fire bars laying at the bottom of it. The surface exposed to heated air or flame from the furnace was 117.8 square feet. The whole of the furnace was enclosed in a casing, except the bottom and the side next the boiler: and the space between the furnace and the casing was 3 inches in the clear, and kept constantly filled with water, and there was a pipe from the side of same which communicated with the underside of boiler, and another pipe was fixed at the top of it, which conducted the steam from it into the boiler. The cylinders had a stroke of 16½ inches, and were placed in a diagonal direction upon each side of the extremity of the boiler, each working a wheel of 4 feet 8½ inches diameter.

The principle of generating steam was by the exhausting power of the chimney, assisted by the impulse of the steam from the cylinders, which escapes from them into the chimney by two pipes, one on each side.

It weighed only 4¼ tons, and the evaporating surface was three times the extent of the former engines, which weighed upwards of 7½ tons, and it attained a speed of 29 miles an hour, and an average velocity of 14¼ miles an hour.

It was soon after found, that by constructing engines of greater size, the increased evaporating powers would make ample amends for the additional weight, and a strong desire was accordingly manifested of having heavier engines, on the Liverpool and Manchester Railway, but owing to the rails not being sufficiently strong to carry them, they were found objectionable, and there was accordingly a constant struggle between light and heavy engines for some time, but the line being now re-laid with heavier rails they are exclusively used. The Locomotives in general use at the present time weigh from 9 to 13 tons, and are mounted on 6 wheels, which are frequently coupled to increase their power of adhesion.

The power of a locomotive engine is different at different velocities, and it cannot be estimated in the same manner as other engines, by taking the actual force upon the piston, and the velocity of its motion, as it is very difficult to ascertain the effective pressure of the steam upon the piston, in consequence of its often differing very considerably from that in the boiler, and on account of the large amount of resistance of the waste steam, owing to the great velocity with which the piston moves; the only true method of determining the power of a locomotive is therefore by experiments. The power of a modern locomotive engine, having 12 inch cylinders and an 18 inch stroke of piston, is about 38 or 40 horse power at high velocities, and 70 to 80 horse power at a slow rate of speed.

N.

O.

P.

Q.

R.

RAILWAY, or RAILROAD.

An improved description of roadway of modern invention.

Railways were first used in the collieries in the North of England, and were originally formed of wood, horses being exclusively used upon the same for many years, and very little attention was bestowed upon the gradients; the horse was therefore required to exert himself according to circumstances, perhaps to the utmost of his power for a short distance, after which he might not be required for some time, and it was customary to unhook the horse at very rapid descents, allowing him to follow after, the gravity being sufficient to propel the waggons. Accidents were very common upon the runs, although there was a break or convoy to check the waggons, but they were frequently prevented acting in wet or damp weather, owing to the steepness of the planes; to obviate which ashes were strewed over the rails, which assisted the working of the break, notwithstanding the works were sometimes completely stopped. Thus, if a sudden shower occurred when a train was descending a very steep plane, it let them down at a fearful velocity, and the ropes which were drawn across the railway to stop them were frequently broken. These railways generally descended in the direction of the delivery of the goods conveyed upon them, and the waggons were easily drawn back when empty. The gross load upon the wooden rails was about 2 or 3 tons, but upon the introduction of iron rails a horse took nearly double, the velocity of the train down the inclined planes was therefore much increased, which is supposed to have suggested self-acting inclined planes.

Upon the application of steam to mechanical purposes generally, its action was applied to railways, and it was first used upon the ascents only, a rope being extended from the steam engine, and made fast to the waggons, whereby they were drawn up, and this principle was afterwards introduced upon the remaining portions of the line. At length locomotive engines were invented and employed upon the more level parts, and the fixed engines were confined to the inclined planes.

Public lines of railway, as the London and Birmingham, are generally made 33 feet wide in cuttings and 30 feet on embankments, the difference being caused by a drain (18 inches wide) which is required in the former, but not in the latter case; the surface of the ballasting or road is laid a little convex, to carry off the water. The width between the rails is 4 feet 8½ inches upon the principal railways throughout the kingdom, as the London and Birmingham and Grand Junction Railways, (the space between the rails is 7 feet upon the Great Western Railway); and the space in the centre between the trackways is usually 6 feet, although it is made the same as the width between the rails upon some lines, as the Newcastle and Carlisle and the Leeds and Selby Railways.

The following Table shews the average expense of working the Liverpool and Manchester Railway, from the year 1831 to 1834.

Heads of Charge.		Merchandise per ton per mile.		Passengers.		Aggregate cost per ton per mile.	
		Useful load or of goods.	Gross load.	Per passenger per mile.	Per ton per mile gross.	Useful load or of goods.	Gross load.
		d.	*d.*	*d.*	*d.*	*d.*	*d.*
*Locomotive power - - -		0·55	0·36	0·27	0·73	0·73	0·51
†Maintenance of railway - -		0·307	0·233	0·085	0·233	0·307	0·233
Coaching	Upholding carriages	- -	- -	0·054	0·146	0·082	0·058
	Conducting Coaching	- -	- -	0·104	0·282	0·158	0·111
	Duty on passengers	- -	- -	0·071	0·216	—	—
Carrying goods	Upholding waggons -	0·227	0·159	- -	- -	0·094	0·067
	Conducting traffic -	1·08	0·76	- -	- -	0·463	0·324
General expenses - - -		0·354	0·248	0·091	0·248	0·354	0·248
Total cost - -		2·518	1·760	0·675	1·855	2·188	1·551

This Table does not include the cost incurred, in laying down new rails where required, as such contingencies would not be likely to occur on another line, neither the interest paid for capital, or the cost of cartage at each end of the line.

The annual cost of private railways is much less, as will be seen by the following Table, which shews the expense of working a line adapted for the conveyance of heavy goods, or for a mixed traffic, where the latter is such that a maximum effect can be produced in the conveyance of heavy goods, without interruption to the general traffic of the line, and where the goods are generally carried in one direction only, the carriages having to be brought back empty in the other direction, deduced from the cost upon the Stockton and Darlington, the Seaham and Clarence, and other railways.

Heads of charge.	Cost per ton per mile.	
	Useful load.	Gross load.
	d.	*d.*
Locomotive power, or haulage - - - - - -	0·380	0·191
Maintenance of railway - - - - - - -	0·208	0·104
Upholding Waggons, including loading and unloading coals	0·265	0·133
General expenses - - - - - - - -	0·100	0·051
Total cost - - -	0·953	0·479

* The cost of Locomotive power differs according to the locality of the railway. The London and Birmingham Railway Company have contracted for their motive power at 0·05*d.* per ton per mile for goods, and 0·25*d.* per mile for passengers.

† The average expence of maintenance was £422. 13s. 6*d.* per mile.

PAGE.

RAILWAY, or RAILROAD *(continued.)*

The following Table gives the comparative average cost of conveying goods and passengers by Locomotive Engines on railroads.

Rate of speed in miles per hour.	Resistance per ton in lbs.	Cost of haulage, per ton per mile.	Cost of carriages per ton per mile.	Cost of conveyance per ton per mile.	Charges for conveyance per ton per mile.	Remarks.
		d.	*d.*	*d.*	*d.*	
8	8·5	0·375	0·19	1·065	1·065	Export coals.
					1·566	Lansdale coals.
12	8·5	0·5	0·227	2·138	3·5	General merchandize.
20	8·5	0·25 per passenger.	0·206	0·675 per passenger	1*d.* to 1½*d.* per passenger.	—
		1·73 per ton		2·855 per ton	12·37 per ton	—

RETAINING WALL.

A wall used for the support and maintenance of a body of earth, when circumstances render it inexpedient to slope the same gradually down.

Retaining walls are sometimes used where land is valuable, and are battered on the outside face, from 1 inch to 1½ inches to the foot; the greatest degree of batter (which is usually curved) being given to the foot of the wall.

Counterforts are generally carried up at the back of the wall, and piers in the front of it.

RIVER.

A natural water channel communicating with the sea. In order to preserve a proper depth of water for the barges navigating rivers, it is found necessary to employ artificial means upon same, by removing all obstructions, and by erecting various works, also by sustaining the banks on each side, which latter also protects the adjacent country from inundation.

ROADS.

The most general means of communication from one place to another.

The ancient Roman military roads always took hilly ground, in preference to level, for the sake of commanding the country by towers of defence on the summits; but the great desideratum in laying out of modern roads, is to obtain the most level as well as the shortest line of country, some attention being paid to the material afforded by the country for the proposed works.

A road, generally speaking, should ber aised 3 or 4 feet above the surface

Roads (*continued.*)

of the ground, in order that it may have the benefit of the sun and wind, as well as to allow for drainage; and it should always have an inclination within 1 in 100, (longitudinally) by which the water will run off into the various water courses. Steep inclinations on a road both impede the passage of coaches, and are exceedingly dangerous, and alternate rises and falls increase the distance. The inclination of turnpike roads, when near towns, should not be less than 1 in 40, and the steepest inclination upon other parts should not be within 1 in 30 under any circumstances, and for parish roads 1 in 20 to 24. The smoother the surface of a road can be got the better, provided it remains hard, as it offers the least resistance; thus, pavement forms the nearest approximation to a railroad.

Roads should be of uniform width throughout, (about 30 feet) and of a convex shape, to carry the water off into the side ditches, which should be kept perfectly clear of obstructions, and led to the natural water courses. Proper mitre drains should be constructed under the road, and filled in loosely with large flints, to carry the water into the side drains, independant of the drains at the junction of the footway with the road, the water from which should be conveyed to the side ditches by culverts.

It is now generally admitted, that the best way of forming a road is according to the system pursued by the late Mr. Telford on the Holyhead Road—viz: by laying down a foundation of regular close set pavement, having the broad part of the stones securely placed upon the bottom of the excavation. A covering or coating of broken stone or gravel is then laid over the whole, from 6 to 9 inches thick, according to circumstances.

Mr. Macneill stated, in his evidence before a committee of the House of Commons, in 1830, that the expense of improving any of the present roads, allowing none of the slopes to be less than 1 in 40, would cost from £600. to £1500. or £2000. per mile, according to circumstances.

The following Table shews the cost of conveying goods and passengers on Turnpike Roads, with the comparative expense of the same upon Railways, both with horses and with locomotive engines.

Description of traffic.	Rate of travelling in miles per hour.	Turnpike roads (with Horses).			Railways (with Horses).			Railways (with Locomotives).			
		Force of traction in lbs. per ton.	Cost of haulage per ton per mile.	Cost of conveyance per mile.	Force of traction in lbs. per ton.	Cost of haulage per ton per mile.	Cost of conveyance per mile.	Rate of travelling in miles per hour.	Force of traction in lbs. per ton.	Cost of haulage per ton per mile.	Cost of conveyance per mile.
Heavy goods (stage vans)	$2\frac{1}{2}$	73	3*d.*	8*d.* per ton	8·5	0·56*d.*	1·65*d.* per ton.	8	$8\frac{1}{2}$	0·375*d.*	1·065*d.* per ton.
Light goods (vans or light carts)	4	73	4·5*d.*	12*d.* per ton	8·5	0·9*d.*	$3\frac{6}{7}$*d.* per ton.	12	$8\frac{1}{2}$	0·5*d.*	3·5*d.* per ton.
Passengers and parcels (stage coach)	9	83	0·7*d.* per passenger, 10*d.* per ton.	3*d.* per passenger 3*s.* per ton	8·5	0·25*d.* per passenger, 2·24*d.* per ton.	1*d.* to $1\frac{1}{2}$*d.* per passenger, 1*s.* 3*d.* per ton.	20	$8\frac{1}{2}$	0·25*d.* per passenger, 0·73*d.* per ton.	1*d.* to $1\frac{1}{2}$*d.* per passenger, 12·37*d.* per ton.

S.

SHEET PILING.

A row of timbers driven into the earth side by side, which are sometimes grooved and tongued together, and used for protecting foundation walls from the effects of water.

Sheet piling is always employed next rivers, canals, &c. and good clay should be well punned in at the back of the piles next the wall.

Sheet piling is usually supported and secured at the top by guide piles and walingpieces.

SHEEVES OR FRICTION ROLLER.

Small wheels, generally made of cast iron, and much used in connection with inclined planes and fixed engines, for the purpose of receiving the rope, thereby diminishing the friction of same, which would otherwise be very great.

The friction rollers used upon inclined planes are from 10 to 15 inches diameter, having their peripheries hollowed out to receive the rope, and are usually fixed about 8 yards apart; the axles rest upon a metal box or socket, which is well bedded in the ballasting, sometimes they are fixed upon stone blocks. When the plane is curved laterally, as some parts of the Euston Square plane on the London and Birmingham Railway, they are fixed in a slanting position, and at different degrees of inclination, according to their situation in the curve, a strong stay-bar being attached at the back of each, which enables them to support the pressure of the rope. Wooden friction rollers and frames are used on the Whitby and Pickering Railway.

An Inclined Plane on a double line of railway is usually worked by an endless rope, and a large metal sheeve or wheel is fixed at the end to pass the rope back again upon its ariving at the bottom, which runs between the flanges on each edge of the periphery of the wheel. This method is applied on the Euston Square plane; the sheeve and tackle being fixed beneath the level of the rails, and in a diagonal direction, and set with masonry and brickwork, it therefore does not form any obstruction, being entirely concealed from view.

PAGE.

SPIRIT LEVEL. (See LEVEL.)

SPOIL, or SPOIL BANK.

Surplus excavation, which is laid by the side of a line of railway, or other the work, to save the expense of removal, and which occurs when the amount of cutting upon the line exceeds that of the embankments.

,,	Its technical signification	101
,,	On London and Brighton Railway	138
,,	Described, and considered objectionable	78, 128

SPRINGS.	On Great Western Railway, observations on	145

STAGE COACHES.	On roads between London and Birmingham: Stamp Office returns .	40
,,	Number of passengers; Fares; Speed; Parcels, &c.	36—43

STAMP OFFICE RETURNS.	Of coaches between London and Birmingham	40
,,	,, London and Bristol	158

STATIONARY OR FIXED ENGINE.

As applied to railways. A steam engine, employed upon the inclined planes, and sometimes upon the other parts of a railway, to drag the carriages.

Where an incline is too great to be overcome by the gravity of the meeting trains, owing to the traffic being both ways, recourse is had to a fixed steam engine, and where it is necessary to pass a steep hill, inclined planes are sometimes made on each side up to the summit, upon which an engine is fixed. In all cases of inclined planes worked by fixed engines, their inclination should be sufficient to enable the empty waggons to descend, and pull the rope after them by gravity, which would thus be in readiness to return with the train passing up.

Some of the private railways in the North are laid out in a suitable manner to be worked by stationary engines throughout; which are fixed in regular succession, and at certain distances, reciprocating with each other. This plan was partly recommended by Mr. J. Walker and Mr. J. U. Rastrick, Civil Engineers, in their celebrated Report to the Directors of the Liverpool and Manchester Railway, in 1829, on the subject of the best motive power to employ upon this line; but locomotive engines were not at that period in the same state of perfection they are at present. The principal objection to the adoption of fixed engines is owing to the great friction arising from the rope, also the inconvenience of same where passenger trains are required to be conveyed. There is not much difference in the expense between fixed and locomotive engines.

The Durham and Sunderland line is entirely worked by fixed engines; upon which the charge for conveying coals is precisely similar to that upon the Stanhope and Tyne line, where locomotives are used, viz. 1·13*d.* per ton per mile; the charges for the same upon the Seaham and Clarence Railway, which is worked by locomotives, is only 0·75*d.* per ton per mile.

(See INCLINED PLANES and FRICTION.)

PAGE.

STONE.

An aggregation of several hard mineral substances, insoluble in water. Stones are named either according to their chemical constituents, physical properties, or from their external appearance, or the names of the places from whence quarried.

Stone for engineering purposes should possess *Strength*, or the power of resistance in every direction; also *Hardness*, or the power of attrition, which enables it to resist blows; and *Durability*, that it shall not be affected by any natural agents, as the atmosphere, water, heat, and frost.

There are three classes of stones, (although some stones partake of all) viz:—

The *Silicious*, which are generally the least liable to decay, comprising granite, sand stone, &c.—

The *Argillaceous*, which comprehends basalt and nearly all slate stones. Stones of this class, although excessively hard when in their beds, are not suitable for building purposes, as upon their being quarried and removed, they are soon acted upon by the atmosphere.

And the *Calcareous*, which is a very plentiful and valuable class, comprising all lime stones, from marble downwards; it is the principal ingredient of all cements: and the most celebrated statues of antiquity being formed of calcareous stones bear proof of its great hardness and durability.

(See SAND STONE, GRANITE, PORTLAND STONE, BATH STONE, SLOPES, and SOIL.)

STONE BLOCKS. (See BLOCKS.)

STOP PLANKS (on CANALS.)

It is necessary to erect dams, at certain distances from each other, on the line of a canal, (except in cases where the space between the locks is very short) to prevent the loss of water that might arise from an accident, and for other purposes.

This is usually done by contracting the water way at such points, and carrying up wing walls from below the bottom of the canal. And vertical grooves are made in the face of the masonry upon each side, corresponding with each other, for the insertion of the stop planks.

T.

A term applied to trees after they are felled. The trunk of a full grown tree presents three distinct parts, viz. the *Bark*, or exterior; next to which is the *Sap*; and the *Heart*, which is the most essential part, forming the centre of the tree. A period of full three years should elapse after the felling of a tree before making use of it for building purposes, during which period it should undergo the process of drying, by being sawed into various thicknesses, as may be required, and properly piled.

Oak is a most durable and tough wood, and much used for all ground purposes, as sleepers, planking, &c.

Foreign fir is also much used in this country. Memel, Riga, and Dantzic timber being considered the best.

(See KYANIZE.)

The power of draught required to overcome the friction or resistance of a road, canal, or railway, and which depends partly upon the contraction of the vehicle to be propelled along the same.

(See ROAD, TRAMROAD, RAILROAD, and CANAL.)

The amount of tractive power necessary to overcome the resistance upon a road, railroad, or canal.

TRAM OR PLATE RAILWAY, TRAMWAY, or TRACKWAY.

A description of roadway prepared to receive the wheels of carriages, whereby the transit of the latter is much facilitated.

Trackways form the nearest approximation to railways, and were introduced into this country about the year 1600; they were originally constructed of timber, the transverse sleepers being of oak or fir, from 4 to 6 inches square, 5 or 6 feet long, and laid about 2 feet apart. The longitudinal beams or rails were generally of sycamore or larch, laid across the former, being secured thereto by pins, and were from 4 to 6 inches square, and laid in about 5 or 6 feet lengths; and this description of line formed what was called a *single way*. When two longitudinal beams were laid one upon the other, it was called a *double way*, the which constituted a great improvement upon the former.

PAGE.

TRAM OR PLATE RAILWAY, TRAMWAY, OR TRACKWAY (*continued.*)

Plates of wrought iron, 2 inches by $\frac{1}{2}$ an inch, were sometimes laid upon the surface of the beams, at any sharp curves or steep planes, to receive the wheels of the waggons, which were secured by counter-sunk bolts.

Cast iron plate rails were introduced in place of longitudinal timber rails in 1767; they are generally made about 3 inches wide; an upright ledge, 3 inches high, being cast on the surface of same, upon the inner side, to keep the wheels on the tracks; and they are generally in about 6 feet lengths, and secured to the sleepers by pins.

There are several modifications of cast plate rails; some have a circular flanche or web on the outer edge, projecting downwards, which increases their strength, but wrought iron plate rails, which were first introduced in the year 1824, are now generally employed.

There is a Tramway from Wandsworth to Croydon and Merstham, formed of plates of cast iron, $4\frac{1}{2}$ inches wide and 1 inch thick, and laid in 3 feet lengths: the plates have an upper vertical guide flanche 2 inches high, and fish-bellied lower flanche on the other side. The guide rails are 4 feet apart, and the space between each line is 5 feet; the plates are bedded on stone blocks, and fastened down by iron spikes driven into wood plugs, which are let into the blocks vertically. Horses are used upon the line; the usual load of a horse being about 4 tons, the waggons weighing 1 ton.

There is also a Trackway at Glasgow, part of which is laid at 1 in 20, upon which a horse can drag 4 tons, and the amount of repairs upon it is very trifling. The trams are 8 inches wide, 2 inches thick, and are made in 3 feet lengths.

Tramways are sometimes constructed of stone. The Tramway along the Commercial Road, London, is formed of blocks of granite, 16 inches wide and 12 inches thick, and in 5 or 6 feet lengths, the space between the trams being filled in with paving When originally opened, in good order and free from dust, (dust increases the friction upon a tramway from $\frac{1}{4}$th to $\frac{1}{5}$th.) the friction upon it did not amount to more than 12 lbs. per ton, but the waggons having since worn it in ruts, it has consequently increased.

Some of the American railways are constructed of granite or hard stone sills, with flat bars of iron laid on same to diminish the wear and tear, which plan has been found to answer very well.

(See RAILWAY and EDGE RAILWAY.)

TRUCK

As applied to railways, a stage or platform running upon wheels, and used upon railways for the conveyance of ordinary stage coaches and carriages, which are placed thereon. The mails, and most of the coaches remaining in the line of the several railways, are thus conveyed, the passengers and luggage remaining in their places.

TUNNEL.

A subterraneous passage or gallery bored through the earth, for the passage of a canal, road, or railway. They are generally formed by sinking vertical shafts and then commencing operations each way, upon arriving at the proper level. The sides and top of the excavation are supported by timber centering and shores, the brickwork and earthwork being carried simultaneously, or as nearly so as possible; it is usually carried forward in 20 feet lengths, and when the brickwork of a length is completed the centering is struck, and pushed on further for another. The excavation is formed as near the size of the tunnel as possible, and the space between the extrados of the arch of the tunnel and the excavation is carefully filled with earth and well rammed, and if any of the timber should be difficult to withdraw it is allowed to remain. The excavation is conveyed up the shaft to the surface of the ground by a horse gin which is fixed at the top.

It is found necessary to set a cast iron curb in the crown of the tunnel under the shafts to support the same.

Tunnels are sometimes worked by horizontal shafts or galleries.

They are also formed in cuttings similar to bridges, the ground being shored up on each side, and again covered over with earth upon the completion of the brickwork, technically termed *Open tunnels.* An inverted arch is unnecessary in this description of tunnel, although always formed at the bottom of those of the former description.

There are several tunnels upon the London and Birmingham Railway, of

W.

THE PLATES

REPRESENT DETAILS OF

ONE OF THE FOUR LOCOMOTIVE ENGINES, (No. 244)

BEING CONSTRUCTED FOR THE

PARIS AND VERSAILLES RAILWAY, ON THE LEFT BANK OF THE SEINE,*

BY

MESSRS. R. AND W. HAWTHORNE,

Civil Engineers, Steam Engine Builders, and Manufacturers of Machinery generally,

OF NEWCASTLE UPON TYNE.

Plate 1. Side Elevation of Locomotive Engine and Tender.

2. End Elevation of Locomotive Engine.

3. Plan taken above top of framing, with section of Fire and Smoke Box.

4. Longitudinal section, shewing reversing Gear, &c.

5. Transverse section taken through Fire Box.

6. Transverse section taken through Smoke Box.

* Kindly communicated to Mr. Williams, the Publisher, by the Inventors.

Diameter of Cylinders . . . . 12 inches.
Steam way to Cylinder . . . . . length $6\frac{3}{8}$ ins., width $1\frac{1}{4}$ ins.
Length of stroke of Piston . . . . 18 inches.
Boiler . . . . 8 feet long and 39 ins. diameter,
„ containing 121 Tubes . . . . . $1\frac{5}{8}$ exterior diameter.
„ Length of same . . . . 8 feet,
„ and presenting a surface exposed to contact
with heated air, of . . . . 428·40 square feet.
Fire Box . . . . length 30 ins., width $41\frac{1}{4}$ ins.
Height above the bars . . . . $40\frac{3}{4}$ inches.
Area of Fire Grate . . . . 8·59 square feet,
presenting a surface exposed to radiant
caloric, of . . . . 46·16 square feet.
Quantity of Fuel contained in fire box, to the
height of the lowest row of tubes . . . 1360 cubic feet.
Diameter of Chimney . . . . 14 inches.
Six wrought iron Wheels, upon Mr. Hawthorne's arrangement:—
viz. One pair of Driving wheels . . . . . . $5\frac{1}{2}$ feet diameter.
Two pair of Supporting wheels . . . . $3\frac{1}{2}$ feet diameter.
Weight of Engine, without water in the boiler . 8 tons. 12 cwt.
Do. with water in the boiler . . 10 tons.

' The most essential improvement in this Engine is in the working of the slide valves, which will be readily understood by a reference to the Longitudinal Section and Plan, (Plates 3 and 4.)—The motion is communicated from the centre of the Connecting Rod, where the slide A A, working in the guide B B, is attached. The upper arm of the latter communicates with the arm C, which is fixed to the weigh bar D, and upon this weigh bar the double ended lever E is fixed, which gives motion to the slide by the rod F. The weigh bar G is worked by the arm H, which is connected to the reversing handle I by the side rod K. Another arm L is also fixed to this weigh bar, which being attached to the rod F by another rod M, lifts it alternately from one pin to the other; (the latter being situated at the ends of the double ended lever E,) and the motion of the slides is thereby reversed.—The reversing gear is represented by dotted lines on the Longitudinal Section.

This new arrangement entirely removes every description of Eccentrics, and substitutes in their stead a combination of levers, at the same time simplifying the machinery, by which the lead of the slide may be raised at pleasure, and the steam applied at the proper time to produce its utmost effect to the propelling power; a desideratum which cannot be obtained equally satisfactory by the eccentric motion, from the constant spring and closing of the apertures through which the steam passes to the Cylinders. The present improvement is calculated most effectually to obviate this, and consequently to increase the real effective force of the engine.

The cranked axle has three inside bearings, attached to strong longitudinal bars, firmly fixed at the ends, to the fire and smoke boxes. The outside frame of this Engine, only differs from those now in general use by having the Pedestals cut out of the solid plates, instead of being attached to the frame by bolts and rivets.

The method of applying the Steam, or working the Slide valves, is extremely ingenious, and appears to be very perfect; the completion of so important an improvement must be of no small degree of satisfaction to the talented Inventors, whose several Improvements connected with Locomotive Engines are well known; the System of working the Slides by the four fast Eccentrics, instead of two loose ones, was also first introduced by them in the Comet Engine, (the first Locomotive run upon the Newcastle and Carlisle Railway,) which has been fully appreciated, and consequently adopted by the most eminent Engineers of the present day.

THE END.

London: Printed by J. Davy,
Queen Street, King Street, Long Acre.

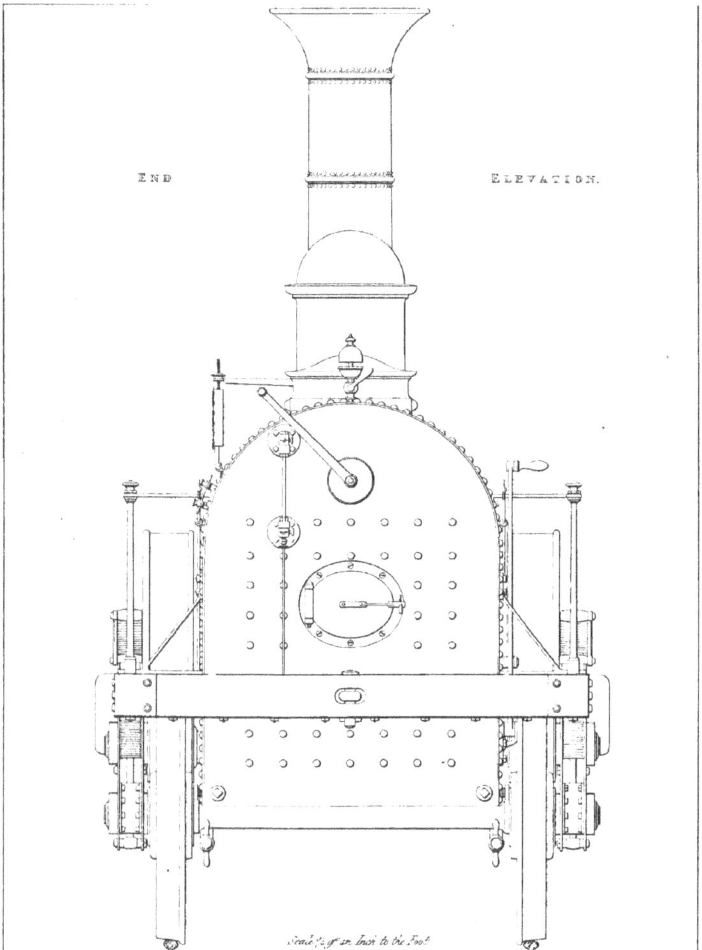

John Hodgson Jun. Del

John Williams Library of Fine Arts, 106 Gt Russell Street

W Kelsall Sc

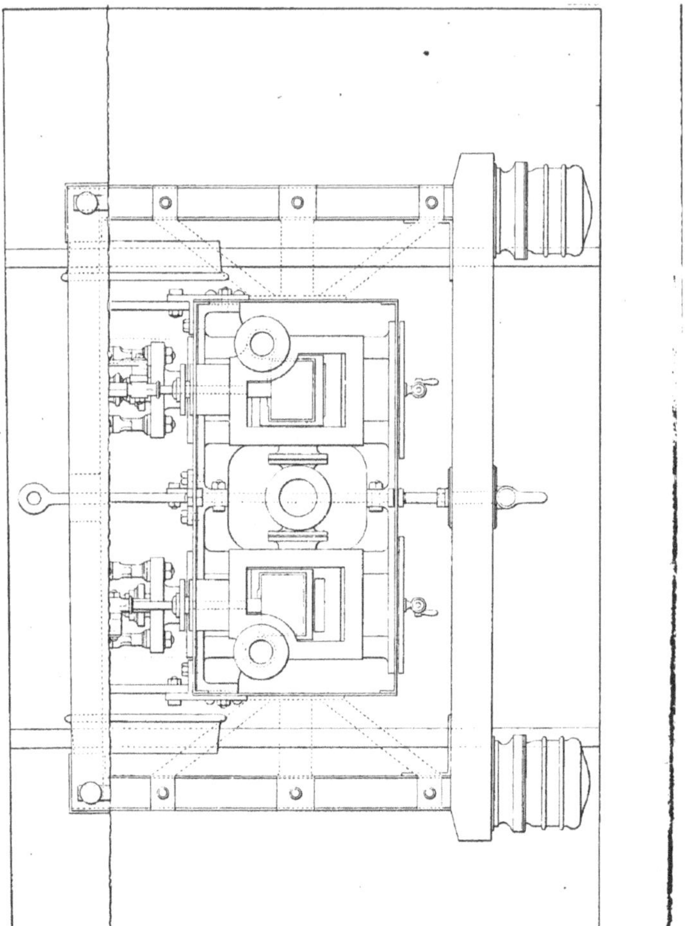
John Hodgson Jun Del
9
10
W Kelsall Sc

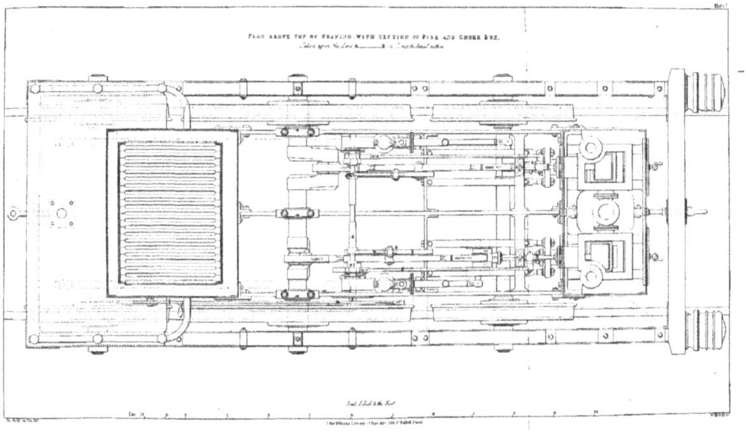
PLAN ABOVE TOP OF FRAMING, WITH SECTION OF FIRE AND SMOKE BOX.
Scale 1 Inch to the Foot

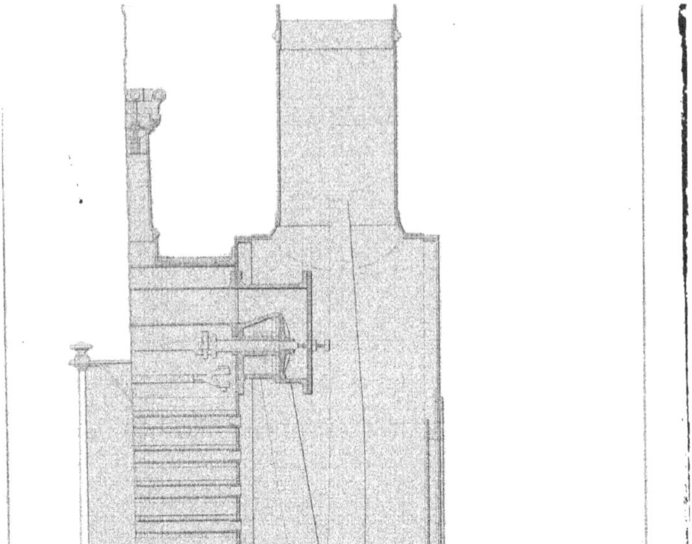
A
B

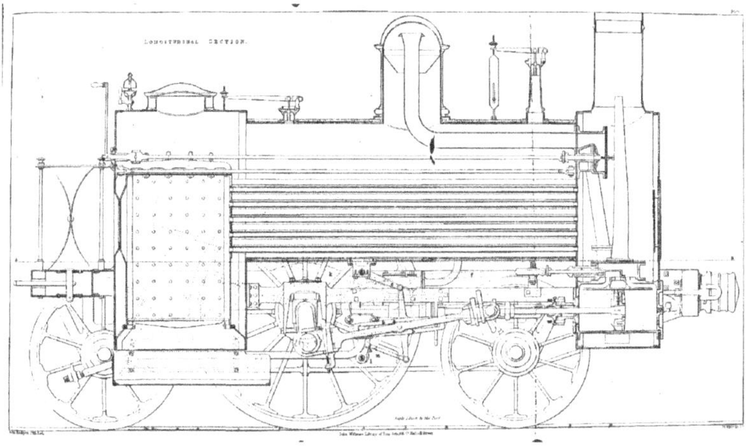
LONGITUDINAL SECTION.

TRANSVERSE SECTION.

Taken through Fire Box

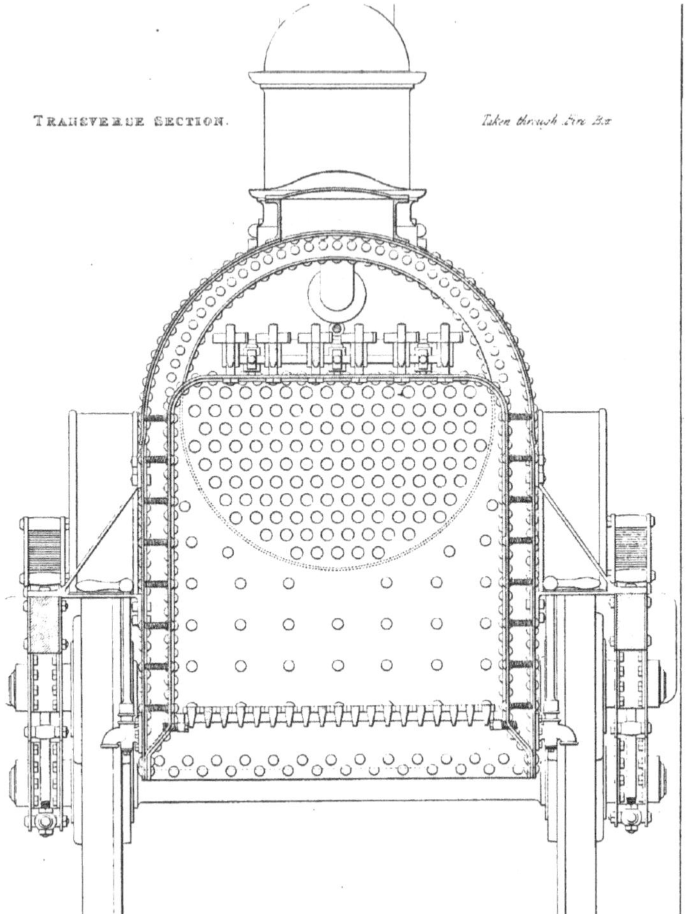

Scale 1 Inch to the Foot.

Ins 12 6 0 1 2 3 4 5 6

John Williams Library of Fine Arts, ... Street

n Hodgson Jun. Del

W Kelsall Sc.

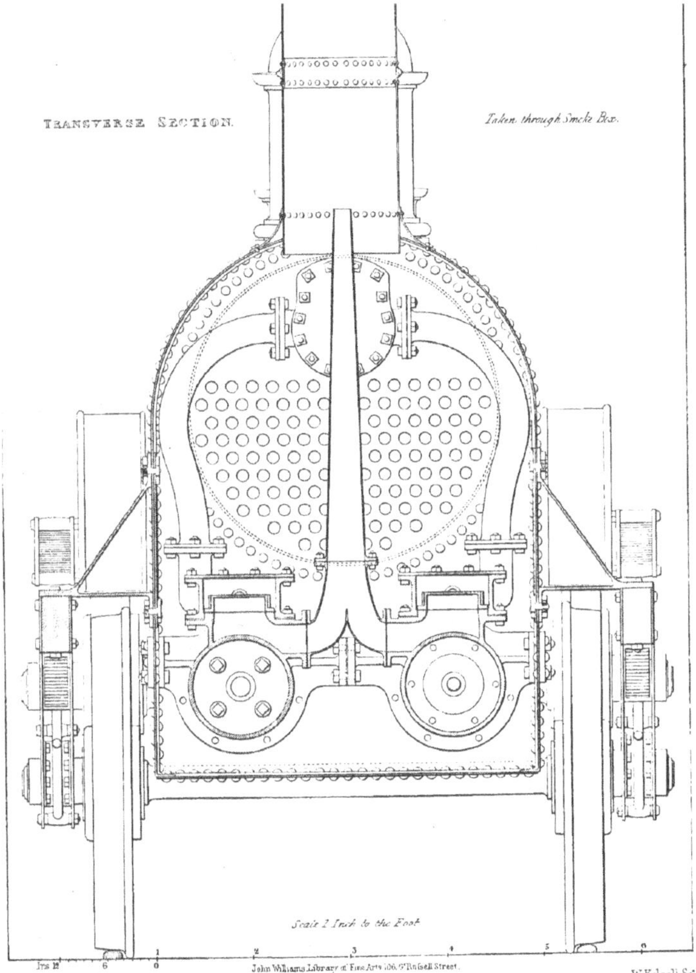

John Hodgson Jun. Del.

John Williams Library of Fine Arts 106, Gt Russell Street.

W. Kelsall Sc.

Spotted something that isn't quite right?

Let us know and we'll fix it for future reprints. And as a thank you for your diligence, we'll send you a free Sothis book of your choice.

publisher@sothis.press

This page intentionally left blank

Except of course for that line above. And this one.
And whilst we have your attention, perhaps you'd
like to follow us on Instagram @sothis.press. Go on.

Made in United States
Cleveland, OH
06 November 2026

47791724R00231